Application of Systemic-Structural Activity Theory to Design and Training

Ergonomics Design and Management: Theory and Applications

Series Editor
Waldemar Karwowski

Industrial Engineering and Management Systems
University of Central Florida (UCF) – Orlando, Florida

Published Titles

Application of Systemic-Structural Activity Theory to Design and Training
Gregory Z. Bedny

Applying Systemic-Structural Activity Theory to Design of Human–Computer Interaction Systems
Gregory Z. Bedny, Waldemar Karwowski, and Inna Bedny

Ergonomics: Foundational Principles, Applications, and Technologies
Pamela McCauley Bush

Aircraft Interior Comfort and Design
Peter Vink and Klaus Brauer

Ergonomics and Psychology: Developments in Theory and Practice
Olexiy Ya Chebykin, Gregory Z. Bedny, and Waldemar Karwowski

Ergonomics in Developing Regions: Needs and Applications
Patricia A. Scott

Handbook of Human Factors in Consumer Product Design, 2 vol. set
Waldemar Karwowski, Marcelo M. Soares, and Neville A. Stanton

> Volume I: Methods and Techniques
>
> Volume II: Uses and Applications

Human–Computer Interaction and Operators' Performance: Optimizing Work Design with Activity Theory
Gregory Z. Bedny and Waldemar Karwowski

Human Factors of a Global Society: A System of Systems Perspective
Tadeusz Marek, Waldemar Karwowski, Marek Frankowicz, Jussi I. Kantola, and Pavel Zgaga

Knowledge Service Engineering Handbook
Jussi Kantola and Waldemar Karwowski

Trust Management in Virtual Organizations: A Human Factors Perspective
Wiesław M. Grudzewski, Irena K. Hejduk, Anna Sankowska, and Monika Wańtuchowicz

Manual Lifting: A Guide to the Study of Simple and Complex Lifting Tasks
Daniela Colombiani, Enrico Ochipinti, Enrique Alvarez-Casado, and Thomas R. Waters

Neuroadaptive Systems: Theory and Applications
Magdalena Fafrowicz, Tadeusz Marek, Waldemar Karwowski, and Dylan Schmorrow

Safety Management in a Competitive Business Environment
Juraj Sinay

Self-Regulation in Activity Theory: Applied Work Design for Human–Computer Systems
Gregory Bedny, Waldemar Karwowski, and Inna Bedny

Forthcoming Titles

Organizational Resource Management: Theories, Methodologies, and Applications
Jussi Kantola

Application of Systemic-Structural Activity Theory to Design and Training

Gregory Z. Bedny

CRC Press
Taylor & Francis Group
Boca Raton London New York

CRC Press is an imprint of the
Taylor & Francis Group, an **informa** business

CRC Press
Taylor & Francis Group
6000 Broken Sound Parkway NW, Suite 300
Boca Raton, FL 33487-2742

First issued in paperback 2017

ISBN-13: 978-1-4822-5802-8 (hbk)
ISBN-13: 978-1-138-74780-7 (pbk)

Contents

Preface

Currently, every complex human–machine system includes a computer as a critically important means of work. However, an operator's interaction with a computerized system cannot be reduced to only performing computer-based tasks. When working with such systems, an operator often monitors and, if needed, enters a control loop to override an automated system response to situational events. For example, manual control is used in the more complicated stages of flight such as takeoff and landing and during system failure. Current developments in the production domain do not eliminate manual components of work. An example of such type of work is complex assembly tasks in contemporary manufacturing. These tasks do not require significant physical efforts but rather use complex technological tools and perform high-precision motor activity in combination with cognitive actions. This makes the performance time of motor components of activity and its relationship with precision an important topic in ergonomics and engineering psychology. Virtual reality techniques become increasingly popular for determining the efficiency of task performance and training in contemporary industry. This book concentrates on discussing such type of work when an operator performs various tasks in highly automated technological systems and interacts with various displays and controls. It also includes consideration of certain aspects of analysis of computerized tasks. At the same time, manual components of work in contemporary industry are also considered.

Cognitive approach considers work analysis and design from information processing perspectives. It focuses on analyzing the constraints that cognition imposes on human performance (Vicente, 1999). With this viewpoint in mind, researchers start task analysis by describing the subject's cognitive characteristics. Cognition is considered as a system of cognitive processes and is presented as an information-processing model. However, cognition is not just a system of cognitive processes that can be described as a sequence of specific stages. It is also a structure that can be presented as a system of cognitive actions and operations that are basic elements of mental activity. Thus, it can be described as a cognitive structure that has a complex relationship with the environment (configuration of equipment, human–computer interface, and social reality). This structure is not a static system. It has dynamic characteristics and is organized depending on the specifics of external conditions and strategies of the subject's activity. Such characteristics of cognition are entirely ignored by the cognitive approach. As Kuutti (1997, p. 19) correctly pointed out, this approach uses experimental laboratory-oriented classical psychology methods that are unable to penetrate the human side of the interface. In practice, the cognitive approach is replaced by experimental procedures similar to the black box approach or input–output analysis. Information-processing models are not task specific. They cannot describe the cognitive structure of activity during task performance due to the fact that cognitive psychology has no clearly developed terminology. Such terms as overt actions, cognitive steps, goals, self-regulation, and strategies can be interpreted differently by various professionals in the field. All of this makes it impossible to describe the flexible cognitive and behavioral structure of activity during task performance. SSAT overcomes these limitations in cognitive psychology and considers cognition as a process and as a structure. The cognitive approach is a parametric method of study that should be accompanied by the systemic approach developed in SSAT.

The book presents new theoretical approaches and principles to study human work from the perspective of systemic-structural activity theory (SSAT). This theory provides a unified framework to ergonomics and work psychology. It can be useful for engineers and economists who work in such fields as equipment design, efficiency of human performance, safety, and training. The book is important because it overcomes the traditional separation between cognition, behavior, and motivation using a systemic approach to the analysis of human work activity. The conceptual apparatus of this approach differs significantly from existing approaches outside of activity theory. The book starts with comparative analysis of general, applied, and systemic-structural activity theory. The author has taken into account the difficulties faced by professionals when they try to use activity theory and presented basic concepts such as goal, relationship between goal and motives, mechanisms of anticipation, vector motive → goal, and cognitive and behavioral actions. The concept of task is considered from the SSAT perspective. Here, for the first time, activity theory terminology is examined not only from a psychological perspective but also from a neurophysiologic perspective.

There is no unified and standardized psychological terminology that can be applied to ergonomic design. The same terms carry different meanings when utilized by different authors. The terminology used in activity theory in recent studies is clarified, and the terminology used in some new psychological approaches that utilize activity theory as a starting point in their development is compared and refined.

Section II has five chapters. In Chapter 5, knowledge and skill acquisition processes are discussed from the SSAT perspective. SSAT offers the following basic principles of learning: unity of cognition, behavior, and motivation; learning as a complex goal-directed self-regulative process that integrates conscious and unconscious levels of self-regulation; learning process as a transition from a less efficient to a more efficient strategy of activity performance; and the ability of a learner to achieve the same goal by various strategies. Learning curve analysis is considered as an efficient tool for studying various tasks in the production environment. The shape of the learning curve is affected by the nature of the task, the idiosyncratic features of the trainee, and the training method. Task analysis during its acquisition is known as a genetic method of study in activity theory. The essence of this principle is to study various psychological phenomena during their development. Cognition and structure of activity in general is studied based on the analysis of the sequential stages of its formation, reconstruction, and dialectic genesis. Thus, studying the acquisition process is a very useful method, which is illustrated by applying it to the analysis of various types of task, including computer-based tasks. In this section, the data in learning by observation are presented.

In Chapters 6 and 7, the data on the design of human–machine system are discussed. The term *design* takes its roots in engineering and cannot be reduced to experimentation as it is done in cognitive psychology. The main objective of design is to create an appropriate documentation that describes a designed object. For example, in manufacturing at the analytical stage, various drawings of the same object are developed and accompanied by related quantitative assessment. The next stage includes creation of a prototype and experimental evaluation of a designed object. The obtained data are further used for the adjustment of the models (drawings and quantitative analysis), and the cycle can be repeated. The engineering design process can be presented as follows: nonformalized (qualitative) stage → formalized stage (drawings and calculations) → experimental stage (experimental evaluation of a prototype). The application of similar methodology to ergonomic design of complex human–machine systems is considered in this book.

Special attention is paid to designing flexible tasks that are performed by an operator in semiautomatic and automatic systems where a complex combination of cognitive and motor components of activity prevails. Similar to engineering design, in SSAT any design involves qualitative and analytical stages. In this section, attention is paid to the first three stages of design and especially to the qualitative stage of analysis that considers human activity as a goal-directed self-regulated system. The process of self-regulation is described as different stages of information processing. These stages have loop structure organization. Each stage is called a function block, because it performs a particular function in activity regulation with forward and backward interconnections between different function blocks. The application of the concept of self-regulation to the analysis of the pilot's activity in emergency situation is considered along with the relationship between time study, error analysis, and design. The formation of professional pace during task performance is discussed. Chaper 9 demonstrates the application of queuing theory to human error analysis, offering quantitative methods for the assessment of human performance in time-restricted conditions.

Section III contains three chapters and describes the basic quantitative analytical method of task evaluation from the SSAT perspective. This method includes the quantitative assessment of task complexity in human–machine systems. Task complexity is a psychological characteristic of task that determines cognitive demands for task performance. The relationship between task complexity and task difficulty is discussed. A critical analysis of some methods of task complexity evaluation outside of SSAT is presented. It is demonstrated that these methods ignore the complex structure of activity that unfolds in time as a process. As a result, such methods utilize incommensurable units of complexity measures. In this section, we demonstrate how the task complexity evaluation method developed in SSAT can be used for the analysis of production operations and evaluation of tasks in highly automated human–machine systems. Task complexity evaluation is an important tool for the optimization of task performance and for equipment design. Complexity evaluation of such tasks has some specifics in comparison with task complexity evaluation of computer-based tasks. In Chapter 12, we demonstrate how task complexity evaluation can be used for safety analysis. This is a totally new method in this area of study. This approach allows detecting errors not by observation or experiment but by just utilizing models of human work activity. This can support more effective safety analysis and help predict possible errors in the early stages of a design process. The developed approach to task complexity and safety evaluation is important for ergonomics, work psychology, engineering, and economics.

It is well known that the effectiveness of all quantitative analytical methods in psychology and ergonomics has been quite limited to this point. In contrast, as described in this book, methods are brought to the level of practical application. This book has the potential to significantly affect the development of work psychology and ergonomics and can be useful for industrial engineers, computer professionals, designers, and specialists in safety and training. SSAT can be considered as one of the main approches to the study of human performance and work design. This book also provides the reader with state-of-the-art information in SSAT and demonstrates its application to task analysis, design, and training.

Author

Dr. Gregory Z. Bedny worked as a professor at several Ukrainian universities, and after arriving in the United States taught at Essex County College in New Jersey. He has now retired and is a research associate at Ergologic, Inc. He earned his doctorate degree (PhD) in industrial organizational psychology from Moscow Educational University and his post-doctorate degree (ScD) in experimental psychology from the National Pedagogical Academy of Science of the Soviet Union. Dr. Bedny is a board-certified professional ergonomist (BCPE). He also is an honorary academician of the International Academy of Human Problems in Aviation and Astronautics in Russia and an honorary doctor of science of the University of South Ukrainian. He has been awarded an honorary medal by the Ukrainian Academy of Pedagogical Sciences for his achievement in psychology and his collaborative works with Ukrainian psychologists.

Dr. Bedny is the founder of the systemic-structural activity theory (SSAT). SSAT is a high-level generality theory or framework that is the basis for unified and standardized methods of studying human work. He authored a number of original scholarly books and multiple articles in this field. He applied his theoretical study in the field of human–computer interaction, manufacturing, merchant marines, robots systems, work motivation, training, fatigue reduction, etc.

Section I

Overview of the Basic Concepts in Activity Theory

1

Activity Theory and the Vector Motive → Goal as Its Basic Concept

---

1.1 General, Applied, and Systemic-Structural Activity Theory: Historical Overview

Activity theory (AT) is a psychological grand theory or framework that has a long history of development in the former Soviet Union. Developments of the theory are associated with the works of Rubinshtein (1959), Vygotsky (1956), Leont'ev (1978), and others. AT has appeared at a time when behaviorism had been developed in the United States. Usually the emergence of AT is explained by the acceptance of Marxist philosophy and the political climate in the country at that time. However, in our view, there is another reason for the emergence of AT, a psycholinguistic one. It was shown in psycholinguistics that the specifics of language have an impact on human thinking. Human beings exist not only in a material and social world but also in a world of utilized language and specific historical conditions (Sapir, 1956). The specificity of language can be, to some extent, an important and independent factor in the development of human thought (Carroll, 1963).

Considering the emergence of AT from a historical perspective, one should pay attention not only to Marxist ideology but also to the role of Russian language in the development of this theory. Activity and action are the most important terms in this theory. These terms are not purely psychological and are used in ordinary Russian language. Moreover, apart from psychologists, these terms are widely used by engineers and economists studying the efficiency of labor. We believe that a better understanding of the origin of these terms allows for a better understanding of the theoretical foundations of AT.

In Russian language, the term *behavior* (*povedenie*) has some similarity with such concepts as cultural mentality, cultural behavior and manners, certain lifestyle, and adequacy of interaction with others according to existing social norms and society regulations.

The term *behavior* is also used to describe what animals do in various situations (animal behavior). The term activity is not used when talking about animal behavior. In the Russian language, we use this term when studying the behavior of humans (work activity, learning activity, games in children's activity, etc.). The term *behavior* is used to describe cultural aspects of human interaction with others or to analyze animals' behavior. All this has greatly contributed to the fact that AT versus behaviorism has been developed in Russia. We can similarly trace the emergence of the term *action*, which was originally associated only with external motor activity. Initially, it was considered as a part or an element of motor activity. Later, the concepts of *mental action* and the basic characteristics of cognitive and motor actions arose in psychology. The term *activity* (or *deyatel'nost'*) integrates the mental and motor components of human behavior. A critically important term for this component of activity is action (*dejstvie*). Initially, the term activity (*deyatel'nost'*) and in relation the term action (*dejstvie*) are used in everyday language and are associated with external, motor human behavior. This led to the fact that these terms were applied by engineers and economists in the study of labor efficiency in the former Soviet Union. Only later was the concept of cognitive actions introduced in general psychology. Thus, activity (*deyatel'nost'*), as a psychological concept, integrates the cognitive and behavioral components, which consist of smaller units—cognitive and behavioral actions. A critical feature of AT is its relation to consciousness. According to Rubinshtein (1957), activity is not merely an external behavior; it is also inextricably linked with internal mental components of activity and consciousness of abstractions from a concrete situation that allow an individual to anticipate the sequences of their situation and provide insight into mental processes that guide conscious and volitional behavior. The unity of consciousness and behavior becomes a major principle of AT. Motivation and conscious goal begin to play an important role in AT.

AT was formulated due to ideological reasons and some historical specificity of the development of the Russian language. The founder of AT was Rubinshtein, who made the first publication in this field (Rubinshtein, 1922/1986). However, significant contributions in the further development of this theory were made by Vygotsky (1956, 1978), Leont'ev (1978), and others. Zinchenko (1995), who worked at Moscow State University, wrote, "In 1922 Sergei Rubinshtein transplanted the philosophical category of 'activity' into psychological soil." Activity, actions, conscious goal, and motivation acquired a deep theoretical and psychological meaning in AT. Prior to that, psychology had been defined as a study of mental experience based on introspective analysis. The latter approach considered psychology as a science that studies external human behavior (stimulus–response psychology). These approaches consider mental activity and behavior as two independent areas of study. In contrast, AT integrates cognition and behavior. Mental activity is not considered an internal independent mechanism but is to be tightly connected with external behavior.

This theory plays a significant role in studying human learning and in school psychology.

AT, which is derived from the works of Vygotsky (1971, 1978), Rubinshtein (1957, 1959), and Leont'ev (1978), is useful in work psychology and ergonomics as a general theoretical and philosophical background for studying human work. However, it is not well adapted for studying human work.

In the 1970s, leading scientists who studied work psychology realized that general AT is a useful philosophical framework, but since it cannot be directly applied to the study of human work in contemporary industry, they started developing the applied activity theory (AAT). Among those were Bedny (1987), Gordeeva and Zinchenko (1982), Galactionov (1978), Kotik (1974), Konopkin (1980), Landa (1976), Platonov (1970), Pushkin (1978), Zarakovsky et al. (1974), Lomov (1966), Zavalova and Ponomarenko (1980), and others. At that time, a new direction in AT had emerged, which is now called AAT. Cognitive psychology played a significant role in the development of AAT at that time. One of the advantages of cognitive psychology compared to general AT was an attempt to study cognitive processes in a more detailed manner. Cognitive psychology introduced such basic concepts as sensory memory and short-term and long-term memory into psychology (Atkinson and Shiffrin, 1968, 1971). Utilizing the partial report method, Sperling (1960) showed that information in the sensory register remains only at a very brief period of time. At the same time, it was demonstrated that sensory memory holds incoming information long enough for further processing. It has been found that a mechanism of repetition is required for preserving information and transferring it into long-term memory. Later, Waugh and Norman (1965) have proposed a model in which three consecutive blocks of information transformation were presented: sensory registers, short-term memory with limited memory capacity and verbal repetition as a way to save the information, and long-term memory with a capacity to store a large amount of information. Sternberg (1969, 1975) developed the additive factor method based on which he confirmed the existence of processing stages. The main intent of additive factors is to define the existence and distinctions of different information processing stages. He proved that these stages are of a very short duration and should be measured in milliseconds. Various block models of information processing that consist of a sequence of stages have been introduced into psychology. Such models have some limitations when applied to the analysis of mental processes, but they open additional opportunities for the analysis of such processes.

All these methods have been accumulated and elaborated in AAT. Human information processing in AAT is usually described as a certain stage of information transformation. However, activity specialists paid more attention to the fact that the specifics of each stage depend on the nature of the task being performed by a subject, past experience, motivation, etc. The contents of the considered stages are also determined by the goal and motives of activity. When analyzing microstructural models of cognitive processes, scientists focused on the role of object-oriented actions in the development of mental operations. In contrast to cognitive psychology, where the main focus of study is memory function, AAT also pays attention to the function of thinking in the process of information transformation during a short period of time. In the study of mental processes, as opposed to cognitive psychology, attention was paid not only to the internal transformation of information but also to the interaction of a subject with an external material world and social factors.

In general AT, a predecessor of AAT, various scientists carried out studies of voluntary and involuntary memory when subjects performed memorization tasks. These scientists focused on analyzing the effect of motives, goals, and individual properties of subjects on their memorization process, etc. (see, e.g., Zinchenko, 1961). However, at that time in general activity theory, nobody studied the short-duration stages of information processing. Scientists in AAT started combining cognitive psychology methods with the AT ones only after the emergence of cognitive psychology. Scientists in AAT demonstrated that an important aspect of an operator's activity is the development of a mental model of reality. This model influences the correct interpretation of a situation and the efficiency of an execution (Konopkin, 1980).

Zinchenko and his colleagues, following cognitive psychologists, started studying short-term memory in the framework of activity theory (Zinchenko et al., 1980). Such concepts as sensory memory, short-term memory, microstages of information processing, etc., were transported into AAT from cognitive psychology. Methods of studying short duration stages of human information processing with some modifications were named microstructural analysis (Zinchenko and Vergiles, 1969). Microstructural analysis is explicitly based on the methods and ideas generated in cognitive psychology. In AAT, the perceptual process is divided into four stages: detection, discrimination, identification, and recognition. The transformative operations at these stages include filtering information, converting information, information identification, and encoding and repetition.

The most representative fields of applied study in AAT are aviation, semiautomatic systems in manufacturing, automatic systems associated with the remote control of various technological processes, and software design (Galactionov, 1978; Ponomarenko and Lapa, 1975; Ponomarenko and Zavalova, 1981; Zarakovsky and Pavlov, 1987 etc.). Several works in AAT of leading scientists that work in aviation have been presented for the first time in a special issue of *Theoretical Issues in Ergonomics Science* (TIES) (Bedny, 2004). They give a general idea about studies in this area of research.

The psychological methods of the study emphasized the analysis of the activity structure. The structural concept emphasizes the relationship between the individual elements of activity. Motor actions and mental processes are the main components of analysis that regulate their execution. Decomposition of activity was usually carried out on technological criteria, as there was no clear distinction between the technological and the psychological units of analysis. In AAT, much time was devoted to the development of mathematical methods to formalize the study of ergonomics. However, the practical effect of these methods, with some exception, was quite limited. It should be noted that AAT is a set of different concepts and approaches developed by various authors. These approaches do not offer the standardized terminology and clearly defined units of activity analysis. AAT has not developed unified and standardized research methods that are needed to address design issues.

Therefore in this book, we describe the systemic-structural activity theory (SSAT) approach (Bedny and Meister, 1997, Bedny 1981, 1987). This theory has been originally presented independently in the work *The Psychological Foundations of Analyzing and Designing Work Processes* (Bedny, 1987) and has been further developed in the United States and published in several monographs (Bedny and Karwowski, 2007; Bedny and Meister, 1997) and numerous articles. This book does not just integrate different data from various publications in this field but presents totally new data. At present, this theory provides a unified framework for studying human work and ergonomic design.

In the general AT, systemic analysis has not gone beyond the general philosophical discussion and has not been brought to the level of practical application. In AAT, only some aspects of systemic analysis have been applied. Systemic analysis as an interdisciplinary field should not negate the need for the development of proper systemic psychological methods of activity studies. Implementing such an approach becomes possible when activity can be described as a complex structure that evolves over time. The creation of such an approach is possible only when we can develop methods of analysis for describing activity as a systemic-structural entity where cognition, behavior, and motivational processes are considered as systemic organization. In this case, activity is described as a system consisting of subsystems and smaller elements that are in specific relation and interaction with other elements of activity. Activity is considered as a logically and hierarchically organized system. The transition from the general philosophical discussion to its practical application is not that easy. Existing methods of systemic analysis of activity are important and useful from the theoretical and practical points of view. However, they are fragmented and cannot substitute a unified and, to some extent, standardized approach to a systemic analysis of work activity.

Cognitive psychology treats cognition only as a process, making it difficult to study human behavior as a system. In the sequential process of human activity, only a particular slice of activity specific for this period of time can be realized at any given moment. Nevertheless, our past and future activities influence what we do at present. We believe that the principles of systemic-structural description of activity are determined by the methods and tools that allow describing it as a system. It means that the same object can also be presented as a system depending on the methods used. Hence, we need to create standardized methods, research procedures, and operations to facilitate the development of various types of models of the same object in the form of a system and structure.

Further, we will concentrate our efforts primarily on the application of the SSAT approach to task analysis in ergonomics, work psychology, and labor economics and study human-computer interaction. These are the main areas where the application of the SSAT analysis of activity is the most important. It raises a number of issues that need to be addressed such as development of standardized units of activity analysis and the language of its description, selection of stages and levels of analysis, development of methods for constructing models of activity, analysis of their relations, and identifying the relations between qualitative, formalized, and quantitative research methods. We believe that SSAT is the most promising general theoretical framework in this regard (Bedny, 1987; Bedny and Karwowski, 2007; Bedny and Meister, 1997).

SSAT views activity as a structurally organized self-regulated system, rather than an aggregation of responses to multiple stimuli, or a linear sequence of information stages as it is described in behavioral or cognitive psychology. Furthermore, it views activity as a goal-directed rather than a homeostatic self-regulative system. Such system is considered goal-directed and self-regulated if it continues to pursue the same goal under changed environmental conditions and can reformulate or formulate the goal while functioning. Activity is a goal-directed self-regulated system that integrates the cognitive, behavioral, and emotional-motivational components. This allows us to analyze and design activity as a flexible and adaptive system. A subject utilizes various strategies for goal achievement during task performance. This eliminates contradictions between constraint- and instruction-based approaches to design, which were described by Vicente (1999).

Cognitive approach, which is presently a dominant one, treats cognition and behavior as a process, making it difficult to study activity and behavior from a systemic-structural perspective. The notion of process does not allow describing activity as a structure. The introduction of standardized and unified units of analysis in SSAT helps to describe activity as a structure that unfolds over time. One can extract from the same activity different structures as independent objects of study, depending upon the purposes of a study. Each of these objects of study can be represented as an independent system. Consequently, we may have different representations of the same activity. Dividing activity into distinct elements and components of activity to construct a holistic activity is an important method of the system-structural analysis of activity.

According to SSAT, activity may be presented as a system that consists of heterogeneous, structural elements, composed of different units that allow for the representation of activity in terms of different models describing the same object of study. The description of activity as a multidimensional system significantly increases the applicability of this approach to the study of human work.

Cognition is not merely a process or a mental picture of the world but is also a system of mental actions and operations intimately related to external actions (Bedny et al., 2000). As in physics, where light has both wave and particle characteristics, in the SSAT, cognition is understood as both a process and a system of actions or other functional information processing units. Thus, cognition incorporates both process and structure. Hence, cognitive task analysis used in ergonomics invites integrating activity principles (Bedny et al., 2008). The basic elements of activity do not exist in isolation; rather they function as a system. Since the 1970s, early childhood development (ECD) systems have rapidly expanded throughout the world (Cole and Maltzman, 1969). This volume demonstrates the application of developmental psychology to education. This field concentrates on education and children's development in various environments, including third world countries (Vargas-Baron, 2013). SSAT, which is derived from the general and applied activity theory and is closely related to the cultural–historical theory of mental development of higher mental functions suggested by Vygotsky (1960, 1978), can contribute to the psychological aspects of study in this field (Bedny et al., 2012). SSAT does not consider a learner as an information processing system but rather as a subject who actively interacts with the situation by using various strategies in achieving the goal of activity. This is especially important for learning and development in various social–cultural environments. New data in SSAT are presented in this book. In this chapter, we give a general introduction into AT, AAT, and SSAT.

1.2 Study of Predictive Mechanisms in Activity Theory

The brain has the ability to not only reflect the real situation but also design the future situation. Thanks to this, the organism not only reflects the surrounding reality but also constructs the future in mental plane and provides adequacy of behavior or activity to the expected events. In psychology, the ability of the brain to predict future events is denoted by such terms as anticipation or expectation. In the English language, the words *anticipation* and *expectation* are used synonymously. In the psychological literature, the word expectation is preferred. For example, in the textbook *Introduction to Psychology*, Hilgard et al. (1979, p. 595) gave the following definition: "Expectation—an anticipation or prediction of future events based on past experience and presented stimuli."

For a long period of time, psychologists did not pay enough attention to the human ability to predict future events. The reason is that behaviorism dominated the first half of the twentieth century. The study of anticipation usually was limited to the analysis of classical conditioning as anticipatory learning (Rescorla, 1972; Zener, 1937). Expectation as an important psychological mechanism of behavior was considered by Tolman (1932). He contrasted the concept of purposive behavior to the traditional view of existing behaviorism. In the analysis of the triadic schema *sign–gestalt–expectation*, he paid attention not only to link the stimulus and rewording response. According to him, motivated and purposeful behavior is performed in a specific way in accordance with an existing cognitive map or a mental picture of the external environment. Expectation is critically important in developing such predictive mechanisms. In the second half of the twentieth century, the notion of expectation also linked to Miller et al.'s (1960) work that is based on the concept of human behavior, which functions in a similar way as a cybernetic device where the concept of conscious goal is missing. Bruner (1964) and Piaget (1952) started to pay attention to expectation in developmental psychology. In cognitive psychology, the problem of expectations also attracts attention (Lindsay and Norman, 1972). In the latter case, this problem was studied first of all in such area as recognition and attention. Different aspects of this problem under *expectation* were discussed in engineering psychology (Wickens and Hollands, 2000).

In the Russian language, the word *expectation* (*ozhidanie*) has a more narrow meaning than anticipation. One of the most important distinguishing characteristics of the concept of anticipation in AT from the concept of expectation is the fact according to which anticipation is closely related to the conscious goal of activity. Of course, the simplest forms of anticipation may occur at an unconscious level. In our opinion, this type of predictive mechanism should be designated as expectation. Such predictive mechanisms are more important for animals. Therefore, in Soviet psychology, the term anticipation has a broader meaning. Anticipation is understood as the ability of humans to predict the possible changes of the situation and the results of human behavior and actions, which in most cases includes consciousness.

The purpose of this chapter does not include a detailed analysis of the problem of expectation or anticipation in English publications. We want to draw attention to how the problem of anticipation has been studied in AT and primarily in AAT and SSAT. The problem of anticipation in AT has received considerable attention because the activity is considered as a goal-directed system. The goal can be seen as a form of anticipation. Moreover, the goal itself has an impact on the development of the process of anticipation. Anticipation is an important aspect in the formation of a mental model of the situation and in particular in the formation of those aspects that relate to the forecasting of the development of the work situation. Self-regulation of activity is always done with mechanisms of anticipation (Bedny et al., 2014). Work activity without prediction is impossible.

The problem of anticipation is considered not only in psychology. Such scientists as philosophers, physiologists, and others are involved in the study of this problem. The works of some outstanding Russian physiologists are important in AT. Anokhin (1962, 1969) introduced the concept of *afferent synthesis,* Bernshtein (1966, 1996) conceptualized this phenomena as *model of required future,* and Sokolov (1969) utilized the notion *the neural model of the stimulus.* All of them have similar meanings. All of them are associated with certain aspects of anticipation. We consider these mechanisms to further analyze the process of self-regulation at the neurophysiological level (Bedny et al., 2014).

Studies of these authors have contributed to a number of actual psychological researches on the psychological level of anticipation. These works had a particularly big impact on the study of anticipation in AAT. Here we can distinguish the work of such scientists as Zavalova and Ponomarenko (1980), Lomov and Surkov (1980), Zabrodin and Chernishev (1977), and others.

Lomov and Surkov introduced the idea of multilevel structure of anticipation. Depending on the specific tasks performed by the operator, the authors identified subsensory, sensory-motor, perceptual, and verbal-thinking levels of anticipation. The proposed levels have a certain conventional meaning. However, their consideration has a practical value. From the standpoint of the theory of self-regulation in SSAT, it is not so much on the levels of anticipation but simply on various aspects or type of anticipation. These types of anticipation are determined by dominant mental processes that are involved in anticipation. Different types of anticipation involve the use of different strategies of activity. Thanks to the process of self-regulation, various strategies of activity in various manners include different psychological mechanisms in the process of anticipation. The development of strategies of task performance is impossible without the prediction of future events. For example, while a subject performed a psychophysics task, it was discovered that he or she utilized the criterion of the decision, which can be changed. This criterion of decision is always associated with the prediction of success. Otherwise, the choice of a decision criterion becomes meaningless. In this task, we can talk about the dominance of perceptual mechanisms of self-regulation. According to Lomov and Surkov (1980), this task is important in the perceptual level of anticipation. Such concept as situation awareness (SA) also includes anticipation components. SA includes such function as projection of the status of situation in the near future (Endsley, 2000). It should be noted that transition from one type or level of anticipation to another is not a linear process. This is a complex process, which is based on feedforward and feedback connections between the different mechanisms of self-regulation. Each mechanism of self-regulation should be considered as various combinations of cognitive processes. Thanks to self-regulation, the type of anticipation changes, is corrected, and becomes more specific.

In order to more accurately analyze the process of anticipation, we introduce some corrections in suggested by Lomov and Surkov's (1980) principles of classification of anticipation. When we consider the earlier level or type of anticipation, we have to note that these levels are not strictly isolated. The considered type indicates what psychological functions take a leading role in the formation of anticipation. For example, in some cases, the leading role in anticipation is visual perception, and in other cases, memory or thinking. Another distinguishing feature of our classification of the types of prediction mechanisms is that we preserve the term expectation. We distinguish two types of not fully comprehended or unconscious expectation: the subsensory type and the set.

The subsensory type of expectation is the level of unconscious neuromuscular tuning of movements. This type of expectation provides tonic and tonic-postural effects associated with the implementation of the forthcoming motor actions and movements. This mechanism of prediction is associated with the functioning of nervous and neuromuscular systems. Expectation of this type takes place in microintervals of time. It is included in the regulation of unconscious movements. This predictive mechanism for the first time was noticed by Bernshtein (1967). It also includes ideomotor acts.

We also relate the set to the specific type of expectation. It operates as a tendency that defines the directness of human activity in order to achieve involuntarily or voluntarily formed goals. The concept of the set was introduced in 1930 by Uznadze (1961) in the former Soviet Union. A stimulus situation interacts with prestimulus motivation, and as a result, a set emerges. The subject may or may not be aware that the set is formed. There is a tendency to react in a specific way to a particular situation. From this, behavior cannot be inferred directly from a stimulus situation. The individual's expectations are included in the broader context of anticipatory mechanisms. A system of expectation and anticipation is linked with the human set and goal of activity. When the goal of activity is altered, expectation is also changed. Initially, an unconscious set can later be transformed into a conscious goal and vice versa. This process is tightly connected with mechanisms of self-regulation (Bedny and Karwowski, 2007).

There are more complex types of prediction, which we distinguished from expectation, that we designate as anticipation. For the first type of anticipation, we can relate the sensory-motor type.

It plays an important role in the performance of sensory-motor reactions, reactions of tracking a moving object, and so on. In this case, a system of visual-motor, auditory-motor, and other types of coordinative predictive systems is developed. At this level, the type of anticipation is an important context of activity. The sensory-perceptual type of anticipation is a more sophisticated one. Pribram (1971) showed that the distortion of the temporal part of the brain violates the perception of context. Behavior occurs only in the present that has no past and future. The sensory-perceptual type of anticipation strictly depends on the specific task. The effectiveness of this type of anticipation depends on the full and timely flow of information to our senses. If the same level of control of activity is transferred from the sensory level to a higher level, for example, imaginative, the accuracy of anticipation is reduced. This was discovered during analyses of various sensory-motor tasks and in particular in the analysis of the tracking tasks.

The next type is perceptual anticipation. This type of anticipation is characterized by the further complication of the integration of mental processes that shape our anticipation. Images that are stored primarily in our short-term memory are important for this type of anticipation. They are activated and modified as the process of perception of objects, and some of the images are constructed directly as perception. For example, thanks to this kind of anticipation, more effective strategies can be developed for finding information in the perceptual field, called primary images.

An example of a perceptual type of anticipation is a situation that requires visual extrapolation of the trajectory of the objects and determining the point in space when these objects meet. Usually, in such situations, information about the trajectories of moving objects deteriorates or disappears completely. The subject must mentally extrapolate possible movements of objects. In performing this type of task, there is complex interaction of perceptual, mnemonic, and cognitive processes. However, the leading role in carrying out such task belongs to perceptual information.

In some cases, such types of anticipation require the development of perceptual hypotheses.

The spatial-temporal type of anticipation is important in the process of control of various types of transportation systems. In these types of task, the operator is on a moving system and he or she is involved in the prevention of a collision. A practical example demonstrates the importance of such type of anticipation. The department of occupational safety Ministry of Merchant Marine asked one of the authors of the book and his colleague to conduct an expert analysis on a major catastrophic shipping accident involving a collision between a freighter and passenger ship in the Black Sea, which took place in 1984 (Bedny and Zelenin, 1988). As a result of this collision, 400 people perished. This accident had multiple causes, including a social one. However, we would like to draw attention to a cause that has, in our opinion, a special meaning. Nobody paid attention to the analysis of this aspect of catastrophic shipping accident. We paid attention primarily on the behavior of the captain of the freighter who was responsible for the considered accident. Examination showed that he was an experienced seaman. He was highly motivated to quickly arrive at the seaport. It was discovered during the analysis that the captain's mental model of situation was inadequately developed with respect to balancing space and time components of a task. The distance of 4 miles between the ships was perceived by the captain as substantial, despite the fact that the physics of vessel movement required approximately 1.5 miles to halt the ship's movements. This means that the spatial-temporal type of anticipation was an important component of the mental model of task. In forecasting a possible collision, it is important to convey the distance between ships into existing reserves of time to prevent collision. Such skills were not developed enough during training on marine simulators.

Anticipation that involves imagination suggests using the secondary types of images, which are stored not only in the short term but also in the long term. At this level, complex images of a situation outside of direct perception are constructed. Such complex images are performed anticipatory functions. Perceptual images are primary images that arise due to the direct impact of stimuli on the sense organs. Secondary images can be developed outside of direct perception. The perceptual type of anticipation works closely with the imaginative type of anticipation. In these types of anticipation, the role of the goal is significantly increased. The interaction of these types of anticipation is essential for the formation of a dynamic mental model of the situation and the conceptual model of activity. Imaginative models help to predict the development of the situation in the future. The leading role in imaginative anticipation belongs to the visual system. The image may not match with the display input and the operator may lose awareness of the real situation. An adequate image allows the operator to see mentally beyond the equipment display into the actual situation. The anticipative imagination can be analyzed from two aspects, as a tool for the comprehension of reality (cognitive function) and as a regulator of activity (regulative function). Therefore, the image is not only a specific stage of cognition but also a regulator of subject actions (Lomov and Surkov, 1980).

The considered type of anticipation is important in deciphering tasks and reading topographic maps. In this type of task, visual images play the leading role. Based on them, the cipher officer performs all the necessary mental operations. These images are of two types: The first type of images, called reproductive images (*predstavlenie*), is based on the mechanisms of memory and the second type of images, called creative images (*voobrazhenie*), involves creative processes. These images are created by subject. The transformation from the first to the second type is important in considered tasks. Anticipative operations are involved in such tasks.

Rubakhin (1974) showed that in deciphering tasks these images may be constructed as the image of a real area (pictures) or image schema. The first type of images usually covers small areas of the land and the other covers large areas. The first type of images is more detailed and the latter has a more general character. The relationship between these images depends on the stage of task performance.

The relationship between the part and the whole in imaginative process is another criterion for the classification of images. There are isolated or elementary images, combined images that consist of elementary images, and complex or holistic images.

One of the essential systems of mental operations with images includes transferring information about a previous familiar situation to a newly considered current situation when performing a deciphering task. Such transformation includes extrapolative predictive operations that are based on the past experience. In the process of gaining experience, logical and calculative operations are carried out with the support of images that are automated and the subject is no longer aware of it. It becomes possible to visually assess the relationship between the imagined and the real object. Understanding the previously described process has been used in developing training for deciphering task performance.

Thinking is the most important mechanism for foresight. This is anticipation at the verbal-logical level. This type of anticipation includes a system of verbal-logical actions. Thanks to such logical actions, the subject can promote and verify various hypotheses.

Verbal-logical actions are combined with imaginative actions. As a result, the dynamic mental and conceptual models become more specific and precise (Bedny and Karwowski, 2007; Bedny and Meister, 1997). In general, such anticipation acquires a leading role in the formation of goals, planning, and programming of activity. In the process of activity, a person can manipulate the image of the object of activity and the image of the situation. The process of verbalized descriptions of imaginative components of tasks helps to transfer the unconscious aspects of anticipation into the conscious level. The process of goal formation is the most important type of anticipation. We consider this type of anticipation in the following chapter.

Anticipation at the verbal-logical level involves manipulation of images, among which are particularly important visual images. This type of anticipation plays a leading role when the operator interacts with the complex information presented by displays, or information presented on the screen of a computer. For example, information on the computer screen should match the dynamic mental model of the situation. Manipulation of data on the screen, extraction of information from human memory, and comparison of data with the extracted information are performed through conscious actions or unconscious operations. Hence, effective interaction of the user with the computer depends on his or her timely preparation to interpret the information and promote hypotheses about appropriate solutions. Information on the computer screen is extremely dynamic. Therefore, it is important for HCI to provide information on the possible consequences of intended actions.

In our analysis of the problem of anticipation, we would like to draw attention to the following factors. It is more correct to speak not about anticipatory levels but about the types of anticipation. Usually each type of anticipation includes several different cognitive processes. Their structural relationship in various types of anticipation is different. The type of anticipation is determined by the leading or dominant role of cognition in the anticipatory process. All types of anticipation are closely interrelated and one type of anticipation can transform into another type. Anticipation appears as a particular aspect of task performance. It is important in the formation of a mental model of the task, the conceptual model, goals, SA, and so on.

An analysis of the earlier material makes it possible to distinguish two types of forecasting: the one identified as *expectation* and the second as *anticipation*.

To expectations we include predictive mechanisms that are not connected with the conscious goal of activity and occur in a short period of time. Such types of prediction are not clearly understood by the subject. To anticipation we attribute such predictive mechanisms, which are included in a goal-directed activity, and predictions are understood by the subject. Each type of prediction includes a certain structural combination of mental processes, under the leadership of one of them.

There are two types of expectations: the subsensory type and the set. Anticipation can be of four types: sensory-perceptual, perceptual, imaginative, and verbal-logical. We can present predictive mechanisms of human activity in the following way:

1. Types of expectations: (a) subsensory and (b) set
2. Types of anticipations: (a) sensory-perceptual, (b) perceptual, (c) imaginative, and (d) verbal-logical

The more complex types of anticipation are the last two (imaginative and verbal-logical). These two types of anticipation are involved in more broad areas of human work. The imaginative type plays an important role in forming the ability to anticipate potential actions. The verbal-logical or thinking type of anticipation helps to plan activity as a whole, by integrating other types of anticipations. The involvement of the thinking process in the manipulation of images leads to the creation of new images and the development of new versions of hypothesis and anticipations. The imaginative and logical components of anticipation have a complex interrelationship. For example, the choice of action can be based on logic, but the method of performance of actions may be based on imagination. Very often the operator can receive information in the conceptual-logical form, which is then transformed into the imaginative form. The image is not static but dynamic. By performing imaginative and thinking actions, the subject transforms the image and therefore the mental model of the situation. Very often in any given moment, only some components of the mental model can be comprehended by the operator. The components of the mental model that are unconscious in certain conditions may become conscious in other conditions. Consideration of anticipation as an independent psychological problem can help further study the issue and, therefore, more effectively use the obtained data in practice.

Anticipation is an active process. It assumes, in some cases, very complex mental and practical activities, analysis of results of activity, the formulation and reassessment of verbalized and nonverbalized hypotheses, and so on. Various strategies of gnostic explorative activity are the bases for promoting hypotheses. Exploration can be a combination of internal or cognitive and external or behavioral actions. The ratio of these two kinds of actions in the activity as a whole can vary. Sometimes explorative activity can be purely mental. Based on the feedback, a person can evaluate the formulated hypothesis. Hence, exploration is an example of a self-regulative process (Bedny and Meister, 1997).

From the activity self-regulation analysis perspective, not only the cognitive but also emotionally evaluative and motivational aspects of activity are involved in the formulation of a hypothesis and in anticipation processes in general. A self-regulative process includes conscious and unconscious levels of activity regulation (Bedny and Karwowski, 2007). The unconscious level of self-regulation is involved in the formulation and selection of hypotheses and *set*. A hypothesis can be considered as a probabilistic model of possible solution. It can include potential goals of activity and mental representation of possible development of events. Such hypotheses are called orienting hypotheses. There are also instrumental hypotheses that are responsible for forecasting the methods of goal attainment.

1.3 Goal Concept in Systemic-Structural Activity Theory

The concept of a *goal* is used in various fields of science and practice. This concept is used in psychology, cybernetics, engineering, management, philosophy, and so on. It can be used in the analysis of living, nonliving, and organizational systems. In psychology, the goal concept is utilized for analyzing the behavior of humans and animals. It is interpreted differently in various fields of psychology and disciplines of science.

In this section, we will consider this concept primarily from the perspective of systemic-structural AT. Since the category of activity is considered as a specific form of behavior that is uniquely human, the concept of goal described in this theory applies only to humans. A goal is one of the most important anticipatory mechanisms at the psychological level of activity regulation. It predicts outcomes of our own activity. A goal is a psychological model of a desired future that results from our own activities. During an ongoing activity, a goal can become more specific and corrected if needed. If the subject totally changes the goal, it will be a new activity. In SSAT, the human goal is the cognitive component of activity, which includes conscious components. Without awareness of the goal, there is no goal in human activity. At least partly the human goal should be verbalized.

The goal can be understood as a model of desired future. It represents the most complex form of anticipation and includes verbal-logical and imaginative components. The relationship between these components varies depending on the specific goal. When we perform the activity based on an externally existing specific sample, the goal includes perceptual components. The goal of the activity may be modified based on the analysis of activity of other people. In this case, the subject forms his or her own goal based on the analysis of his own activity and the activity of others. Evaluation of activity of others and his or her own activity is the source of information for correcting the subject's own goal. In these complex processes of goal formation, it is necessary to understand the goals and motives of other people.

Goal-directed activity is formed in labor. This becomes obvious by analyzing human labor in a historical perspective. Rubinshtein (1959) wrote that for the work activity goal, directness of actions is the main manifestation of human consciousness, which is fundamentally different from unconscious *instinctive* in its basic animal behavior. A person interacts with an external situation in two different ways. The first way is a reactive behavior, and the second one involves goal-directed human activity. The second way to interact with the situation is actualization of purposeful, consciously, and voluntarily performed activity. This human activity should be distinguished from human reactive behavior. Reactive behavior can be considered as a lower level of activity performance.

Involuntary reactions and voluntary, goal-directed actions are basic components of human work. Highly automated actions prompt reactive behavior. For example, an alarm sounds at a nuclear control station when certain parameters exceed the limit. This is a signal for the operator to take a specific, highly automated action. However, this is still a meaningful and purposeful action (not a reaction) because it has a corresponding specific goal or desired future result. Elimination of the task's goal and goals of separate actions during task performance reduces human work activity to a chain of reactions or responses. Such behavior is passive and can be triggered by outside stimulation. Each new stimulus initiates a new reaction. Generally, such behavior entirely depends on external, environmental stimulation. AT psychologists that study human work cannot agree with such interpretation of human work behavior.

The goal in AT is a conscious image or logical representation of a desired future result. An image of a future result, when a subject is not directly involved in achieving this result, is not a goal of activity. An imaginative, future result emerges as a goal only when it is a consequence of the subject's own activity. For example, a student may know that an excellent grade requires 4–5 h of preparation. Such knowledge does not create a goal unless the student is motivated to achieve the desired grade. Hence, only when a conscious image of a desired future result joins with motivation and active student preparation for the exam is this future result transferred into a goal of activity.

We also need to distinguish between the *overall or terminal goal of the task* and *partial or intermittent goals of actions and subgoals of a task.* In AT, there are energetic and informational (cognitive) components of activity that are interdependent but still different. A goal is the cognitive component of activity. Motives or motivation is the energetic component of activity.

A goal cannot be presented to the subject in a ready form but rather as an objective requirement of the task. However, these requirements should be conscious and interpreted by the subject. At the next stage, these requirements should be compared with the past experience and the motivational state, which leads to the goal acceptance process. A subjectively accepted goal does not always match the objectively presented goal (requirements). Moreover, very often subjects can formulate the goal independently. As it can be seen, a goal always assumes some stage of activity, which requires interpretation and acceptance of the goal. So we can conclude that the goal does not exist in a ready form for the subject and cannot be considered simply as an end state to which human behavior is directed.

A goal is understood differently in various fields of psychology. For us first of all, it is important to know how a goal is understood in industrial and organizational psychology (I/O psychology). Therefore, we consider understanding the goal in this applied area of psychology, which studies human work.

For example, Austin and Vancouver (1996, p. 338) defined goal in the following manner:

> We define *goals* as internal representations of desired states, where states are broadly construed as outcomes, events, or processes. Internally represented desired states range from biological set points for internal processes (e.g., body temperature) to complex cognitive depictions of desired outcomes (e.g., career success).

The reasonableness of such definition of the goal is explained in the following way:

> Using this broad definition, we attempt to show that part of the diversity of goal-based hypotheses and vocabulary can be understood more frugally using common concepts.

The *frugality* and psychological factors are different criteria of justification of psychological terms. It is clear that the human goal, which includes the realization of what the subject wants to reach, has nothing to do with this definition. This definition of goal contradicts not only its understanding in AT. This definition is contrary to the understanding of the goal in personality and social psychology (see, e.g., Pervin, 1989). In his later works, Vancouver attempts to introduce the classification of human goals.

Vancouver (2005, p. 329) insists that there are two types of goals. The first type is a goal for the perceptual unit of behavior and the second one is for the internal unit of behavior. For the internal unit, a goal is simply the desired level of errors. In this case, the author's arguments are contrary to all existing data in psychology. A goal cannot be considered as the desired level of errors. There are no external perceptual and internal goals. Perceptual activity cannot be strictly isolated from other kinds of mental activity and from the emotional-motivational process. Moreover, external behavior is closely linked with mental and emotional-motivational processes.

The author also introduced another type of goal. Trying to link the goal with self-regulation, he introduced attainment goal and maintenance goal. The notion of *maintenance goal* from his point of view helps to connect the concept of goal with the concept of self-regulation. The author misinterprets the concept of self-regulation of human activity by reducing the process of self-regulation to elimination of deviations from a so-called maintenance goal. Understanding self-regulation as a process of elimination from *maintenance goal* is a homeostatic principle of self-regulation. However, human activity is not limited to the elimination of errors that deviate from the standard. Such simple tasks are usually accomplished by technical systems. Typically, if a situation deviates from acceptable limits, the performer formulates a new goal and therefore a new task, which help to eliminate the deviation. At the next stage, he or she performs a logical system of actions aimed at achieving this new goal. Thus, the introduction of *maintenance goal* is totally unfounded. Moreover, self-regulation cannot be reduced to elimination of errors. According to Vancouver (2005), if a subject makes an error, it gives him or her an opportunity to eliminate it. Then another error is made and corrected. Human behavior cannot be reduced to a process of error elimination or moving from one error to another. In ergonomics, there are such notions as error, mistake, failure, range of tolerance, acceptable level of deviation, and so on. Vancouver demonstrates his complete misunderstanding of the concept of error in psychology. There are a lot of data in psychological literature about errors (see, e.g., Bedny and Mester, 1997; Kirwan, 1994; Norman, 1981; Reason, 1990; Senders and Moray, 1991).

To substantiate his arguments and theories, Vancouver utilized the following hypothetical example (Vancouver, 2005, p. 305). In maintenance context, a widget maker performs the following task. He or she has to monitor the state of the shelves in his store. He has a goal to keep the shelf full. When customers purchase widgets, the widget maker must make more, but only enough, to fill the shelf. So the customers are considered to be a source of disturbance of the variable (state of the shelf). A customer disturbs the variable and produces an error. In the attainment context, the widget maker has a goal of producing the required number of widgets to fill the shelf and to keep it full. Each workday begins anew with zero widgets made and ends when the goal is reached. Further, the author writes (Vancouver, 2005, p. 315): "as customers purchase widgets, an error is created between the goal and the widget maker's perception of the state of the shelf."

The presented ridiculed example demonstrates that the author does not understand the meaning of production process, work process, task, errors, etc. First of all, we need to emphasize that a widget maker performs incompatible functions. When a worker produces a widget, he or she can perform a number of production tasks. Transportation also can include a number of tasks and take time. A customer cannot wait for the completion of the production process. The same person cannot be responsible for production, storing, and selling. Customers do not disturb businesses. The main goal of the work process is to sell a product, not to keep shelves full. In this example, a widget maker has multiple goals that are not compatible with time. Moreover, it is not reasonable to fill the shelves after a sale of each widget. Only when the number of widgets approach a minimum required quantity does a worker formulate a task (therefore a goal of the task) to go into the production room and bring the widgets to fill the shelf. When the number of widgets becomes lower than the minimum required, resulting in the inability to timely serve the customer, such events can be considered as errors. In other situations, a lower number of widgets can be considered as a permissible level of deviation. Bringing the widgets into the shelves is a particular stage of the work process that might include a number of tasks. If there are a number of tasks, hence there are a number of tasks' goals. Each task includes a number of cognitive and behavioral actions that also have their goals. Hence, self-regulation cannot be considered as elimination of so-called disturbance and errors. The self-regulation process allows not only correction of errors but also their prediction and prevention. Self-regulation happens even when there are no disturbance and errors as considered in the previously described example. Vancouver reduces the self-regulative process to elimination of errors, which is the result of disturbances. Our activity is a self-regulative system. Self-regulation is a complex process that regulates the entire activity and the term *maintenance goal* is not an accurate one. Disturbances include danger, unanticipated events, and emergencies. Subjects have to improvise and adapt to the contingency of such disturbances. Because of the disturbances, the self-regulation process becomes more complex and strategies of task performance change. There are strategies utilized in normal work conditions, in dangerous situations, or in other disturbances and transitory strategies where a subject transfers from an existing strategy to a new one. The process of self-regulation is the foundation for all these strategies, which always involves goal formation. We will consider some examples of self-regulation in a pilot's activity during the performance of a variety of tasks in emergency conditions in other chapters.

Analysis of Vancouver's publications demonstrates that there is currently no clear understanding of goal, task, self-regulation, and other important concepts that are necessary for task analysis. I/O psychologists who study human work cannot use such primitive examples even in their theoretical discussions. One would not find this level of examples in ergonomics where such specialists as psychologists, engineers, computer scientists, etc., work together.

It should be noted that in AT, the goal of a desired future result of performance can be represented as an image and as verbal or symbolic descriptions. The goal also has various levels of specificity or precision and can be clarified and become more precise in the course of activity. However, if the goal is completely changed, the achievement of this new goal will mean that the subject was involved in a new activity. The image of the future result becomes the goal of the task only during interaction with the motives of activity, which determines directness of the activity to achieve the goal. Depending on the motives with which the goal is connected, it acquires a different personal sense. The goal acts as a cognitive component of activity and motivation as an energetic component. In view of the fact that the activity is often polymotivated, it is possible that the subject relates to the same goal differently in various circumstances. For example, if there are conflicts between positive and negative motives in the process of approaching the goal, the acceptance or the formation of goals and hence the task performance in general is conducted in a conflicting motivational background. This conflict of motives can be changed by the process of goal attainment.

In such areas as cognitive psychology, social psychology, and motivation theories, the notion of a goal is very amorphous and undetermined. The goal emerges as a motive, as a cognitive entity, as a characteristic of personality, as state of the system, etc.

In AT, the concept of goals is closely linked to the concept of task. The goal of the task determines the integrity of activity during task performance. The goals of individual actions during task performance are of particular importance in the analysis of individual actions and in the formation of a task performance program. Awareness of the goal of the task as one of its most important characteristics is ignored outside of AT. All these drawbacks are eliminated by goal interpretation in AT. The goal of the task and its relation with motivation is central to the task analysis in AT.

It is also important to consider the relationship between the goal and consciousness. In various areas of psychology, the goal is seen as a conscious or unconscious element of human behavior or activity. For example, Austin and Vancouver (1996) wrote that goals are not limited to a conscious level. Indeed, if we are talking about low-level goals, which are specific to biological systems, these goals can be unconscious. However, in AT, the human goal is always conscious to some degree. This is not to deny the existence of the low-level unconscious anticipatory mechanisms of behavior. But in psychology, there are other terms for low-level goals such as purpose (Tolman, 1932), anticipatory results (Anokhin, 1969), required future (Bernshtein, 1967), and neural model of a stimulus (Sokolov, 1969). These unconscious anticipatory mechanisms, as well as conscious predictive mechanisms (human goals), manifest themselves at different levels of activity performance. All of these predictive mechanisms direct human activity. However, only human beings reach the more complex conscious level of reflection of the desired future result during their own activities. This level of reflection always ends with conscious goal.

In AT, it is possible to distinguish various types of goals that may have a different organization and relationship. For example, each task has its final goal, the achievement of which is the completion of the task. However, an individual needs to perform the logical sequence of actions that have their own goals in order to achieve the goal of the task. Goals of actions often are formed involuntarily. They can be conscious within a short period of time and quickly forgotten. The goal of the task can be formulated more consciously and stored in memory for a longer time. There are proximate and distal goals. The proximate goals can be achieved in a relatively short period of time. The distal goals becomes a long-term one. Progress toward a distal goal requires achievement of a number of intermediate goals. Goals may have various levels of difficulty and significance.

There are also potential and actual goals. In analyzing the goals that are connected with past experience, it is useful to distinguish potential and actual goals. Potential goal is not a real goal. Potential goals are information that are kept in memory and associated with existing needs. An actual goal is actualized in memory thanks to a higher level of needs potentially associated with information that can become a goal. Needs might potentially change over time due to the activation of associative memory connections (Zarakovsky and Pavlov, 1987). The external situation can trigger the activation of these associations. Potential goals are not conscious. If potential needs exceed the certain intensity threshold, the potential goal is transformed into an actual goal. Such goal formation is unconscious, and after formation of the goal, the performance program is triggered almost automatically. The formation of a required level of intensity of potential needs and the level of intensity of the motivation associated with it also depends on the nonconscious feeling of the importance of the situation, the feeling of danger, and the subjective assessment of the difficulty of the attainment of a future possible, not clearly defined result (Bedny, 2006). The more significant such feelings are in general, the higher the intensity of the need-motivational components of activity.

A described method of goal formation should be distinguished from a voluntarily one that is often associated with a willing process that involves goal formation. An involuntary goal formation process is more typical for the formation of the goals of separate actions and applies especially to habitual actions. If we are talking about a task's goal, such a goal is often formed arbitrarily. Voluntary goal formation is particularly important to the study of human work. A subject accepts or formulates a goal of task based on analysis of the situation, existing requirements, guidelines, and past experience. When considering the process of goal acceptance or formation, it is essential to evaluate goal significance (Bedny and Karwowski, 2006, 2007). In extreme situations, the role of volitional processes is elevated. For example, in case of danger, the operator can utilize volitional efforts to suppress the motive to escaping the danger and to increase the importance of social motives aimed at suppressing fear and rescuing people. In all these examples, such concepts as significance, motivation, difficulty, etc., are considered as functional mechanisms of self-regulation (Bedny et al., 2014).

Many scientists are missing the fact that the goals of the activity and behavior in general have not only hierarchical but also logical organization. Most goals have no subordinated relations with the other goals. Their consistent achievement may be defined arbitrarily by the subject. Austin and Vancouver (1996) reduce behavior and relationship between goals only to the hierarchical organization. Simon (1999) introduced the concept of *span of control*. If the level of goals' subordination exceeds the span of control, such hierarchical system has limitations in being efficiently regulated. Moreover, Austin and Vancouver use term *low-level goals*. However, in AT, they are not considered as goals because these goals are not conscious ones.

Since the goal is interpreted in different ways in various fields of psychology and directions of sciences, this concept is utilized by specialists in work psychology and ergonomics inadequately. Some scientists suggest eliminating this concept completely.

The ambiguity of interpretation of the goal in psychology and the impossibility of its effective use in applied studies lead to the fact that some authors try to eliminate the goal concept from the research or replace it with a more appropriate concept.

For example, Diaper (2004) attempts to substitute the concept of goal with what he calls the forward scenario simulation (FSS) process. This author mixed goal with motives because goal in cognitive psychology is not distinguished from motives. In Diaper's theoretical substantiation of the concept of goal by the FSS process, he wrote that this approach is a multitheological versus existing single-theological approach that utilizes a single desired future state of the system. According to SSAT, the author mixes the anticipatory stage of activity when a subject can formulate multiple potential goals and hypotheses about the state of the system with the final stage of a goal formation process when a subject chooses one goal from a number of potential goals. A goal formation process is followed by the development of a number of instrumental hypotheses and selection of one of them. The goal sometimes is formulated in ambiguous terms. For example, what happens if I perform this particular course of actions? A subject can check his or her instrumental hypothesis mentally or practically, abandon this hypothesis, and formulate a new one. If the subject formulates a new goal of task, it leads to a completely new task. Moreover, the user can formulate not only the final goal of the task but also a number of intermittent goals that depend on a selected strategy of task performance, which is a subgalling process of breaking down the overall goal of the task into smaller goal-directed steps. This process to some degree is similar to the means–ends analysis that was introduced by Newell and Simon (1972).

In SSAT, it is treated as a goal-directed self-regulation process. In computer-based task, the final goal of the task very often is formulated by a subject and the goals and strategies of task performance are not defined in advance. Subjects progress from the anticipatory stage of activity to the executive stage of activity while the overall goal of the task gradually becomes clearer. Anticipatory and executive stages of activity include complex exploration of possible strategies of task performance. The user can mentally operate with various elements of the task presented on the screen and camper and combine these data with information in memory. As a result, the same external situation is constantly changing in the user's mind. This phenomenon has been called gnostic dynamic (Pushkin, 1978). The self-regulation process with its mental transformation, evaluation, and correction of the situation is the basis of gnostic dynamics (Bedny and Karwowski, 2007). This process utilizes not only conscious mental actions but also unconscious mental operations. There is a complex relationship between these conscious and unconscious components of activity that can sometimes be transformed into each other (Bedny and Meister, 1997) when a user eventually selects a subjectively accepted goal and possible strategies of its achievement. Hence, Diaper's (2004, p. 17) unrelated goals are simply hypothetical motivational factors.

In applied cognitive psychology, which is a base for engineering psychology and ergonomics, the concept of goal is practically unused. For example, in the basic textbook of engineering psychology written by Wickens and Holland (2000), we only discover the term *intended goal* when authors considered feedback in modelling human information processing stages.

We can formulate the following basic conclusion. Goal is the basic concept of activity. Goal is a conscious image and logical representation of a desired future result. Our activity during task performance is always goal directed. A goal integrates all components of activity into a system. A goal cannot be presented to the subject in a ready form but rather as an objective requirement of the task. However, these requirements should be conscious and interpreted by the subject. At the next stage, these requirements should be compared with the past experience and the motivational state, which leads to the goal acceptance process. A subjectively accepted goal does not always match the objectively presented goal (requirements). Moreover, very often subjects can formulate the goal independently. As it can be seen, the goal always assumes some stage of activity, which requires interpretation and acceptance of the goal. So we can conclude that the goal does not exist in a ready form for the subject and cannot be considered simply as an end state to which human behavior is directed.

1.4 Goal and Motivation in Activity Theory

As we discussed in the previous section, an image or verbal representation of a desired future result does not necessary constitute a goal. The desired future result emerges as a goal only when it is joined with motivation and the subject is involved in the activity for achieving this result. For example, a student has a desire to obtain a good grade. Such desire does not create a goal. This desire is not a motive. Only when the student starts to perform a number of tasks will he or she eventually be able to reach the final goal. Therefore, this imaginative result is transformed into a goal only when a desired future result is coupled with motivation and desires, wishes, etc., become motives. It becomes clear that the goal does not exist apart from the motivation. There are no unmotivated goals. Goals and motives are interdependent, but not the same. They create the vector *motive* → *goal*. This is critically an important mechanism of goal-directed activity.

We view the vector *motive* → *goal* as a source of active and purposeful human activity. The goal cannot exist outside of this vector. However, this does not mean that the goal and motive are the same. Each component has its own specificity in human activity. The vector can be viewed as a two-component subsystem of activity. The vector defines not only the course of activity but also the energetic characteristic of this orientation. With the ability to maintain performance over time, the resistance confounding factor is determined by this vector.

The concept of *spatial vector* or simply *vector* is mainly used in physics and engineering to represent directed quantities. Many physical quantities and direction can be represented by the length and direction of an arrow. Their magnitude and direction can be added to other vectors according to vector algebra. Of course it is important to distinguish understanding of the vector in psychology from its understanding in physics or engineering. Similarly to vector in physics, psychological vector has intensity and direction. Moreover, motivation can be metaphorically represented as a resultant vector. So if we assume that human activity is determined by the interaction of two motives that are in the opposite direction, there is a *conflict of motives*. Directness toward a goal or avoidance of it will depend on the intensity of the appropriate motive. Naturally, the understanding of the vector in psychology does not have the exact meaning as in physics, where the vector has physical units of measurements. Motive and motivation in psychology are hypothetical constructs. Hence, the vector *motive* → *goal* is a qualitative psychological concept. It has only some similarity with the vector in physics. However, in principle, it can be used in psychophysical measurement methods by which it is possible to depict the psychological intensity of a vector. It is more difficult to precisely depict the psychological orientation of the vector *motive* → *goal* in relation to each other. Despite these limitations, the vector *motive* → *goal* pretty accurately reflects the essence of the phenomenon. It reflects the intensity of activity directness to achieve a certain goal and the ability to sustain this directness. In the absence of such a vector, goal directness of activity disappears. Therefore, the vector *motive* → *goal* should be considered as a basic concept of AT. Specifically, the vector *motive* → *goal* most clearly manifests the interaction of energetic and informational components of activity. We can outline the following critical points of this important concept: Motive and goal are not the same; motives are energetic and goal cognitive mechanisms; the more intensive the motive is, the stronger is the subject's desire to reach the goal; the vector *motive* → *goal* gives activity a goal-directed character and sustains this directness in time, etc.

In AT, the goal and the motives are considered as interrelated mechanisms. However, even in AT, where these two concepts have fundamental meanings, some leading scientists admitted mistakes in their analysis. For example, Leont'ev (1978) wrote about the possibility of a shift in the motives to the goal of activity. According to Leont'ev in this case, the motive and goal match. Leont'ev (1981) stated the when a person realizes a conscious motive, the goal might coincide with the motive. However, in this situation, the vector orientation of activity is lost. Lomov (1986, p. 173) also claims that under certain conditions, the motive slides into and coincides with the goal. However, this contradicts the notion of a vector. Such sliding of the motive into the goal leads to a situation when the vector is transformed into a point and activity goal directness is lost. Information and energy are interdependent but not the same. Why a person performs the action and what he or she wants to achieve during performance of the action are not the same. For example, a person wants to drink water. On the table, there are different tableware. A person moves his or her hand to the glass, not to the pot. In this situation, the person is keenly aware of his desire to drink water. The need for water was transformed into a conscious motive. In this situation, the goal of the motor action (grasp the glass) and conscious motive (I want the water) is not the same.

Freud (1916/1917) was the first one who introduced the concept of energy into psychological studies. According to Freud, people are complex energetic systems. Energy is necessary for the functioning of mental processes. The idea that energy is an important component of mental functioning has been borrowed from biological sciences and physics. At the same time, the interaction between energy and informational aspects of human functioning has not been described in detail in Freud's work. Today in psychology, such concepts as energy and information are clearly defined. Information processing is a cognitive function. The energy concept that derives from the neural system electrophysiological function has been utilized in studies of emotional-motivational aspects of behavior or activity. Although in AT these two concepts are distinguishable phenomenon, they are considered as tightly interconnected. There are several types of informational-energetic interconnections in activity (Vekker and Paley, 1971). The first type was shown in the psychophysical studies where it was discovered that an increase in the intensity of external stimuli gives rise to sensory qualities. Another type of interconnection is related to the reticular activating system of the brain, which plays an important role in controlling the state of arousal and awareness. For example, Kahneman's (1973) view on attention includes the concept of energy. This model demonstrates that the success of cognitive processes depends not only on the physical characteristics of information presented to the subject but also on the level of activation of the neural system. A third group of interconnection is linked to the emotional-motivational components of activity and to the specificity of information processing. These groups are interdependent. Currently, the cognitive approach dominates in ergonomics, when a person is considered as a pure information processing system that picks up information from an external situation and from memory, because the relationship between cognition and energetic aspects of activity is not sufficiently studied. However, interpretation of information depends on the emotional-motivational stage of a person. Moreover, information cannot be transmitted without energy.

Motivation is an intentional or inducing component of activity, which is tightly connected with human needs. A need is an internal state of individuals that is less than satisfactory and produces a feeling of a desire for something (Carver and Scheier, 1998). Some needs are biological in nature. Other needs such as achievement and power are secondary or psychological needs. Human needs are a result not just of biological evolution but also of an experience acquired through human culture.

If needs are connected with the goal, they become motives of activity. Motives are an inducing force that catalyzes the person's desire to reach the activity goal. So needs operate through another construct called motives. Motives derive from needs but they are closer to our activity or behavior. For example, a need for food as a physiological state can be transferred into a motivational state called hunger. This state is experienced cognitively and affectively. Motives are also influenced by external events that sometimes are called *press* (Murray, 1938). These external stimuli create a desire to obtain or avoid something. If somebody received recognition, this can trigger his or her own motives for further recognition. Hence, internal needs and external press can activate motives to engage in a particular kind of activity to achieve a conscious goal. The relationship between internal needs, external press, motives, and goal is presented in Figure 1.1.

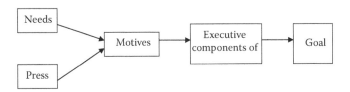

FIGURE 1.1
Relationship between needs, press, motives, and goal in the activity structure.

Motivation is another important construct that includes diverse motives that have a hierarchical organization. Some motives can be more important than others. Some motives can be conscious, semiconscious, or even unconscious. The relationship between motives is typically dynamic and can be modified during activity. The goal very often can be induced not by one but by several motives. Activity result can coincide with the goal or deviate from it and a subject has to correct his or her actions or activity strategy (Bedny et al., 2014).

Deviation of the result from the goal can be considered as an error if this deviation is outside of the subjective criteria of success that is not always the same as the objective one. Sometimes the activity result that deviates from its goal can be useful to the subject when it is a desired accessory result.

The principle of the functioning vector *motive* → *goal* can be summarized as follows. When a person formulates a goal, motivational tension is set up. Once the subject reaches the goal, existing motivational tension is relieved. Motivational tension is analogous to the forces that direct our activity toward the goal. Activity can be habitual and does not require significant motivational forces to reach a goal. A person can be unaware of these motivational forces. Motivational forces become capable of automatically activating human activity. A person can be aware of the goal of activity without being conscious of the motivational factors. A conscious goal-directed activity is a voluntary process. A latent or potential goal can be activated by a situation and arise immediately in our consciousness—which is the involuntarily goal formation process. In human activity, the goal formation process is more often a voluntary conscious process. In some cases, the conscious formulation of the goal can arise as an independent task. In contrast, automatic human behavior can be triggered outside of goal awareness. Motivational tendencies can be activated directly by the environment. As a result, automated behavior response can be triggered by the environmental situation without consciousness.

There is another aspect of motivation that is associated with the role of emotions in the motivational process. According to some authors (Zarakovsky and Pavlov, 1987), emotions reflect the relationship between our needs and the real or possible success in their satisfaction.

Motivation is distinct from needs, wishes, desires, intentions, etc., through their link with the goal of activity. Not only human needs but wishes, desires, intentions, etc., are transformed into motives when they are connected with a goal.

A person's motives can be divided into two groups: sense formative and situational (Leont'ev, 1871). Sense formative motives are relatively stable and determine a person's general motivational direction. Situational motives are connected with immediate ongoing activity and the performance of a specific task. As a result, situational motives are more flexible. The sense formative motives are connected with personality and may be more important in the selection of people for different jobs. The situational motives are involved in task performance. Therefore, they are more important in task analysis.

Motivation is always connected with emotions. Some scientists wrote that there is no clear-cut distinction between emotions and motivation and it is an unresolved issue in psychology. However, according to the activity approach, motivation is emotions plus the directness to a specific goal. Emotions can be positive and negative. However, even negative emotions cannot be considered as only disorganizational factors. In some situations, negative emotions can improve performance. Emotion, according to Simonov (1982), is a specific *currency* of the brain or a universal measure of utility. Our feelings (happiness, anger, outrage, etc.) present a measure of quality and level of our need in relation to a possibility of satisfying it.

Zarakovsky and Pavlov (1987) described different functions of emotions in activity regulation. The inducing component of emotions has only one function: to direct the person to achieve a conscious goal. The regulatory aspects of emotions have four functions: switching, reinforcing, compensation, and organization. Switching functions enable the person to concentrate on the activity most closely related to the goal, which has a more subjective value for that particular person. The reinforcing function provides rewards and, thus, reinforces the desired behavior. Compensative functions enable to change emotional tension and therefore to transfer into a higher level of activity regulation. However, this sometimes can result in activity disorganization. The organizational function of emotions contributes to the creation of orienting reaction and reflection of the mismatch between available and required ways of organizing performance.

A presented analysis of the relationship between the goal and motives demonstrates that actions during task performance can be successful and unsuccessful. Each action can be evaluated in relation to the goal of action and in relation to the goal of the task. Evaluation of actions or activity always includes emotional components or subjective significance of the obtained result. From the AT perspective, the task always includes a motivational component and there is no such thing as unmotivated task.

Some authors use terms such as *anti-goal* or *work avoidant goal* (Carver and Scheier, 2005; Pintrich and Zeidner, 2005; etc.). Such goals do not exist in AT. The subject may or may not accept the goal formulated by the instruction. Moreover, in response to the presented goal, a subject can formulate his or her own goal, which contradicts with the objectively presented instruction goal. In AT, the goal is always associated with motives and creates the vector *motives* → *goal*. This vector defines the direction of activity. This vector is directed at achieving the required goal. *Anti-goal*, *work avoidant goal*, etc., are just new goals, and thus, the new vector that determines the directions of activity is formed and the subject formulates a plan for a new type of activity or task. One goal can contradict another, but this is not an *anti-goal*, because such goal is created based on the same mechanisms.

Consider a hypothetical example. There is a family with children. The father is a gambler. Sometimes after receiving salary, father withdraws all the money from the ATM and spends it in the casino. At the same time,

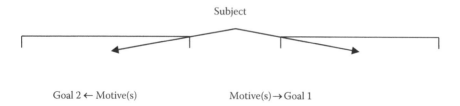

FIGURE 1.2
Scheme for the formation of two opposing goals.

he feels responsible for the family and tries to not take a lot from the ATM machine and keep the rest of the money for his family. In this situation, we can distinguish two types of motivations and two goals. Positive motivation creates the first goal vector *motive(s)* → *goal 1* (keeping money for the family). Negative motivation creates vector *motive(s)* → *goal 2*, which directs the father's activity to spend money in the casino. These two situations are represented by two vectors (see Figure 1.2).

We can see that two vectors have opposite directions. But here we have the presence of two exclusive goals. The second goal has its own vector and cannot be considered as an anti-goal. The mechanisms of goal formation of this vector are the same. The differences lie in the fact that the significance of goals for the subject, motivation aimed at achieving them, will be different.

The process of goal selection according Zarakovsky and Pavlov (1987), with some of our modification, can be presented in the following manner. The long-term memory contains units of memory (engrams) of the intentional type. They can be considered as vectors. We designate these vectors as *need* → *potential goal* engrams of memory. A set of such vectors have internal energy or inducing forces that depend on the intensity of needs. Similarly as in psychophysics, vectors of *need* → *potential goal* engrams may be of a higher or lower level than the existing absolute threshold according to their energetic characteristics. If *need* → *potential goal* engrams are above the existing energetic threshold, then the person starts to become aware of his or her feelings or desires, acting toward achieving the potential goal. There is a possibility that not one but several potential vectors may exceed the threshold level. Which of the vectors will dominate depends on the differences in the engrams' potential of the specific vector. With the aid of psychophysical methods, it is possible to evaluate the engrams' potential of such vectors. Without going into the psychophysical analysis of this issue, we just want to say the following. If at the same time the person has two divergent vectors, *need* → *potential goal*, and both reached sufficiently high activation levels, the subjects may be well aware of them. In such case, we can use the term vectors' *motive* → *potential goal*. Of course such difference in terminology has some arbitrariness. However, this difference can be useful. In the first case, the subject may be unconscious or may not be well aware of the vector. In the second case, we speak about conscious goals or even well-realized motives. If two *motive → potential goal* vectors have approximately similar and at the same time high-intensity potential engrams, there is conflict of motives. When one vector begins to dominate over the other, the person begins to perform work in accordance with the chosen vector. The selected vector, the *need → potential goal* vector, is transformed into the *motive → potential goal* vector, and at the final steps, the predominant vector is transformed into the vector *motive → goal*. The subject starts to act according to this vector.

Materials presented in Sections 1.3 and 1.4 demonstrate that the goal concept is interpreted in different ways in various fields of psychology. In all theories outside of AT, goal and motives are not distinguished. The goal is considered in unity with motivation. For example, representatives of personality and social psychology suppose that the more intensive the goal is, the more such goal pulls human behavior (Lee et al., 1989; Pervin, 1989). In action theory, scientists considered goal in a similar way (Frese and Zapf, 1994). Kleinback and Schmidt (1990) described the volitional process where the goal pushes human behavior.

In general AT and AAT, there is belief that goal and motive are different; however, sometimes motives can move toward the goal and they are integrated (Leont'ev 1972; Lomove, 1986). Then, the vector is transformed into the point and the goal directness of activity disappears. Therefore, according to SSAT, motive and goal cannot be integrated. They are different and at the same time interdependent mechanisms of activity regulation. Motives are energetic and goal is informational.

An incorrect understanding of the goal and its relation to motive when the goal is mixed with motive can be seen from the following hypothetical examples. If a thief spots a hundred dollar bill in somebody's pocket, his or her goal is to pull it out as gently as possible so the victim does not catch him or her. Obtaining this money is his motive but not the goal when he is performing the action. The more money in the victim's pocket and the more the thief needs the money, the more he is motivated to steal. *Obtaining the money* is a simply verbalized description of one possible motive in this example. Another motive may be, for example, the desire to demonstrate his *perfect skills to other craft—brother*. Refocusing attention of the thief on the fact that he will get the money, but not to the goal of performed actions, can lead to inaccuracies of actions and therefore undesirable consequences.

When a computer programmer works with an issue that requires fixing production database, his or her goal is to make the corrections without damaging the sensitive information, while the motive is to resolve the issue at hand. Such a task may be accompanied by great emotional tension, because the error can lead to very undesirable consequences. The more significant the task is, the more the subject is motivated to reach the goal.

The ambiguousness of the term goal in cognitive psychology results in a situation where some scientists suggest to eliminate this concept altogether. For example, a scientist in the field of HCI, Diaper utilized an example in which he attempted to prove his theoretical position according to which the human goal is a redundant concept. He wrote:

> A hierarchy of goals, as used in HTA, consists of multiple related goals, but a person can also perform an action on the basis of unrelated goals. Furthermore, unrelated goals that nonetheless motivate the same behavior cannot be simply prioritized in a list, because different goals have more or less motivational potency depending on their specific context.
>
> For example, a chemical plant operator's unrelated goals for closing a valve might be (1) to stop the vat temperature from rising, (2) to earn a salary, and (3) to avoid criticism from the plant manager. The first might concern the safety of a large number of people, the second is sociopsychological and might concern the operator's family responsibilities, and the third is personal and might concern the operator's self-esteem. These three goals correspond to different analysis perspectives—the sociological, the sociopsychological, and the personal psychological—and there are other possible perspectives as well. Furthermore, people might have different goals within a single perspective.
>
> **(Diaper, 2004, p. 17)**

It becomes clear that the concept of human goal is confusing for HCI practitioners. For example, what is unrelated goal? In AT, unrelated goal does not exist. It is not Diaper's fault that he confused the goal with motives. It can be explained by the fact that a concept of goal is not precisely defined in psychology and ergonomics and goal and motives are not distinguished. Our behavior is polymotivated and can include a number of motives. If the goal includes various motives, according to cognitive psychology, a person pursues not one but multiple *goals → motives* at any given time during the task performance. In SSAT, the goal is only an informational or cognitive component of human activity. In contrast, motives or motivation is an energetic, inducing aspect of activity. Therefore, in the earlier example, the goal of human activity is *closing a valve*. In contrast, the motives, which push the operator to close the valve, may be (1) to stop the vat temperature from rising, (2) to earn the salary, and (3) to avoid criticism from the plant manager. The earlier listed wishes or desires (which are described verbally) in connection with the goal become motives. They create motivational inducing forces to reach one single goal—*closing a valve*. The presented example earlier demonstrates that including motivational components into the goal of a task or actions makes it difficult to perform task analysis in practical settings. Hence, it is not accidental that in their concluding article, Diaper and Stanton (2004) suggest to eliminate the concept of human goal from task analysis. The authors suggest to substitute the concept of goal and the connected motives with their method called the *FSS approach*. As a result, a critically important area of psychology that is involved in studying anticipation, forecasting, formation of hypothesis, goal formation, etc., is entirely ignored.

Some other professionals attempt to eliminate the concept of goal in the HCI field. We will discuss this problem later. In the traditional area of ergonomics, which is based on the cognitive approach, the goal concept is not considered at all.

An analysis of the material in this section brings us to the following conclusion. The vector *motive → goal* is the basic concept of AAT and SSAT. However, according to SSAT, as components of activity, goal is informational and motivation is energetic. Therefore, motives cannot move to the goal, because they are different mechanisms of activity regulation. The vector disappears when motives move to the goal. In such situation, the goal directness of activity is also eliminated. Integration of motives with the goal, as it is done in cognitive psychology, results in a situation where one task has various goals. Task analysis is impossible in this situation. Therefore, experimental cognitive psychologists practically do not use the goal concept in task analysis. This term has for them some common meaning. The traditional interpretation of goal eliminates the concept of the goal-directed process of activity self-regulation. Self-regulation can be considered only as a homeostatic process, which was criticized by various authors in personal and social psychology. This problem will be considered further in the book.

2

Vector Motive → Goal *and Brain Functioning*

2.1 Neuropsychological Mechanisms of Attention and Goal-Directed Activity

In our previous discussions, we demonstrated the difference between the goal and motive(s) by using a psychological approach. Now we will discuss the relationship between the goal and motives from a neuropsychological perspective. It should be noted that in Sections 1.3 and 1.4 we only examined such a relationship from a psychological point of view. The relationship between the goal and motives from a neuropsychological point of view was not considered in activity theory.

We begin our analysis with a brief review of the neuropsychological mechanisms of goal-directed activity. At the first stage of our discussion, we consider the frontal lobe of the brain that is responsible for most of the abstract level of information processing such as selection of words with particular meaning, development of a plan of activity and following through with it, formation of goal of activity and making a reasonable projection about the future, flexible adaptation to the social environment, and selection of information that should be held *online* in memory. According to Luria (1966, 1970), the most general functions of the frontal lobe cortex are temporal organization and temporal integration of goal-directed cognitive and behavioral actions. From this depends their sequence. In other words, it is the temporal organization of human speech, reasoning, and behavior. The human frontal lobe provides *representational processing* that allows the human subject to interpret reality, to understand past events, and, based on it, to predict future events. Representational processing cannot only work with reality but also operates on hypothetical content. This is a dynamic mental model of reality. Mental models can be imagined as dynamically constructed mental scenarios of a variety of situations that unfold in time and space in which the subject, if necessary, may act in accordance with their intentions, goals, and specifics of the environment. This is the mental theater of the human mind. However, it is not a passive Cartesian theater of philosophers but the active cortical theater of human brain functioning (Picton et al., 2002). A person with lesions in this cortex has difficulty in sequencing propositional language and complex mental operations and executing plans of behavior. Temporal integration and sequencing of goal-directed activity elements require cooperation with some other areas of the cortex and subcortical structures. According to Fuster (2002), there are two types of sequences of actions toward a goal: routine and complex.

In routine sequencing, each action leads to the next, in a chain-like fashion, with contingences only between successive actions. In complex sequencing, actions are contingent across time on the plan, on the goal, and on other acts. In performing such complex functions, the frontal cortex performs cortical control of attention. In temporal integration of goal-directed activity, the frontal lobe performs the following functions: regulates attention processes and coordinates them with functioning of working memory, develops preparatory set, and provides monitoring. Working memory is essentially based on the activation of the neuronal cortical network of long-term memory, which is necessary for the functioning of attention. A preparatory set includes anticipatory mechanisms for anticipating signals and actions. The other important role of the frontal lobe of the brain is monitoring. According to the systemic-structural activity theory, human activity is organized and performed based on mechanisms of self-regulation where the concept of feedback is particularly important. Hence, monitoring means feedback and self-regulation of motor actions (Bedny et al., 2014).

Neuropsychology provides significant evidence of the participation of the prefrontal lobe in the cortical control of attention. An electroencephalographic pattern of predominantly high-frequency waves developed over the frontal lobe during a goal-directed activity supports this hypothesis. It was discovered that psychological inertia is caused by a dysfunction of the frontal lobe (Luria, 1966). An individual with prefrontal damage experiences difficulty starting an activity, but once engaged in it, he or she has great difficulty stopping it. Once a person is engaged in the course of a particular activity, he or she has also difficulty deviating from the expected path. Such type of activity is often referred to as perseveration.

Often a dysfunction of the frontal lobe is the cause of a situation when a subject's behavior is triggered by some object in the environment. Such type of human activity or behavior is referred to as an environmental dependency syndrome. This environmental dependency syndrome emerges in different forms. Very often this dependency is a result of a person's individual history of prior injury. The mental deficit may in part be characterized as a disconnection between thought and actions.

Another deficit exhibited by individuals with damage to the frontal lobe is an inability to process presented data in an abstract rather than a concrete manner. It was discovered that persons with frontal lobe damage have problems with abstract conceptualization. They demonstrate an inability to use known information and to make reasonable judgments or deductions about the world. Individuals with frontal lobe damage cannot make attributions about another person's mental state, which can be used for the prediction of an other's activity. A person with frontal lobe dysfunction has trouble demonstrating flexible and adaptive styles of activity (Fuster, 1985). It was discovered that the frontal lobe of the brain plays an important role not only in the initiation of actions but also in enabling novel and adequate actions.

The more important aspects of conscious activity are associated with goal directedness. The ability of a person to organize his or her own activity toward a goal is critically important and has multifaceted aspects. The loss of any aspect of goal-directed behavior can destroy the entire plan of goal-directed activity. First of all, the overall goal of activity (task) must be kept in mind until this goal can be achieved. The person has to keep this goal in mind even if intermittent goals of actions can vary. Activity should be flexible and adaptive in attaining the overall goal of the task. Further, the subject has to keep in mind what he or she already did and what must be done further. Activity is organized according to the mechanisms of self-regulation. Hence, the person should be able to use feedback and correct his or her activity. This means that actions must be flexibly sequenced toward the goal.

One of the basic prerequisites for goal-directed activity is the ability to stay on the task. Individuals with frontal lobe damage cannot monitor their performance. They have a tendency to *wander off task*. If such persons would be asked to perform a specific task, they may start to do it. However, later, they might start to perform some unrelated actions that are incompatible with what they should do (Luria, 1966). This means that such persons start to perform non-task-relevant activities. The other critical feature of goal-directed activity is sequencing. Reaching a goal requires some order of steps of task performance. It was demonstrated that planning abilities rely mainly on the frontal regions. There are two types of memory. One is recognition memory and the other memory is the item order. These two types of memory rely on different neural areas of the brain. The ability to remember the sequence in which information is presented is disrupted by damage to the frontal regions. However, such damage leaves recognition memory unaffected. The ability to remember whether an item was presented before is disrupted by damage to the temporal regions (Milner, 1982). Another important aspect of sequencing of activity steps is the ability to select the more effective strategy of a goal achievement. The subject with frontal lobe damage demonstrates less efficient strategies of the goal attained. Their strategies involve a lot of aimless actions, which cannot be directed toward the goal. All this demonstrates the role of the frontal lobe in planning the sequence of cognitive and behavioral action performance and consciously regulating activity. Selected strategies very often should be modified. Frontal lobe damage makes such modifications more difficult (Delis et al., 1992). It was discovered that a considered area of the brain plays an important role in the generation of strategies and switching of hypotheses (Cicerone et al., 1983). Frontal lobe damage disrupts the controlled processes. Further, it was discovered that the frontal lobe region is more important in controlled processing as compared to automatic processing (Shallice, 1988). This region is important in the functioning of short-term memory. The previously data demonstrate that the frontal lobe region is critical in keeping a goal in mind, directing activity toward a goal, and performing consciously, associated with verbalization activity.

To substantiate the importance of separation of motives from goal and consider them not as unitary but as interdependent yet separate psychological mechanisms, we need to consider in a more detailed manner some neuropsychological mechanisms of attention. Attention is inseparable from working memory, preparatory set, execution of motor actions, and response monitoring. All of these are key to the temporal integration of the entire goal-directed activity. People do not analyze all the information that is available to them at one time. They select particular information in the environment, concentrate on one portion of the information and ignore the rest, etc. All of these are examples of attention functioning. Somehow our brain selects that information which is pertinent and ignores the others.

Three are six neural areas of the brain responsible for controlling attention (Banich, 2004). They are located in different parts of the brain. In the cortex, there are (1) the frontal lobe, (2) the parietal lobe, and (3) the anterior cingulate cortex (ACC). Outside the cortex, there are the following areas: (1) the reticular activating system (RAS) located in the brain stem, (2) the superior colliculus located in the midbrain, and (3) the thalamus that is related to the subcortical structure.

Attentional functions can vary from basic functions as alertness to the more sophisticated functions such as those associated with goal-directed activity. The more sophisticated functions of attention associated with goal-directed activity are performed by such areas as the frontal lobe. To summarize attentional functions of the frontal lobe region in goal-directed activity, we can outline the following. This area of the brain as a mechanism of attention is responsible for processing more abstract characteristics of information such as selecting words with specific meaning and keeping information online in memory; it aids in selecting, initiating, and inhibiting motor responses; and it volitionally directs the eyes to a particular point in space. There are other critically important functions of attention without which conscious goal-directed activity is impossible. These are alertness and arousal that represent the most basic level of attention. It is associated with the term *activation of the nervous system*. The term is also often used synonymously with *energy mobilization*. This term was introduced by Lindsley (1957). According to this author, the main specificity of emotions is associated with different levels of activation of nerve processes. Changing behavioral characteristics such as wakefulness depends on the activation of the nervous system. When we are tired or sleepy, our alertness and arousal are low. At this state of our brain, we cannot efficiently process the required information and perform correctly necessary motor actions. The major function of the RAS is arousal of the cortex and screening of incoming information. The system is also responsible for the regulation of sleep–wake cycles. The cell bodies of the RAS are located in the medulla and the pons, which have different connections to most regions of the cortex. Without these functions, the subject is unable to extract information from the environment and select an adequate response. The reticular formation consists of a number of small neural networks with varied functions.

Interaction of the reticular system with the cerebral cortex of the brain can be presented in the following way.

Information from each receptor comes into the cerebral cortex, where a formal analysis of information messages plays a leading role. Simultaneously on collaterals, this information goes into the reticular formation, where its specific meaning is largely lost. The quantitative aspects of information stimulation play a leading role as well. This system is very sensitive to the total number of impulses coming to it (Bloch, 1966). The more the number of impulses is in the reticular system, the more is the activation of the reticular formation. Activation of the reticular system is related to arousal or widespread activation of the brain. The RAS provides an attentive state of the brain that is known as tonic arousal and contributes to sustained attention. Violation of these functions when the RAS is damaged results in a coma state. The individual loses consciousness and awareness of the outside world and becomes unresponsive to external stimuli. In addition to tonic arousal, the RAS is involved in alerting the brain for receiving information and making required actions. This means that the RAS has important functions in goal-directed activity.

The frontal lobe of the cerebral cortex has the most intimate connection with arousal processes in such conscious activity. Cortical awaking and therefore consciousness depend on the level of activity of the RAS. From this follows that external stimulation can directly influence cortical information processing and through the reticular formation indirectly influence the intensification of information processing. The reticular formation system consists of many functional subdivisions. For us, the influence of this system on the cerebral cortex is important, because it is responsible for the activation of the level of consciousness. Because of that, it is also named as the *activating system of the brain stem*. All data demonstrate that this system is closely linked to the control of awareness and attention. Such functions as abstract thought and reasoning and organization of behavior in time and space toward future goals are associated with wakefulness and frontal lobe functioning (Goldman-Rakic and Raric, 1984). Emotional states such as anger, excitement, and anxiety are also associated with increasing activation of the reticular system. The reticular formation system also includes cell groups that are concerned with the regulation of sleep.

The reticular formation is a nonspecific nervous system. It does not convey a meaningful message into the cerebral cortex. The frequency-impulsive code of this system in the cerebral cortex of the brain only has a general activating effect on its functioning. The role of the frequency-impulsive code of the reticular formation in the functioning of the cortex can be explained by an example, which is used only for illustration of the major idea.

Assume that the operator should perform important vigilance tasks during the night shift. His or her partner becomes ill. After the day shift, the operator must work the night shift. For such a situation, a special chair was designed. It vibrates at certain periods of time with some frequency. The vibration protects the operator from falling asleep. The vibration of the chair does not convey meaningful information or massages. The vibration of the chair has one main function—protecting the operator from falling asleep—or it maintains the operator's level of alertness. The real information is transferred from the appropriate equipment and processed in the cerebral cortex.

Certainly the relationship between the reticular formation and the cerebral cortex is much more complex. There is evidence that the value of some important information can be interpreted by the reticular system at an unconscious level. For example, this system can be involved in learning to ignore repetitive, meaningless stimulations while remaining sensitive to others.

The other region of the brain that is important in our discussions is the cingulate cortex.

It performs interface functions between the subcortical and cortical regions of the brain. This area is particularly important in the functioning of attention during selection of appropriate responses and performing cognitively demanding or complex tasks. It was discovered that the anterior cingulate cortex (ACC) becomes activated especially in situations when selection of an appropriate response is difficult for the subject. For example, this can be a task that requires attentional control over response selection when a person has to use effort to inhibit an automatic and habitual response for producing an appropriate action. Activation of a considered area of the brain is observed not only when there is a need to select between conflicted responses. This was also discovered when a person selects the correct response in demanding and complicated tasks (Petersen et al., 1988). In neuroimaging studies, it was discovered that the more difficult the task is, the more likely is the activation of the anterior cingulate (Paus et al., 1998). The close relationship of the ACC with the RAS reflects the need for increasing arousal and attention as the task becomes more complex and cognitive demands increase. Moreover, activation of a specific part of the ACC directly depends on the cognitive demand of the task rather than any other type of stimulation that can capture attention. For example, activity of the considered area of the brain is not increased when given a noxious stimulus that grabs attention. However, when a complex cognitive task without any noxious stimulus was presented, activation of some part of the ACC was observed. These data are particularly important in task complexity evaluations. They demonstrate the role of such areas of the brain as the RAS and the ACC on attention processes during the performance of a task with different levels of difficulty for a subject.

We considered frontal lobe functioning in conscious goal-directed activity and associated with it some attentional mechanisms.

Let us briefly consider three other areas of the brain responsible for controlling attention below the cortex that do not involve verbalized processing of information. A region such as the parietal lobe is important for the visual and spatial aspects of attention. However, this region is not activated by highly demanding tasks, which includes verbalization functions (Wojeiulik and Kanwisher, 1999). At the same time, verbalization is important in goal-directed activity. Another region of the brain that is involved in attentional process is the superior colliculus area, which is involved in automatic moving of the eyes in a particular position—an involuntary process of eye shifting. In such a process, the eyes jump involuntarily in a particular position. Such saccades take about 120 ms. When the superior colliculi are damaged, such saccades are extinguished. The other types of saccades are performed under voluntary control and take about 200–300 ms (Schiller et al., 1987), which are not affected by damage of the superior colliculus area of the brain. However, they are disrupted by damage to part of the frontal lobe, which is called the frontal eye field.

The other area of the brain that is involved in attention process is the thalamus. Some of the nuclei of the thalamus are involved in the modulation of arousal of the cortex. The other important function of this area of the brain is gating or filtering of sensory information that impinges upon the brain. Sensory information goes into the brain through the thalamus.

It was discovered that the thalamus is more engaged when filtering of information is required (LaBerge and Buchsbaum, 1990). Analysis of the last three regions of the brain demonstrates that there is no evidence of the importance of these areas in verbalized goal-directed activity.

We considered the relationship between neural mechanisms of the brain and the formation of goal-directed activity. Analysis of the presented material is important in considering the relationship between motive and goal from a neuropsychological perspective. This issue will be discussed in Section 2.2.

2.2 Relationship between Goal and Motives from Neuropsychological Perspectives

Before we examine the relationship between motive and goal from the standpoint of neuropsychology, it is necessary to clarify the relationship between such basic concepts as needs, motives, and emotions. In activity theory, they differ from a functional point of view. However, they have a very close meaning in neurophysiology and neuropsychology. The concept of motive is tightly connected with the concept of emotion. Both are related to energetic mechanisms of activity regulation. According to some American psychologists, there is no clear-cut distinction between these notions (Hilgard et al., 1979, p. 329). However, in activity theory, these notions are clearly distinguished. Reykovski (1979) wrote that motivation is emotions plus the directness of activity to a specific goal. Emotional behavior is expressive and not oriented toward a certain goal. In activity theory, emotions are considered as a form of reflection by subject relations to the objects or phenomena of reality, due to their conformity or nonconformity with the needs of humans (Smirnov et al., 1962, p. 384). Activity theory emphasizes two aspects of emotions and feelings: (1) The aspect of reflection is that the emotions and feelings are a reflection of the values of a specific form of objects and phenomena for the subject, and (2) the aspect of relation lies in the fact that people do not passively, but actively interact with the environment and learn it and at the same time reflect their attitude toward objects and phenomena of reality. There are two types of emotional processes. In cases where the subcortical region of the brain dominates emotional processes, we talk about emotions. If the cortex also plays an important role in the functioning of emotions, we talk about feelings. This is specifically human emotions, which are primarily social in nature.

Motives are associated with various needs. However, needs do not directly generate activity. They merely create dispositions directed toward their satisfaction. Activity is derived from those cases in which the person consciously imagines a specific goal. Achievement of the goal can satisfy the subject's needs. The factors that determine a person's effort to reach the goal are motives. Reflection of the values of the objects and phenomena of reality for the subject in the process of goal-directed activity is provided by emotions. Thus, the needs and emotions experienced during the activity can serve as motivational factors. From a physiological point of view, needs, motivations, and emotions are connected with the functioning of virtually the same brain mechanisms. In this regard, we shall analyze the relationship between the neural mechanisms of emotions and the brain. This is stipulated by the fact that emotional and motivational processes are closely related and most data in this field discuss relationship between emotions and regions of the brain. Analyzing the relationship between emotions and their neuropsychological mechanisms, we at the same time analyze the relationship between emotional-motivational processes and their neural mechanisms.

To prove from neuropsychological perspectives that goal formation and emotionally-motivational processes should be considered as separate mechanisms, it is sufficient to analyze the relationship between different parts of the brain that are involved in the formation of goals, emotions, and motivations. If the formation of the goal and emotional-motivational processes involved different parts of the brain and their structural relationship is different, then we have grounds to assert that the goal cannot be considered as a unitary mechanism that combines information (cognitive) and energy (emotion and motivation), which are different components of activity. From this follows that such concepts as vector *motive* → *goal* can receive not only psychological but also neuropsychological bases. To solve the problem, it is not necessary to perform a detailed analysis of the physiological mechanisms of the brain that are involved in goal formation and emotionally motivational processes. It is sufficient to analyze some basic aspects of brain function during the formation of goals and generation of emotions and motives of activity. In other words, we want to make a point that the vector *motive(s)* → *goal* has not only a psychological but also a neuropsychological foundation.

This issue is fundamentally important in understanding work activity as a goal-directed self-regulation process.

The relationship between the neural mechanisms of the brain and the formation of goal-directed activity has been examined in the previous section. In this section, we consider in a simple manner the brain mechanisms of emotions and motivation. The two-factor theory of emotion suggests that emotional experience is the outcome of physiological arousal and the cognitive attribution of the cause for that arousal. Cortical and subcortical areas of the brain play an important role in the formation of a particular type of emotion in humans. According to this theory, emotional experience is the physiological arousal or state and the cognitive attribution of causes for that state. The physiological state and the causal attribution determine the emotional experience (Schachter and Singer, 1962).

Many emotional processes unfold outside of consciousness. However, in human beings, emotions can be significantly changed by our appraisals of the situation.

The major brain areas that are responsible for our emotions include the limbic system, which contains among others such regions as the amygdala, hippocampus, hypothalamus, and ACC. There are also various subdivisions of the prefrontal cortex (PFC) and posterior regions of the brain. The last includes the parietal lobes and posterior cingulate cortex. We consider some of these regions for illustration.

We begin our discussion with brain regions that belong to the noncortical regions. These regions are involved with more automatic or subconscious aspects of emotions. For example, when a person is in a dangerous situation, his or her body has to mobilize its resources and quickly perform some type of protection actions.

At present, the amygdala is identified as a critically important region in the study of emotions, which plays a significant role in the formation and evaluation of fear. Based on the thalamo-amygdala route, the fear response can be quickly developed. Case studies of people with amygdala damage demonstrate a loss of ability to emotionally react to aversive visual and auditory stimuli (Aggleton and Young, 2000). There is also a second and more complex pathway in fear evaluation for humans. This route includes the thalamo-cortico-amygdala pathway. It includes some cognitive mechanisms in the formation of fear. At the same time, we can note that the amygdala seems to be especially involved in the representation of unconscious aspects of emotions.

Such region as the hippocampus plays an important role in processing contextual information and its emotional evaluation. The next region is part of the subcortical system and is called the hypothalamus. It is also involved in the quick analysis of emotional information. This region of the brain helps to prepare the organism to approach a dangerous situation or to withdraw from it. For example, it involves an increase in blood pressure, when the subject is confronted with a threatening stimulus. For example, Olds and Milner (1954) demonstrate that rats exhibited electrical stimulation of the hypothalamus by pressing a lever to deliver electrical self-stimulation of the hypothalamus. Similar data were obtained in a study of self-stimulation in humans (Heath, 1972). There is evidence that this area of the brain is involved in emotionally motivational processes. The ACC of the brain is involved in a variety of emotional reactions. This region performs reciprocal dynamic function between emotions and cognition. It is responsible for shutting down certain cognitive systems. In contrast, this region can be involved in increasing the cognitive effort and concentration of attention and suppressing strong emotions. There are some data according to which the ACC can be involved in the reciprocal relationship between emotion and cognition. For example, when there is strong emotional tension, this area can shut down certain cognitive systems of the brain (Bush et al., 2000). In general, this region plays an important role in evaluative self-monitoring along an affective dimension (Luu et al., 2000). Another subcortical region of the brain involved in emotion is the retrosplenial cortex, which is part of the posterior cingulate cortex. This area is involved in evaluating the emotional and motivational significance of ongoing stimuli and events. This is particularly interesting because in the psychological model of self-regulation of activity, there is a functional mechanism that is responsible for the evaluation of the significance of an element of a situation (Bedny and Karwowski, 2007; Bedny and Mejster, 1997).

One important mechanism of motivation is the dopamine system, which depends on the functioning of the brain's ventral tegmental area (VTA). The dopamine system works as a motivational mechanism (Nicholson, 1997). Studies have indicated that such types of activities as risk-taking behavior, gambling, and motivated work behavior increase the release of dopamine in the VTA (Carlson, 2004). The amygdala, hippocampus, cingulate cortex, etc., are also involved in motivational processes in the same way as in emotional processes (Franken, 2002; Taylor et al., 2004).

We described the subcortical regions of the brain in emotionally motivational processes. Simplifying, we can state that the subcortical regions enable humans to quickly develop an emotional-motivational response. However, an emotionally motivational reaction can be mediated by cognition. In such situations, emotionally motivational processes can be developed more slowly. The more important cortical area responsible for this type of reaction is the frontal lobe region. However, it is not a unitary region. It has a prefrontal region that can be further divided into areas that can be important in emotion. These are the orbitofrontal prefrontal cortex (OFC), the ventromedial prefrontal cortex (vmPFC), and the dorsolateral prefrontal cortex (DLPFC). All of them have their specific purpose. In a brief manner, we consider some functions associated with emotion. The OFC area is involved in the cognitive evaluation of emotional processes. For example, this area is involved in evaluating reward and punishment contingencies (Thut et al., 1997). In a more recent study (Gray et al., 2002), it was found that the DLPFC is also involved in the integration of cognition and emotion. We do not discuss neuropsychological mechanisms further. We simply attempt to illustrate that in the formation of human emotions, the interaction of the cortical and subcortical areas of the brain is important. Interpretation of a situation and the physiological state of an organism are important aspects of emotional regulation of human behavior.

Even a short analysis of neuropsychological mechanisms of human emotion demonstrates that the frontal region of the cortex is important not only in the formation of a goal, planning of activity, and decision making but also for the interpretation of an emotional state. Such interpretation requires not only analysis of the physiological state but also analysis of the situation. Hence, interpretation of emotions is, first of all, a cognitive process. From this follows that in stressful situations, the frontal lobe performs a dual function. On the one hand, this region is involved in the regulation of goal-directed activity. On the other hand, the same region is involved in the interpretation of an emotional state. It should be noted that this interpretation can be subjective and not always correct. Some cortical and subcortical areas can be in a dynamic reciprocal relationship during the performance of emotional tasks.

Hence, comparison of functioning of the brain in goal-directed and emotionally neutral activity and functioning of the brain in an emotionally stressful situation demonstrates different brain structures that are involved in these two activities.

It is important to compare brain functioning in goal-directed and emotionally neutral activity with goal-directed activity in a stressful situation. In a goal-directed and emotionally neutral situation, the frontal lobe plays a particular role. However, in providing an active state of this region and maintaining the required vigilance and arousal of this area, a subcortical area such as the RAS is critically important. The more difficult the emotionally neutral task for the subject, the more RAS is activated in the frontal lobe. At a particular level of difficulty, activation reaches such a level when emotions emerge. Hence, initially, an emotionally neutral task can transform into an emotional task. This cognitive characteristic of the task as difficulty is an emotions-producing factor.

The brain works in a different way when a person performs a relatively simple but very dangerous task. In this situation, the role of the cingulate cortex imparts motivational significance to receive information. Its functioning becomes critical in developing strategies of selecting information.

Here, neuropsychological data supplement psychological studies in the self-regulation of activity, where a mechanism that is involved in the evaluation of significance is important (Bedny and Karwowski, 2007; Bedny et al., 2014).

According to Freud's (1916/1917) psychoanalytical theory, our activity or behavior is determined by inner forces and impulses, often operating below the level of consciousness. The concept of unconscious motives is an important notion for him. At present, most psychologists agree that unconscious motives may exist. However, it is more correct to think in terms of degrees of awareness of different motives. The driving forces of different motives, including social motives, always exist in human activity. However, we often are precisely aware of them. Here we pay attention to the fact that a subject may be unaware of the energetic factors of activity together with the emotionally motivational aspects during performance of a routine activity. These data allow us also to conclude that during performance of an emotionally neutral and relatively simple task at the certain period of time, a person can be unaware of the motivational forces that direct activity toward the goal. However, a motivational factor is always present, because an unmotivated goal-directed activity does not exist.

We limit ourselves by considering only the most basic mechanisms of the brain involved in goal-directed activity and emotionally motivational processes. There are some other mechanisms of the nervous system that can also be considered when we discuss this topic. For our analysis of the concept of the *vector motive → goal*, this material is sufficient for the required conclusions.

Although different psychological functions are controlled by a network of interacting brain regions with overlapping functions, goal-directed activity is provided by a brain structure that significantly differs from the brain structure that is involved in emotionally motivational reflection of reality. In goal-directed activity, the frontal lobe of the brain plays a dominant role. The most general function of the frontal lobe is the formation of the goal, the temporal organization of behavior or activity, monitoring of speech, and reasoning. Humans with lesions of this cortex have difficulty executing plans of activity, sequences of propositional language, and complex mental operations. The human frontal lobe provides *representational processing* that allows the human subject to interpret reality, to understand the meaning of the situation, to fit the possible to the actual and the past to the future, and to regulate goal-directed activity in general.

In contrast, emotionally motivational components of activity are more related to subcortical functions than a conscious goal-directed activity with a low level of emotionally motivational activation. In the activation of the cerebral cortex during performance of goal-directed activity, the reticular formation plays the leading role. Moreover, in an emotionally motivated activity, such regions of the brain as the anterior cingulate, amygdala, hypothalamus, and hippocampus play the leading role.

The cortical regions of the brain play a different role in goal-directed activity, with slightly expressed emotional and motivational components of activity in comparison to highly emotionally motivated activity. In the latter situation, the cerebral cortex of the brain, in addition to conscious regulation of goal-directed activity, performs interpretational functions of physiological processes associated with emotionally motivational components of activity in a particular situation. Thus, data in neuropsychology confirm that the goal of activity is a cognitive component, and emotionally motivational mechanisms are energetic components of activity. These components are interrelated but not identical. All these give us an opportunity not only from psychological but also from neuropsychological perspectives to prove that such concepts as goal and motives should be considered as separate and at the same time interdependent mechanisms. They are best described as a vector *motive* → *goal* that consists of two interrelated but different components from a neuropsychological point of view. A comparison of neuropsychological data and mechanisms of self-regulation of activity demonstrates that emotions and motivation are intimately involved in the development of strategies of selection of information and its processing. The vector *motive* → *goal* reflects the distinction and unity of cognitive or informational and energetic components of activity. Even in a situation where the same regions of the brain may be partly involved in the formation of goals, and the formation of emotional-motivational states, their role in these processes and their relationship are different. All these lead to the conclusion that the vector *motive* → *goal* as a psychological concept has also received neuropsychological justification.

3

Concept of Task from the Systemic-Structural Activity Theory Perspective

3.1 Task Structure and Its Basic Characteristics

The task is the basic component of human work. Changing any equipment characteristics influences the method of task performance. Therefore, we can evaluate the usability of equipment based on task analysis procedures. The same equipment can be used by utilizing different methods of task performance. Some of these methods are more efficient than others. Hence, efficiency of performance also can be evaluated based on task analysis. Task analysis makes an important contribution to the development of training (Patrick, 1992). By analyzing the tasks that are being performed by workers, we can estimate the safety of the equipment. In this chapter, we will consider the task concept in psychology and ergonomics. Ordinarily, the work is broken down into specific tasks or production operations. All tasks and production operations should be specified and, where possible, standardized. At present, even standardized methods of work are frequently very flexible and complex. From systemic-structural activity theory (SSAT), flexibility of task performance means that the method of task performance is not rigid and should be designed based on existing constraints. Moreover, any design process should be performed based on existing constrains. This aspect of design will be discussed in Section 6.1. Using various types of equipment, man transforms the object of work, making in it premeditated changes and transforming it to a finished product. At the stage of mechanization when manual components are quite substantial, such terms as production operation or task can be used. Professionals with technical backgrounds prefer the term production operation. In ergonomics, the terms production operations or tasks can be used synonymously. At the stage of automation or the semiautomatic type of work when a person performs the transformation of information, the state of remotely controlling objects or processes, and so on, the term task is used.

For analysis of tasks or production operations, the more relevant is such terms as equipment, means of work and tools. This terminology is interrelated and can be understood in comparison with each other. In addition, the concept of the tool is connected to the term introduced by Vygotsky, which he described as a mental tool. The latter is of fundamental importance in activity theory. There are differences among equipment, means of work, and tools in the engineering field. The worker operates with machine or equipment. For example, in order to cut the blank, the worker needs to install a cutting tool. Depending on the nature of the production operation, the worker may change a tool. The worker sets different tools, depending on the specifics of the blank and produced parts. Here, we consider a machine tool. There are also hand tools, which are used in manual work, for example, a hammer or a saw. Compared with the concept of equipment, there is a more general term—a means of work—which is a combination of any physical equipment and tools used by a worker for transforming material objects, physical and chemical processes, or information. In this sense, a personal computer is not a tool. Computers are a means of work. Hand tools are directly manipulated by work. Human beings use different tools for transforming objects from one state or position to another. Vygotsky considered a variety of images and sign systems, which can be external and internal, as mental tools. The computer is a means of presentation or creation of different artificial tools for the users. Moreover, the computer as a means of work not only presents to subjects a variety of tools but also creates different artificial objects that can be modified by the user. Computers are new means of work mediating human interaction with the external world by creating artificial objects and tools required for the performance of computer-based tasks. In traditional task analysis, the tool has a technical meaning. In activity approach, the tool also has a different meaning. It is used as a psychological tool for mental components of activity. This important concept demonstrates that the subject can utilize mental actions and mental tools in a similar way as external tools with material objects. Such concepts as material and mental actions and material and mental objects and tools are important concepts in task analysis from an activity theory perspective. Therefore, Kaptelinin's (1997) statement that the computer is a tool that mediates the interaction of human beings with their environment is not correct. Computers as a means of work, which creates artificial objects and tools, are necessary to support human beings' mental activity.

The existence of the same equipment and tools, object of work, and technological completeness characterizes a production operation or tasks. Production operations according to the technological principle can be divided into smaller standardized technological units. Production operations may be studied from the technological frame, or from an activity perspective. In the first case, the leading figure is the production engineer or related professionals. In the second case, a human factors specialist or work psychologist is called for.

In semiautomatic and automatic systems, the worker's demands during task performance are primarily cognitive in nature. The tasks can be with different degrees of freedom of performance.

Analysis of the material presented earlier shows that there are two main groups of tasks performed by the worker. The first group includes the traditional production tasks or production operations. The simplest of them are performed at the stage of mechanization of work, and the more complex ones are performed at the stage of automation and centralization (Vicente, 1999). At present, both types of tasks are used in industry. The basic characteristics of such types of task are described in ergonomics and industrial engineering (Barness, 1980; Sanders and McCormick, 1993; Wickens and Hollands, 2000).

In ergonomic literature, we have not seen a clear definition of what constitutes a task in the psychological or ergonomic sense. Each author gives his or her definition of the task, using terminology that is not clearly described. We present as an example the definition of the task that was given by Vicente. This definition is more relevant to activity theory: "Task—Actions that can or should be performed by one or more Actors to achieve a particular Goal" (Vicente, 1999, p. 9).

It immediately raises questions of what authors understand under the term *actions*, how they are classified and extracted from the flow of behavior that unfolds over time, what is a goal, and so on. Normally, all these terminologies have simply common sense. For example, in SSAT, there are cognitive and behavioral actions (Bedny and Karwowski, 2007). Without understanding the concept of cognitive and behavioral actions, it is impossible to describe the structure of the task. The structure of activity during task performance is determined by the spatially temporal and logical organization of the elements of activity. The main elements are the cognitive and behavioral actions and psychological operations of activity. The last are elements of cognitive or behavioral actions. Analysis of the relation between the structure of activity and the configuration of the equipment is the most important principle of ergonomic design in SSAT.

In SSAT, each task in the work process is regarded as a situation-bounded activity that is directed to achieve the goal of the task under given conditions. According to SSAT, the task is an overall goal of situation-bounded activity that should be achieved in particular conditions. This definition of the task is based on the data of the general activity theory (Leont'ev, 1977; Rubinshtein, 1959). For example, Leont'ev (1981, p. 63) gives the following definition of task: "the task is the goal given under curtained conditions." The goal of the task does not exist in a complete form. Only requirements are given objectively in a specific situation. For example, requirements may be presented to the subject in the form of special instructions.

These requirements should be transformed by the subject into a subjectively accepted goal at a later stage. This is the goal interpretation and acceptance stage of activity. Hence, there are requirements given in particular conditions and the task emerges after the transformation of these requirements into a goal. In some cases, the goal of the task, and hence the task, is formulated by the subject independently. In this case, we speak about the formulation of a goal of the task or a task entity independently. This is the goal formation or task formation stage.

Acceptance or formulation of the task goal is closely associated with the subjective mental representation of the task. The mental representation of the task determines its performance. From the standpoint of self-regulation, the mental representation of the task is seen as a process of formation a mental model of the task. The mental model of task includes not only verbalized but also nonverbalized components and conscious and unconscious elements.

Any task includes both the subject's activity and the material components of the task, with all the elements of activity during task performance being organized by the task goal. There are also goals of separate actions, and goals of subtasks, which integrate several actions. Actions have logical organization. Cognitive and behavioral actions are basic elements of activity during task performance. In the production environment, there are tasks that can be formulated by workers independently. Such tasks are particularly important for the stage of automation and centralization and specifically for computerization of human work. Therefore, there are objectively given or subjectively formulated tasks. However, even in cases when the task assignment is a result of an external given instruction, we need to distinguish *objectively given* task and *subjectively accepted or formulated* task. This distinction is important, not only in the case of studies carried out directly in the production environment, but also in laboratory conditions. In psychology, it is common practice to consider that an objectively given goal or a goal formulated by instruction tasks is uniquely interpreted and accepted by the subject. However, it was shown that in the goal formulated by rigorous instructions, the goal and task very often do not coincide with the subjectively accepted goal of the task and the mental representation of the task in general. The goal acceptance or goal formation stages are important aspects of task analysis. These stages can be a source of error when the worker performs various tasks. Thus objectively, there are only requirements of tasks. When the task requirements are accepted by the individual, these requirements become of the individual's personal goal of the task. This is an important step in task performance and task analysis. This step is omitted not only in ergonomics but also in experimental psychology.

Whatever is presented to the subject for the performance of the required actions constitutes the conditions of the task. Task conditions include the subject's past experience and such material components as instructions, means of work in given conditions, raw material, and input information. These conditions also determine the possible constraints on activity performance. The raw material, or input information, is considered to be the object of activity. What is actually achieved (finished product, output) is the result of the activity. The vector *motive* → *goal* determines the directness of activity during task performance.

Any task includes an initial situation, intermittent situations, and a final situation. By associating the notion of a *situation* with stages of task performance, it becomes possible to study how the structure of a task changes during different stages of performance and how many basic transformational stages are required. The more complex the task is, the more possible the stages of task performance are. Computerization significantly alters the specifics of performed tasks. When a subject directly interacts with a computer during task performance, this becomes a specifically human–computer interaction (HCI) process, which includes logically organized computer-based tasks. The only major difference being that the computer is now the dominant means of work.

We will now briefly summarize the main features of task.

Each task in the work process is regarded as a situation-bounded activity that is directed to achieve a goal of the task under given conditions. Any task includes both the subject's activity and the material components of the task, with all the elements of activity during task performance being organized by the task goal. It is only when the objectively given or subjectively formulated requirements of the task are accepted by the subject as a desired future result that they become the goal of the task.

A continuum of possible tasks can be imagined as skill-based tasks from one side and complex problem-solving tasks from the other side. We determine skill-based tasks following Rasmussen's terminology of *skill-based behavior* (Rasmussen, 1983). Skill-based tasks can vary in their complexity. Any task includes some problem-solving aspects. When a task requires conscious deliberation about how to accomplish the goal, it becomes one of problem solving.

Problem-solving tasks can be divided into two major groups: skill-based tasks and problem-solving tasks. Skill-based tasks are performed in a rapid automatic way with minimum concentration. The simple production operations are examples of such tasks (Bedny and Karwowski, 2007). According to Russian terminology, such tasks require automatically performed actions—called *naviky*. However, there is the second level of skills, called *umeniya*, which consists of an individual's ability to organize knowledge and the first level of skills into a system and efficiently use it to perform a particular class of tasks or solve a particular class of problems. When a task requires conscious deliberation on how to accomplish the goal, it becomes one of problem solving. Hence, all tasks can be presented as a continuum. From one side there are skill-based tasks and on the other side there are problem-solving tasks. Problem-solving tasks can be divided into two major groups: algorithmic and nonalgorithmic (Bedny and Karwowski, 2007). Algorithmic tasks are performed according some logic and rules. Algorithmic tasks can be divided into deterministic and probabilistic types. In deterministic-algorithmic tasks, workers perform simple *if-then* decisions based on familiar perceptual signals. Each decision usually has only two outputs. For example, "if the red bulb is lit, then perform action A; if the green bulb is lit, then perform action B." Algorithmic tasks completely define the rules and logic of actions to be performed and guarantee successful performance if the subject follows prescribed instructions. Deterministic-algorithmic tasks can be compared with rule-based tasks according to Rasmussen's terminology.

Probabilistic-algorithmic tasks involve logical conditions with the possibility of three and more outputs, each of which possess a different probability of occurrence. This probabilistic element significantly increases the operator's memory workload and the complexity of task performance in general. Probabilistic-algorithmic tasks can also include nonalgorithmic problem-solving components.

The more complex tasks are nonalgorithmic. The latter class, according to Landa (1984), should be divided into three subgroups: *semialgorithmic, semiheuristic,* and *heuristic.* The distinction between these types of tasks is relative, not absolute. We recommend the following major criteria for the classification of such tasks: (a) indeterminacy of initial data; (b) indeterminacy of the goal of the task; (c) existence of redundant and unnecessary data for task performance; (d) contradictions in task conditions and complexity or difficulty of the task; (e) time restrictions in task performance; (f) specifics of instructions and their ability to describe adequate performance and restrictions; and (g) adequacy of the subject's past experience for task requirements.

If the situation or instructions contain some uncertainty resulting from the vagueness of criteria that determine the logical sequence of actions and, therefore, require the subject not only to perform actions based on prescribed rules but also to create his or her own independent cognitive actions that should be performed in order to achieve the required goal, such tasks are semialgorithmic. Because of the inability to remember all possible rules for the performance a probabilistic-algorithmic task and insufficient familiarity with the probabilistic characteristics of the task, this type of task very often becomes semialgorithmic and even semiheuristic.

If uncertainty is even greater and includes some independent solutions without precise criteria, the tasks are semiheuristic. This class of tasks does not only fully determine executive actions but also requires explorative actions for analyses and comprehension of the situation. Semiheuristic problems may include algorithmic and semialgorithmic subproblems. The purpose of the ergonomic design of such tasks is to reduce the degree of objective and subjective uncertainty in problem solving. A significant part of probabilistic-algorithmic tasks and all nonalgorithmic tasks can be considered as knowledge-based tasks according to Rasmussen's terminology (Rasmussen, 1983).

Purely creative tasks are a heuristic task problem. The major criteria of such tasks are an undefined field of solution, indeterminacy of initial data, and indeterminacy of the goal of the task.

The last basic characteristic of the task is the degree of physical efforts. It is particularly important for the evaluation of manual task. Physical efforts also have subjective components as a feeling of physical stress. The proposed classification of tasks suggests that the type of task depends not only on its characteristics but also on the past experience of the performer. If the past experience of the performer is not adequate for the task, even a skill-based task can turn into a problem-solving task. Thus, the major characteristics of tasks are structure, complexity and difficulty, and degree of physical effort. We consider the complexity and difficulty of task assessment in Sections 10.1 and 10.2.

Analysis of the previously presented material demonstrates that the concept of task in activity theory includes goal, motivation, and behavioral and cognitive actions, which can be decomposed into cognitive and motor operations. Goal and motivation greatly influence the specificity of information processing. The overall goal of a task should be separated from the goals of individual actions. In contrast to cognitive psychology, the goal in SSAT cannot be integrated with motives. In cognitive psychology, the goal can have such attribute as intensity. However, the goal as a cognitive mechanism can be precise or imprecise and can include verbally logical and imaginative components. According to the SSAT, the goal is an informational component only and does not have the attribute of intensity. From the SSAT point of view, the more intense the motives are, the more a subject will expend his or her effort to reach the goal. The subject's goal cannot be sufficiently defined at the beginning of task performance and then can be specified or modified during task performance. The goal of the task and motives creates the vector *motive → goal* that gives directness to activity during task performance. The task can be presented by instructions in a ready form. In certain situations, tasks are not well delineated. In accordance with prescribed rules and restrictions, as well as contextual purposes, operators formulate a goal of a task and a task itself. Changes in the situation, conditions, or objectives may lead to the reformulation of a task, rejection of a task, shifting attention to new tasks, etc. This is a self-formulated task. Independently formulated tasks are especially common when operators control automated systems and a lot of unpredictable situations may arise. In such situations, a subject often has to formulate the tasks independently.

Some specialists considered the goal as an end state of the system toward which human activity is directed. This is a cybernetic understanding of the goal. The shortcoming of such an understanding of the goal is that the goal of a technical system is not distinguished from a human goal. The goal of the task cannot be considered as the end state of the system. There are only objectively given requirements of the task, which need to be transferred into a subjectively accepted goal of the task. Even in situations when a task's goal is clearly defined, a person has to interpret, clarify, reformulate, and accept it.

The person formulates intermittent goals, the achievement of which brings him or her closer to the overall goal of the task. The goal is just one anticipatory mechanism of the activity during task performance. Every task has requirements and conditions. Anything that is presented to the worker or known by him or her is conditions of the task. The requirements of the task include the information of what needs to be achieved. When requirements are interpreted and accepted by a person, they become the goal of the task.

The transformational process is a major means by which the goal can be achieved. The task includes the initial situation, the acceptance of the goal, formulated independently or formulated by others, that is associated with motives, the cognitive and behavioral actions required for achievement of the goal, and the elements of the external environment as a basic component of conditions for task performance.

3.2 Computer-Based Tasks in Production and Nonproduction Environments

In the HCI field, various professionals remark that in different task analysis techniques, professionals utilize the same terms that have different meanings. For example, for some of them, such terms as *goal* and *task* are synonymous. Similarly, goal, task, actions, etc., will be different for various professionals, depending on their previous backgrounds (Preece et al., 1994, p. 411). Preece and his colleagues introduced such terms as external and internal tasks in HCI study. "A goal (also called external tasks) may be defined as a state of a system that the human wishes to achieve." And further on the same page, they wrote "We can define a task (or internal task) as the activities required, used or believed to be necessary to achieve a goal using a particular device."

It becomes clear that the authors confused the technological or control processes with the work process during performance of a specific task. Naturally, these components are interconnected in task performance. However, ergonomists or work psychologists concentrate their efforts primarily on the analysis of the work process. Technologist or software engineers analyze the tasks primarily from the standpoint of the technological or control analysis processes. It is natural that these aspects of the analysis of task performance are interrelated. However, there are those aspects of task analysis that are specific to different specialists. For example, technologists or process engineers in manufacturing can be engaged in engineering design or those aspects of task analysis that are specific to their profession. Their familiarity with work psychology or ergonomics can be very limited. Similar in the analysis of the control process, software engineers can be involved in those aspects of task analysis that are not within the competence of specialists in the field of human factors or work psychology.

All of these demonstrate that the notion of *task* is increasingly difficult to use in studies in the computer-based field. In recent work, Diaper (2004) and others discuss different meanings of task and goal in the HCI field. The author draws attention to the fact that in cognitive psychology, such terms as goal and task are ambiguous terms that are not clearly defined in ergonomics and cognitive psychology.

These terms are understood differently and have fundamental meaning in activity theory. Our lives can be conceptualized as a continuing attempt to solve various tasks. The task in activity theory is not only routine work but also a problem-solving process through which a person interacts with the outside world and other people. Such concepts as motives, goal, task, and cognitive and behavioral actions are important and clearly defined in SSAT. Activity during task performance is considered as a process; however, this process consists of a sequence of logically organized structured elements, which includes actions, operations, and other components. The structure of activity is organized based on principles of the goal-directed self-regulation process. This process is a basis for the integration of activity elements and development strategies of task performance (Bedny et al., 2014). Such understanding of task is totally different from its ambiguous interpretation in other fields of psychology and ergonomics. The task concept in SSAT helps to understand how people work with computers in the real world. The central focus of study for SSAT in HCI is work settings where users carry out their tasks. Consideration of activity as a process and structure helps to conceptualize the cognitive aspects of HCI from an activity theory perspective (Bedny and Chebykin, 2013). Thanks to such concepts as cognitive and behavioral actions, the principle of unity of cognition and behavior and cognitive processes become considered in a real-world context.

The task concept becomes very confusing in the study of HCIs in a nonproduction environment. In such an environment, the computer is used apart from the workplace. The computer as an informational means is now present in the everyday lives of people outside of the production environment. In this chapter, we demonstrate that task analysis is a major tool used for optimizing human performance in the production environment as well as in the nonproduction environment.

From the systemic-structural activity theory perspective, task analysis in the HCI field can be defined as the study of work activity strategies in terms of logical organization of cognitive and behavioral actions to achieve the goal of the task that derives from demands of the computerized system. Per this definition, the goal of the task and the goal of the system are not the same. The overall goal of the task should be accepted or formulated by the performer based on the requirements of the system. Furthermore, there is never just one optimal way of task performance. This is specifically relevant for computer-based tasks. Possible strategies of task performance should be developed based on constraint-based principles that are derived from system requirements.

One weakness common to cognitive approaches is the disregard for the emotionally motivational aspects of human work activity in the HCI study. Usually this problem in ergonomics is reduced to the study of emotional or psychic tension of the operator in a stressful situation. Psychic tension is the state of the operator that arises in difficult activity conditions. In the West, this problem is known as emotional stress. This problem is important for reliability analysis. Nayenko (1976) distinguishes between what he calls operational and emotional tension. Operational tension is determined by a combination of task difficulty and lack of available time. Emotional tension is determined by the personal significance of the task. These two kinds of tensions are closely interrelated. Under certain conditions, one of them causes the other.

This abbreviated analysis shows that emotionally motivational aspects of task analysis in cognitive psychology are reduced to a relatively narrow problem of task performance in stressful conditions. This approach does not consider positive emotions in task performance, the relationship between emotions and motivation, the significance of the task for the person, etc. In industrial/organizational psychology, motivation is considered separately from the human information processing approach developed in cognitive psychology. Usually motivation is described from personality, group dynamics, and productivity perspectives.

A functional analysis of activity (Bedny and Karwowski, 2007; Bedny and Meister, 1997) clearly demonstrates that cognitive functions are tightly connected with emotionally motivational factors of activity. People can process the same information differently depending on its subjective significance (positive or negative) for a person. This is why SSAT considers activity as a system, which integrates not only cognitive and behavioral but also emotional motivational components.

Today, computer-based technology opens a new wide area of nontraditional ergonomics design that includes recreational and nonproductive design. Computing technology with increasing frequency is now used for nonproductive purposes. There is a trend to design a human–computer system for recreation activity, education, games, etc. (Karat et al., 2004). In order to improve the design of such systems, we need a good understanding of the following: task, goal, emotionally motivational aspects of human performance, etc. A not sufficiently precise definition of such basic concepts in cognitive psychology is a source of confusion, erroneous terminology, and incorrect task analysis in general.

Recently in ergonomics, the concept of affective design emerged (Helander, 2001). This field of design attempts to introduce in ergonomics pleasure-based principles of task analysis as an important field in ergonomics. Such concepts as emotion, motivation, or aesthetic requirements are particularly important in this design. However, in activity theory, emotionally motivational aspects of human performance were always important. Human information processing cannot be separated from emotionally motivational aspects of activity. A person is not a computer and cannot be considered simply as logical devises. A person is always emotionally related to the information presented to him or her. Hence, emotionally motivational processes always interact with cognition. Interpretation of information depends on emotionally motivational aspects of activity. Pleasure-based design is only one aspect of the implication of the principle of unity cognition, behavior, and emotionally motivational components of activity. In traditional design, this principle is also critical. Conceptualization of the principle of interdependency of cognition, behavior, and motivation can be found in SSAT (Bedny and Karwowski, 2007). In the area of pleasure-based design, we need to know how a user emotionally evaluates a product, an interface, or a system. Application of emotionally motivational aspects of affective design requires development of adequate concepts, terminology, and theory that would consider their relevance to the design process. Design principles and concepts developed in SSAT can be very useful for this purpose. In this section, we considered the concept of task and its main attributes from the SSAT perspective.

There are two main types of activity: object-oriented and subject-oriented activity. The first type is performed by a subject using tools or material objects. The simplest scheme of such activity can be presented as follows:

$$\text{Subject} \rightarrow \text{tools} \rightarrow \text{object}$$

Through the use of mental and external tools, the object is modified in accordance with the required goal. Objects may be either concrete or abstract (mental signs, symbols, images, etc.). The next type of activity refers to what is commonly called social interaction and can be presented as follows:

$$\text{Subject} \leftrightarrow \text{tool} \leftrightarrow \text{subject}$$

Just like object-oriented activity, social interaction begins with subjects' goals, orientation in a situation, etc. However, this kind of activity includes understanding of partners, prediction of their goals, strategies of activity, and personal features. Social interaction includes three aspects: exchange of information, personal interaction, and mutual understanding (Bedny and Karwowski, 2007). Intersubjective aspects of activity can be found even in subject–object activity when there is no direct contact with others. These data are presented in the works of famous philosopher and literature theorist Bakhtin (1979). Social interaction between subject → object can be uncovered through *inner dialogue*.

There is another activity classification: play, learning, and work. All three kinds of activity have similar structure and include goals, motives, and cognitive and behavior actions. Playing and games play an important role in the study of HCI in the nonproduction environment. For our analysis, it is useful to consider the role of playing in mental development. Vygotsky's (1978) work gave the most wide-ranging account of psychological characteristics of the game and its role in mental development. When a child is playing, this fulfills two functions: formation of a child's needs and formation of his or her cognitive functions. The purpose of play is not in achieving some useful result of activity but rather the activity process by itself. However, this fact does not eliminate goal formation and motivational aspects of activity. Actions of a child are purposeful and goal directed. For example, an adult demonstrates how to feed a doll. Children have their own experience of being fed. In spite of the fact that a child cannot really feed the doll, he or she still performs conscious goal-directed actions. Moreover, imaginative aspects of the play become critical when children operate with a variety of objects, which has a particular purpose and is associated with objects' meaning. The child at play operates with the meanings that are often detached from their usual objects (Vygotsky, 1978). For example, a child can take a stick and tell that it is a horse and start performing meaningful actions that have some similarity with a rider's actions. Play is associated with pleasure and develops motivational components of activity. Gradually, the play becomes more and more important and a child develops the rules for his or her play. Subordination to rules makes playing more complex and it is transferred into a game. The imposed rules force children to learn how to suppress involuntarily impulses. Hence, the childhood games are preceding real adult activity.

In the game, the same as in work, the tasks can be formulated in advance by others or formulated by the subject independently. Independently formulated tasks exist even in jobs that have rigorous rules and requirements. For example, during the flight from New York to Moscow (Russia), a pilot performed not only prescribed tasks but also multiple tasks formulated independently depending on the weather, information from the air controllers, etc. Similarly, there are multiple self-initiated tasks during performance of computer-based tasks in productive and nonproductive environments. In this type of task, goal formation and motivational aspects of activity are particularly important. Analysis of games and other types of activity demonstrates that the goal is one of the central concepts in psychology, but in cognitive psychology, it is mixed with motivation, which is one of the main confusing factors in the applied field. For example, Diaper and Stanton (2004, p. 611), criticizing the goal concept in cognitive psychology, wrote:

> The basic idea is that there is some sort of psychological energy that can flow, be blocked, deviated, and so forth. Goals as motivators of behavior would seem to be a part of this type of psychological hydraulics. Given that there is no empirical evidence of any physical substrate that could function in such a hydraulic fashion, perhaps we do not need concept of goals as behavior motivators.

We may agree with such comments when a goal is considered in cognitive psychology. However, in activity theory, the relationship between the goal and energetic components of activity is treated totally different. As we have discussed earlier, goal in SSAT and applied activity theory (AAT) is a cognitive component that interacts with motives and creates a vector *motives → goal*, or more specifically *motivation → goal*. In this regard, it is interesting to mention Klochko's (1978) work where he studied contradictions between task elements (conditions) and emotional tension during task performance. Subjects did not know about contradictions in task conditions and could not report them verbally. It was discovered that galvanic skin response (GSR) was increased when a subject reads the text of the task description related to contradictions. Electric resistance of the skin is an indicator of emotional tension. Hence, emotion reactions emerged as an indicator of contradictions in a problem situation without the subject's awareness. This is just an example that demonstrates a complicated relationship between emotionally motivational and cognitive components of activity. Analysis of relationship between cognitive and emotional-motivational aspects of task performance can be found in Bedny and Karwowski (2004b).

With the development of the computer industry, games became important even for adults. The motivational factor plays a particular role in a game. So we will consider the stages of motivational process in human activity. According to the concept of motivation developed by Bedny and Karwowski (2006), there are five stages of motivation: (1) preconscious, (2) goal related, (3) task evaluative, (4) executive or process related, and (5) result related. These stages, according to the principle of self-regulation, are organized as a loop structure, and depending on the specificity of the task, some stages can be more important than others. Depending on task specificity, scientists should give more attention to some of these stages and their relationship.

The first preconscious stage of motivation predetermines motivational tendencies. This stage is not associated with a conscious goal but rather with an unconscious set that can be later transferred into conscious goal and vice versa. The second goal-related motivational stage is important for goal formation and acceptance of the goal. This stage can be developed in two ways: by bypassing the preconscious stage of motivation or through the transformation of an unconscious set into a conscious goal. When the current task is interrupted and attention is shifted to a new goal, the previous goal does not disappear but is transformed into a preconscious set. It helps a subject to return to an interrupted task, if necessary, through transition of a set into a conscious goal. The third motivational stage is related to the evaluation of the task difficulty and significance, which has been discussed in the previous sections. The fourth executive or process-related motivational stage is associated with executive aspects of task performance. Goal formation, task evaluation (evaluation of task difficulty and its significance and their relationship), and process-related stages of motivation are particularly important for understanding risky tasks, games, and development of recreational computer-based tasks. The fifth stage of motivation is related to the evaluation of activity result (completion of task). All stages of motivation can be in agreement or in conflict.

Let us consider some examples. The relationship between process-related and result-related stages of motivation is important for the production environment. In some cases, the work process itself does not produce a positive emotionally motivational state. This can be observed during the performance of a boring job when the work process–related stage of motivation is negative. In order to sustain positive motivation during such task performance, commitment to the goal (stage 2) and result-related motivational stages (5) should be positive to offset it.

In computer-based games, the process-related stage (4) is critical and should be associated with the positive emotional-motivational state. The result-related stage (5) should vary when positive results are combined with negative results producing a combination of positive and negative emotionally motivational states. At the same time, only a positive result in computer-based games can reduce interest in the game. A simple game without the risk of losing can reduce the positive aspects of the process-related stage of motivation. Hence, the complexity of the task should be regulated depending on the previously obtained results. If the game is designed for children, a possibility to obtain a positive result should be significantly increased. Even in gambling when people can lose their money, some relationship between success and failure is important. The strength of positive and negative emotions during different stages of motivation is also important. This is particularly relevant for the risk-addicted people since manipulation with process- and result-related stages of motivation is critical. Of course, other stages of motivation also should be taken into consideration. In nonproductive tasks, the simplicity or difficulty to obtain a desired result is an important factor. Understanding motivation as a sequence of interdependent motivational stages helps us to create a desired motivational state in the production and nonproduction environment.

Sometimes the goal of the task game is not precisely defined, and at the beginning, the goal is presented in a very general form. Only at the final stage of the game, the goal becomes clear and specific. This is not new. For example, when playing chess, the goal can be formulated only in a very general form *to win* or *to tie the game*. A chess player also formulates in advance some hypotheses about his or her possible strategies that are tightly connected with the goal of the task. For this purpose, a player uses some algorithmic and heuristic rules that he stores in his memory. When a chess player selects a possible strategy, he or she starts to formulate multiple intermittent goals that correspond to this strategy. The selected strategy can be corrected or totally abandoned depending on the strategies of the opponent. A clear and specific understanding of an overall goal is possible only at the final stage of the game, just before a checkmate. Even when a goal of the task is externally given in a very precise form, a subject can reach this goal by using a variety of strategies and various intermittent goals. Therefore, a goal cannot be considered as an end state of the system that the human or machine wishes to achieve as has been stated by Preece et al. (1994, p. 411). A goal of the system and a goal of a person are two totally different concepts.

Let us consider as an example one possible task from everyday life. Suppose one of the authors of this chapter wants to get ready for a formal meeting. He needs to select a tie that matches his suit. First of all, he formulates the goal: "I need to select the most suitable tie." Then he opens his closet door and looks at his ties comparing their colors with his suit, asks his wife if he made the right choice, etc. The goal of this task is not precise at the beginning of the task performance, and only at the final stage of the task, the subject identifies what tie should be selected. However, the goal of the task exists in a general form. If somebody needs to find a tie, he does not look for shoes.

This understanding of goal is radically different from its description as a clearly defined end state of the system as it is considered by some usability engineers. Karat et al. (2004) stated that a task in the production environment has a clearly intended purpose or goal. According to these authors, the

HCI field shifts its focus from the production environment with its clearly defined tasks and goals to the nonproduction field, where the major purpose is communication, engaging, education, game, and so on.

> HCI professionals might say that people use technology because they have the goal of reaching a pleasurable state, but this is awkward and has not proven useful as a guiding approach in design. This is partly because of the difficulty in objectively defining the goal state, and without this, there is not much the field can say about the path to the goal (Karat et al., 2004, p. 587).

Here, the authors mix the goal of the task with the motive and insist that in the nonproduction environment and particularly in games, there are no tasks and goals. However, in our earlier example, we showed that the goal is to reach the desired future result of the game. The motive is to obtain a pleasurable state during a game and some satisfaction after a game. The goal of the game in the nonproduction and production environment can be not precise at the beginning of the task performance. In designing a task in any environment, it is very important to find out how an initially formed goal is gradually clarified and specified further during the task performance. The previously described data in combination with data obtained in SSAT and AT (Bedny and Karwowski, 2007; Pushkin, 1978; Rubinshtein, 1959; Tikhomirov, 1984) and cognitive psychology (Newell and Simon, 1972) can be of use for these purposes. In cognitive psychology, intermittent goals are known as subgoals. The subgoaling process is performed based on the means–ends analysis where the desired subgoal is considered as the end state of the step. This end state is compared with the present state of knowledge. From the AT perspective, these data require some additional interpretations because the subgoals are mental representations of desired future results. Hence, a subgoal is a cognitive and conscious entity. There is also a need for a general motivational state that creates an inducing force to produce this subgoaling process. Comparison of a future hypothetical end state with an existing state is provided by feedback that is performed in a mental plane. This demonstrates that thinking works as a self-regulative process. Moreover, there are well-defined and ill-defined problems. Well-defined problems are those that have a clearly stated goal. Performance of ill-defined problems in more complex situations begins with searching for and forming the goal. A subject promotes hypotheses, formulates hypothetical goals, and evaluates them mentally or practically. Only after that can he or she formulate hypotheses on how to achieve a defined goal. The hypothesis formation process can include conscious and unconscious components. Unconscious hypotheses are not verbalized. They can be performed, for example, in a visual plane (Pushkin, 1978). Some of these hypotheses can be later transferred into a verbalized, conscious plane. There are also hypotheses that are conscious during a short time and then they are forgotten. People are goal-directed systems.

According to Karat et al. (2004), the movement of technology into the home environment and everyday life of people eliminates task-oriented activity.

This can be explained by the fact that concepts of task and goal are not clearly defined outside of activity theory. Moreover, Karat et al. (2004, p. 588) wrote that "the science of enjoyment is not capable to define a goal-directed approach." We would like, first of all, to draw attention to the fact that there is no such science as *enjoyment*. In psychology, this term simply refers to a certain emotional state. Then it should be noted that even in entertainment, the concept of *task* is very important. Our activity or behavior strives toward anticipated goals in the production and nonproduction environment. Analysis of a variety of tasks people perform in everyday life demonstrates that they attempt to break down the flow of activity into smaller segments or tasks that are often self-initiated. Users' everyday life activities cannot be understood without referring to the motivational and goal formation process (Bedny et al., 2011). In contrast to cognitive psychology, in AT, a task always has its desired final goal and motivational forces. Similarly, social interaction, learning, playing, and games always, as any other type of human activity, are motivated and goal directed.

Karat et al. (2004) substitute such a complicated concept as motivation by the term *value* that just has simply in their discussion a common sense meaning. Hence, the classification of HCI systems as communication driven, content driven, etc., is questionable. The authors come up with a new *science of enjoyment*, which in their words is not a goal-directed approach. It is hard to agree with such interpretation of enjoyment. People can enjoy drugs, alcohol, work, sports, etc., depending on the motivational factors. Hence, the study of motivation should be associated with enjoyment. Is there a need for a new *science of enjoyment* when there is psychology and motivation as its branch?

Any technology is just a means and/or tools of human work or entertainment activity. Hence, we need to study the specifics of utilizing such tools or means of work in various kinds of activities. For instance, in order to design a computer-based system of person-to-person communication, such a system should be adapted for social interaction activity providing a means of understanding the partners, prediction of their goal and motivational state, ability to formulate a general goal for members of communication, ability to emotionally interact with each other, etc. Usability engineers should work together with psychologists to improve the design of computer-based tasks in various environments.

Each task in the work process or operational-monitoring process should be regarded as a situation-bounded activity that is directed to achieve a task goal under given conditions. The task includes motivational components and problem-solving aspects. The relationship between an overall goal of the task and the conditions in which it is presented determines the task to be performed. The task formation process is presented in Figure 3.1.

From this figure, one can see that the vector *motive* → *goal* and the task are a result of the subjective reflection of reality, which of course depends on such objective factors as conditions and requirements. There are also intermediate goals of actions during task performance. Usually they are quickly

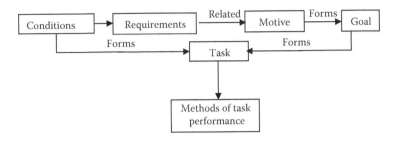

FIGURE 3.1
The task formation process.

forgotten during task performance. From a functional analysis perspective, goal acceptance should be distinguished from goal formation. Goal formation is associated with self-initiated task. The goal acceptance process is connected with a prescribed task (Bedny and Karwowski, 2007).

As can be seen from Figure 3.1, a goal cannot be given in a ready form. The conditions and requirements together with motives (motivation) form an overall goal of the task. Conditions and goal form a task. For any objectively given task, its requirements should be transferred into a subjectively accepted goal. Thus, objective requirements and subjectively accepted goals are never exactly the same. In AT, objective requirements and objectively given goals sometimes are used synonymously. Requirements can be modified and interpreted and subjectively accepted. Analysis of presented data demonstrates that concepts of goal, goal acceptance, goal formation, task formation, etc., are discussed in a totally different way in systemic-structural activity theory in comparison to any other fields outside the activity approach.

A task that has the same objective complexity can be evaluated by subjects as a task with different subjective difficulty. It depends on individual differences, past experience, and the temporal state of the subject. An individual might under- or overestimate the objective complexity. For example, a subject can overestimate a task's difficulty and can reject task performance. In other cases, the subject can accept the task but reduce the quality of performance.

The relationship between difficulty and significance determines the level of motivation. If the task is very difficult for the subject and significant for him or her (subjectively important), the task can be accepted. If the task is very difficult but not significant, it can be rejected.

Strategies of task performance depend not only on cognitive but also on emotionally motivational mechanisms of activity regulation. In particular, they depend on emotionally evaluative mechanisms or factor of significance. Subjects actively select the required information and interpret it depending on the expected goal and personal significance of the task or its elements. Such mechanism as significance is specifically important in task analysis in the nonproductive environment. The other specifically important mechanisms in the formation of strategies of task performance in the non-production environment are mechanisms of subjective evaluation of task difficulty and subjective criteria of success. We consider these mechanisms in more detail in Bedny et al. (2014) where we describe models of activity self-regulations developed in SSAT. Here we will consider this issue in relation to the analysis of the task in the nonproduction environment.

For example, in gambling, the goal is to win money in a risky situation. Achievement of this goal in such conditions is conveyed also with the possibility of losing money. These are two significant aspects of the gambling task. The relationship between these two kinds of significance determines the subject's level of involvement in gambling. Moreover, there is also a possibility that a risk-addicted person will choose a task with a high probability of failure. For example, the task can be perceived as very difficult, but the subject still desires to perform the task. For such a subject, it is important to not only obtain money but also experience the danger. Elimination of danger immediately removes feelings of positive significance in gambling for such a person. Risk-addicted people are usually involved in gambling based on some proportion of positive and negative significance associated with the mental representation of a probability to lose or win money. Feeling some level of risk (to lose money) increases the feeling of positive significance, which in turn triggers inducing components of motivation. After success in a risky situation, risk seekers often get involved in even riskier tasks until they fail. Positive and negative significance factors interact and influence each other. Hence, there are individual differences between risk-addicted and non-risk-addicted people in the selection of activity strategies in risky situations. Some individuals always attempt to be involved in tasks that have some negatively significant factors with various proportions of positive and negative significance. Moreover, this proportion can have different values for different individuals. For some individuals, such relationship is dynamic and depends on their history of success and failures. The previously considered aspects of motivation are important for the development of computer-based gambling tasks and nonproduction tasks in general. All these aspects of motivation cannot be understood without analyzing activity self-regulation (Bedny et al., 2014).

Outside of SSAT, there is no agreed upon understanding of the concept of task and of its main attributes in cognitive psychology and even in AT. Task analysis is a multitude of independent, and sometimes not sufficiently grounded from the theoretical point of view, techniques. There is a reasonable opinion that at this point, task analysis is a mess. Moreover, some practitioners raise the questions about the future of task analysis (Diaper, 2004; Diaper and Stanton, 2004). There are even suggestions to eliminate the concept of task because it ignores motivational forces, or to eliminate this concept just for entertainment systems. Other professionals insist on eliminating the concept of goal in task analysis. All of these examples clearly demonstrate that in cognitive psychology, there is no clear understanding of the goal and task concepts and their application does not bring any value for practitioners.

From the SSAT perspective, any task includes a means of work, materials and mental tools, work processes, and technological or control process. Hence, the task can be studied from a behavioral or technical perspective. Of course these two aspects of study are interdependent. In ergonomics and work psychology, behavioral aspects of task analysis are more important. At the same time, these two aspects sometimes can be considered as relatively independent. For instance, a process engineer can develop a technological process that does not require consideration of behavioral aspects of task performance. However, this independency is not absolute. The means of work, including informational technology, often significantly influences human work activity. One aspect of task analysis in the design process is to find out the relationship between the physical configuration of equipment and the structure of activity during task performance. Task analysis is important for the training process and for safety and also for the development of selection methods for very demanding professions. Therefore, task analysis is critically important for the study of human work from a behavioral point of view.

Cognitive task analysis concentrates its efforts first of all on cognitive aspects of human work and does not pay enough attention to the interaction of cognitive, behavioral, and motivational aspects of human work. Work motivation in industrial/organizational psychology is considered separate from ergonomic design. However, human information processing inherently interacts with emotionally motivational aspects of activity. The more complex and significant the task is for a person, the more mental efforts it takes and the higher is the required level of motivation. Hence, it is important to take into consideration these aspects of activity not only for pleasure-based design but also in traditional design, and not just for pleasure but also for more efficient interaction with technology in its broader sense. All of the previously mentioned aspects are particularly relevant for the study of HCI.

In this chapter, we have considered some basic characteristics of task for the production and nonproduction environment. It was demonstrated that this concept is critically important for the study of any kind of human activity, including games and entertainment in general. In contrast to cognitive psychology, the task in SSAT includes motivational forces and goal as its major attribute. An attempt to eliminate the goal as an attribute of task is incorrect because the goal is not well developed in cognitive psychology where it is often not considered at all, or considered as a combination of cognitive components with a number of different motives. As a result, some practitioners mixed goal and motives and the same task has multiple goals. In contrast, in AT, the goal is a cognitive component of activity and motives are the energetic component. *Motives* → *goal* creates an inducing vector around which all elements of task are organized. Goal is not an end state of the system. From a psychological viewpoint, it is a mental representation of a desired future result, which can be clarified and specified during task performance. If a subject totally changes his or her goal, he or she changes the task itself.

Attempts to eliminate the concepts of task and its main attributes such as goal are a source of confusion and of strange terminology such as *nontask system* and *the science of enjoyment*. From an SSAT perspective, such concepts as task, motives, goal, and interdependence of cognitive, behavioral, and energetic aspects of activity are critically important in the study of human work including computer-based tasks in non-production environment.

Our lives, even in the nonproduction environment, can be conceptualized as a continuing chain of various tasks. For example, people run a variety of tasks to maintain their houses, such as cleaning and washing dishes. These kinds of tasks can be an object of ergonomic studies when working on designing or improving household devices. The design principles for the development of machines, computers, and kitchen appliances are the same. We can evaluate the adequacy of equipment, computer interface, or home appliances only by assessing them in the context of task performance.

3.3 Structure of Production and Operational-Monitoring Processes

Production and operational-monitoring processes are other key concepts for ergonomics and work psychology. Analysis of their structure facilitates a better understanding of the relationship between technology and human work activity. However, these important concepts do not receive deserved attention in ergonomics. The concept of production process can be applied to various types of industry, but its basic characteristics are described in great detail in the manufacturing industry. Hence, we will briefly consider this concept and then review the operational-monitoring process. The manufacturing production process is also called the manufacturing process. Barness (1980) has listed six planning steps of the production process: (1) design of a product (developing drawings of a product that show its shape, size, weight, material, and so on), (2) design of the production process (designing technological process), (3) design of a work method (how an operator is to perform an operation), (4) design of tools and equipment (usually, equipment design is not included in considered steps because it already exists; however, some special tools are often manufactured for a specific production process), (5) design of equipment layout, and (6) determining the time for the production process (machine time and an operator's work activity time). We want to stress the fact that a presented sequence of stages makes it evident that design always precedes manufacturing process. Design is the process of working with information and not with real objects. A designed object should be created based on presented drawings and descriptions. Therefore, a design cannot be reduced to an experimental procedure as is often done in ergonomics. Experiments should be used at the final stage of evaluating a designed object.

A production process can be defined as a sequence of transformations of raw materials into a finished product that usually begins with entering raw materials and proceeds through various steps until the material becomes a finished product, or results are achieved in accordance with the purpose of the production process. Any production process contains three basic elements: human work activity, or work process; means of work; and product (Bedny and Harris, 2005). The means of work are those tools, equipment, and instruments used by subjects in the production process. There are various types of production processes. For example, in manufacturing, there are mechanical production processes, physical–chemical processes, transportation, and control production processes. The structure of the production process is presented in Figure 3.2a as a combination of work and technological processes and operational-monitoring processes as a combination of work and control processes (Figure 3.2b).

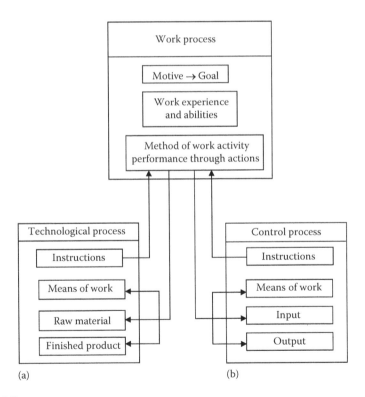

(a)   (b)

FIGURE 3.2
The structure of the production and operational-monitoring processes, (a) the work process and technological process that create the production process, and (b) the work process and control process that create the operational-monitoring process.

Here, the term process is used to emphasize that activity is performed according to some prescription or order. A work process contains a substructure of basic components: *motive → goal* as a vector that depicts directional and energetic aspects of work activity, *knowledge and skills* that demonstrate the relevance of past experience to the work process, *abilities* that are in relation to tasks to be performed, and *work actions* that are organized into a structure and together present a *method of work*. In this scheme, *action* refers to both cognitive and motor action. The presence of the concepts of knowledge and action in the structure of work process also implies the existence of mental tools.

A production process can be described as a sequence of separate steps or production operations (tasks). In engineering work design under production operation, one understands an isolated part of a production process that is performed upon a work object in one workplace by one or several workers. The existence of the same equipment and tools, object of work, and technological completeness characterizes a production operation. Production operations can be divided into smaller standardized technological units based on technological principles. While, in general, the purpose of any production process is the transformation of raw material into a finished product, it is possible to distinguish various types of production processes. For example, in manufacturing, there are mechanical production processes, physical–chemical processes, transportation, and control production processes (Gal'sev, 1973).

Figure 3.2 demonstrates that the production and operational-monitoring processes have two basic components: technological process and work process. All tasks performed by the workers must be adequate to all technological requirements of the production.

A technological process includes instructions according to which a worker performs his or her job and means of production such as equipment, the raw material or initial material object, and a finished product or result. The technological components of a production process are also performed in accordance with some prescriptive technological order. Any production process always involves some standardized requirements on how it should be performed.

Figure 3.2a shows that production operations include human activity, technological components or tools, and a transformed object. An object's nature changes depending on both human activity and specifics of the technological process. Therefore, production operations can be studied in the technological frame or from an activity theory perspective. According to the latter, a task or production operation can be divided into a logically organized system of cognitive and motor actions that can be in turn divided into smaller units or integrated into a combination of actions. In the production process, tasks can have various degrees of freedom of performance such as deterministic or skill-based task or rule-based tasks.

When we consider the production process through the prism of the work activity, the object of analysis becomes the work process. Thus, the production process includes technological and work processes. In the study of work process, the major focus of analysis is work activity during the performance of various tasks. The work process and therefore tasks are always performed according to some prescriptions and constraints. Figure 3.2a demonstrates the interaction of human work activity and the technological process. If the technological process changes, human actions also change. From the perspective of activity theory, a work process may be understood as a combination of various types of work activity during performance of tasks by subjects for accomplishing the objectives of the system.

In some cases, the production operation (e.g., in manufacturing) can be executed outside of in advance designed production process. For example, a mechanic repairman should independently make a new part needed for repair of broken equipment. The production of such items is not described in advance in the special documentation. A highly skilled worker simply uses his or her knowledge and experience to the manufacture of parts. In this case, we are talking about knowledge-based production operation. We use the terminology proposed in English language for describing different types of behavior (Rasmussen, 1983). Performance of knowledge-based production operation requires the second level of skills or *umeniya* that constitutes an individual's ability to organize knowledge and the first-level skills into a system efficiently used to perform a particular class of tasks or to solve a particular class of problems.

The described production process is relevant for the mechanization stage of the production process. At the automation stage when workers perform tasks in automated and semiautomated systems, the nature of the production process significantly changes. The main functions are no longer transformation of physical or material object but transformation of information. Tasks are not only performed based on some procedural rules but also involve problem solving. Rather than controlling power sources directly, an operator uses intermediary control devices. In this type of work process inspection, controlling and monitoring functions predominate, and the motor components of activity are significantly reduced. Another characteristic of an operator's performance with the automated systems is that he or she is required to perceive information from a variety of displays and instrumentation. Work activity of the worker–operator is extremely flexible. In this type of work process, inspection, controlling, and monitoring functions predominate, and the motor components of activity are significantly reduced. Work activity in its external appearance (as motor actions) loses continuity and acquires an episodic character, while at the same time, the role played by the sensory-perceptual and thinking components of activity increases (Bedny and Harris, 2005). Recognition of the specific characteristics of the operational-monitoring process requires some reconsideration of our description of the work process.

As a result, the term production process has been replaced by the term operational-monitoring process and the term technological process has been substituted by the term control process. An operational-monitoring process is defined as a combination of duties essential to accomplish some automated or semiautomated system functions. The notion of the production operation is no longer appropriate. Rather, the task is seen as the basic component of operational-monitoring processes. In this case, the task goal is not always reduced to the transformation of material or physical object but often involves changes in the state of a control object, or the transformation of information. The task then becomes one of problem solving. The operational-monitoring process can thus also be described as a system of logically organized problem-solving tasks.

The structure of an operational-monitoring process is depicted on the right side of Figure 3.2b. In the control process, the components *raw material* and *finished product* are replaced by *input* and *output*. In most cases, the input is the information received by an operator about the initial state of the controlled object. The output is the information about a controlled object following completion of a task. An operational-monitoring process can thus also be described as a system of logically organized problem-solving tasks. In a production process, tasks or production operations are prescribed in advance. In an operational-monitoring process on the other hand, tasks are often self-initiating. In many cases, it is the operator themselves who must determine what has to be done. Although the work is based on certain procedures and rules, it also involves creativity and problem solving. In these cases, the work is organized as probabilistic algorithms and quasi-algorithms, with the operator forming the performance rules based on his or her experience.

Computerization significantly alters specifics of an operational-monitoring process, often leading to greater demands on a task performer. When a subject directly interacts with a computer during task performance, this is an HCI process that includes logically organized computer-based tasks. A schema of such process is basically the same as the one for operational-monitoring processes, the only difference being that a computer is now the dominant means of work. Thus, each task in the work process or operational-monitoring process is regarded as a situation-bounded activity within this process. The main components of the task are a goal of the task and a logically organized system of cognitive and behavioral actions that are directed to achieve this task's goal. In the operational-monitoring processes, the proportion of tasks that is formulated by the operator or the user independently dramatically increases. This means that the proportion of the knowledge-based tasks is increased significantly.

Analysis of production or operational-monitoring processes demonstrates that there are feedforward and feedback influences between human work activity and technological processes. A worker selects required actions for task performance according to existing restrictions, prescribed instructions, and existing knowledge. However, if we change the configuration of the equipment, the structure of activity also changes. Any changes in the physical configuration of an equipment or interactive software affect the activity structure in a probabilistic manner. Such changes cannot be described and accounted for without the concept of human cognitive and behavioral actions. An analytical comparison of activity structure and equipment or software configuration is the central component of ergonomic design. Further, the presented material demonstrates that design is first of all an analytical process that cannot be reduced to experimentation. In this framework, the main component of a unified ergonomic design approach is the comparison of (theoretical) models of activity with the models of equipment or software. Experimental methods are then used in the latter stage of design. Sometimes a specialist can use some prototypes as models. However, theoretical models in such situation are still required.

Very often it is required to determine the most effective work method when a worker uses the same equipment. In such a situation, a specialist is involved in designing the most effective methods of work. Each method may be sufficiently flexible. In this case, we are talking about designing effective strategies of activity during task performance. Analysis of the previously presented material demonstrates that the work process includes a number of different tasks that can be classified according to previously described criteria.

4

Basic Concepts and Terminology: Comparative Analysis

4.1 Analysis of the Basic Terminology in Activity Theory

The success of any theoretical and applied research in psychology, as in any science, largely depends on the proper use of basic terminology and fundamental concepts underlying the theory. Terminology is a major problem of psychology. This is particularly relevant to papers published in English. This language is international in science. Multiple publications and translations from various languages into English are sources of incorrect interpretation of psychological terminology. Moreover, the same terminology in different languages has totally different meanings. This is related also to activity theory (AT) terminology, which in addition is overloaded with Marxist terms. All of these make analysis of the terminology in AT particularly relevant. The other factor that influences the incorrect interpretation of AT's basic terminology is the fact that the same terms in various approaches of psychology have totally different meanings.

Analysis of publications in the field of AT demonstrates that scientists have great difficulty in translating AT terminology and its interpretation. Many of the translations have failed to capture the original meaning of the basic concept of AT (Diaper and Lindgaard, 2008). This could be due to multiple reasons. One of the reasons is that this theory has been developed in very specific sociocultural context. The other reason was that specifics of the language and the associated history of the origin of various terms have not been sufficiently taken into account. It is important to understand how difficult it is to translate AT into English language and adapt this theory for practical application.

In this section, we conduct a comparative analysis of the basic terminology that is used in AT and in other theories of psychology and philosophy such as cognitive psychology, action theory, and praxiology that share some general ideas. The purpose of this chapter is not to identify the advantages or disadvantages of these areas but rather to identify the specifics and differences in terminology used by AT and by previously listed theories.

However, sometimes this requires a comparison of the advantages and disadvantages of various interpretations of the terminology in different theories. The other aspect of the chapter is analysis of the terminology in general, applied, and systemic-structural AT. Without understanding basic AT terminology, it is impossible to use it. Knowledge of AT terminology is particularly important due to the fact that the basic concepts, ideas, and principles of AT are now widely discussed in the field of ergonomics and human–computer interaction (HCI).

We start our analysis with comparison of AT and action theory.

One direction that has close ties with AT is action theory. Action can be described as a purposive or goal-directed human behavior. Humans perform actions in order to accomplish a goal. The tension that motivates an action performance is similar to the forces that tend to close the gap between the initial situation and the imagined future goal state of the situation. Currently, action theory has three versions. Two of them have been developed in Germany and one abbreviated concept has been developed in the United States.

In one German version, action theory considers action as a goal-oriented behavior that also assimilates in it cognition. Frese and Zapf (1994, p. 271) offered the following definition of this action regulation theory: "Action is goal-oriented behavior that is organized in a specific way by goals, information integration, plans, and feedback and can be regulated consciously or via routines." Further, they explained that action theory is a cognitive and behavior-oriented theory that can be applied to the study of human work.

The other German version is motivational action theory (Heckhausen, 1991; Heckhausen and Gollwitzer, 1987). Heckhausen and Gollwitzer mainly described volitional and motivational aspects of action regulation when people attempt to translate their wishes and desires into reality. According to Frese and Zapf (1994), this version of action theory is not utilized in the study of human work. Hence, we are not going to discuss it further. In the United States, Norman (1986) advanced what he called the *approximate theory of action*. In his theory, the term action is used similarly to the way it was utilized by Frese and Zapf, when they considered it as a goal-directed behavior, with a feedback loop.

Specialists who study human work do not always distinguish between described versions of action theories, which led to terminological confusion.

Praxiology is another approach that studies human behavior as a goal-oriented system.

This is a philosophical theory developed by Polish scientist Kotarbinski (1965) who studied human actions in terms of their efficiency.

Praxiology provides an explanation of human purposive action and formulates principles of its efficient and normative performance (Gaspaski, 1984). An action highly depends on the attitude of its subject, an actor or an agent. Kotarbinski treats this theory as the systemic philosophical approach that studies human work. Such basic concepts as agent, action, aim, product, result, system, and efficiency have been developed in this framework. Kotarbinski's theory is an important theoretical framework in studying different forms of the human praxis. The previously listed notions are also important in AT. The concept of action in praxiological terms has some similarity with the concept of activity. The difference between these two theoretical concepts is that one approach studies human praxis from philosophical perspectives, while the other uses psychological perspectives.

Even this short introduction demonstrates that there are different versions of action theories that are not clearly distinguished by some specialists in psychology and professionals in ergonomics. This leads to terminological confusion. Different versions of action theory and AT utilize similar terminology. For example, there are such concepts as activity, action, goal, goal-directed system, subject, result, self-regulation, systemic approach, etc., in various versions of action theories, but they have different meanings in AT. In translated AT publications, this basic terminology is mixed and used synonymously with various terms in action theory. Hence, in the following text, we consider the basic terminology of AT, compare it with the terminology in other related theories, and clarify the basic terminology of AT used in the West.

At present, there is an attempt to use only general AT while disregarding the data obtained in applied AT and systemic-structural activity theories (Bedny and Chebykin, 2013). This aggravates the issues of interpretation of psychological terminology. General AT is a grand theory or a framework. Such theories often are not very well developed empirically and theoretically for the direct application in the study of human work. The degree of specialization of a theory should be adequate for its application. Frese and Zapf (1994) wrote: "It is naive to think of general theories as ideas to be simply applied in work." There is a general consensus between applied AT and SSAT professionals that it is often difficult to use general AT in practice. One cannot expect from the philosopher and psychologist as Leont'ev was to create a theory that can meet all needs required for the study of human work in ergonomics and HCI in particular. The terminology of general AT is not well adapted to study human work.

The next important aspect of our discussion is considering concepts of activity and action from different theoretical perspectives. We start our analysis with consideration of such basic concepts of AT as activity (*deyatel'nost'*) and action (*dejstvie*). In the West, the terms activity and action are often considered interchangeable, but in AT, these terms have totally different meanings. We considered several definitions of activity before. In our discussion, we utilize a definition that was introduced by Bedny and Karwowski (2007). Activity is a self-regulated system that integrates cognitive, behavioral, and motivational components and is directed toward achieving a conscious goal of activity. From this definition, it follows that cognition, behavior, and motivation influence each other and make clear that not only cognition regulates behavior, but behavior also regulates cognition through feedback. It is also important to underline that emotionally motivational processes interact with cognition and behavior through feedforward and feedback. Further, activity is described as a goal-directed system. If we compare this definition of activity with the definition of action presented by Frese and Zapf (1994) earlier, one can see that the concepts of activity in AT and action in action theory have a similar meaning.

German psychologist Hacker (1986) presents the following definition of action: "Action is the smallest unit of behavior that is related to conscious goal." Such definition of action is similar to the understanding of action in AT. What is surprising is that Hacker's definition of action is also utilized by Frese and Zapf (1994, p. 274). These authors did not even notice that their first definition of action (Frese and Zapf, 1994, p. 271) is in conflict with Hacker's definition of action, which they utilize at page 274. In the first case, we are talking about goal-directed behavior, which includes mental processes and external behavior (in Russian language, it is activity). In the second case, we are talking about the smallest element of activity, which includes a conscious goal. This is how action (*dejstvie*) is understood in AT. At the time, this definition was written by Hacker when he lived in East Germany, where Russian AT was popular. Considered earlier, two definitions of action are in conflict. According to the first definition, action is a complex whole that has a conscious goal. According to the second definition, action is the smallest element of the complex whole that has a conscious goal. This example demonstrates how confusing terminology is in contemporary psychology. Frese and Zapf and Norman described action theory from cognitive and behavioral perspectives where action is considered as a goal-directed system.

Sometimes the term *action* is utilized to describe external behavior. At the same time, the goals, operators, methods, and selection rules (GOMS) system utilizes the term action to depict an isolated element of goal-directed behavior, which can be behavioral or cognitive (Kieras, 1993). Preece et al. (1994, p. 411) consider an action as a task that involves no problem-solving components. In AT, actions are elements of tasks that can include some cognitive and problem-solving aspects where actions should be distinguished from the task. That is, the term action in the West has two meanings: goal-directed behavior or separate smallest element of goal-directed behavior.

Hence, activity in AT is understood as a goal-directed system that integrates cognitive and behavioral actions. Actions are only elements of activity and cannot be mixed with activity. In action theory, the term action has a similar (but not exactly the same) meaning as the term activity in AT. Human activity is always conscious because it is directed to achieving a conscious goal. Actions as elements of activity are also directed to achieving their own conscious goals. The major type of activity is work activity. It should be noted that in English language, the term *activities* very often is utilized. However, this term is used in a non-AT sense to mean some bit of work people do (Diaper and Lindgaard, 2008). This term does not imply conscious goal and the integration of cognitive, behavioral, and motivational components of activity based on the mechanisms of self-regulation (Bedny and Meister, 1997).

A general AT also does not have a clearly developed terminology, which would allow the effective use of this theory in the study of human work. Leont'ev (1978), which is considered as one of the founder of AT, emphasizes on the excessive dynamism between actions and operations and on the lack of clear boundaries between them. Action during skill acquisition can transfer into an operation and become a component of a more complex action. At the same time, when conditions change, an operation can return to the level of conscious action. Kuutti (1997, p. 32), who attempted to discuss this problem, rightly noted that excessive flexibility of the basic concepts, according to Leont'ev's view, makes it impossible to create a general classification of what activity, action, and operations are.

Despite these comments, Kuutti (1997, p. 33) tried to demonstrate a relationship between activity, actions, and operations based on Leont'ev's work. In one of his examples, he suggested that building a house is an activity, but fixing a roof is an action. Such interpretation does not make much sense. Any person with a technical background understands that *fixing a roof* and *building a house* are examples of a production process that consists of a technological process and a work process (Bedny and Harris, 2005). They could be further divided into production operations or work tasks and described from technological and work activity perspectives. In the production processes such as *building a house* or *fixing a roof*, the technical operation associated with hammering nails is widely utilized. Let us consider this example. Such example was utilized by Miller et al. (1960) in their famous analysis of the test–operate–test–exit (TOTE) unit, which has its similarity with the concept of motor action in AT. Hammering a nail is considered completed when the nail is flush with the roof board.

Suppose a worker has to fix a roof by hammering 10 nails. Suppose that in order to hammer a nail until it is flush, the worker has to perform in average five strokes with a hammer. The worker raises the hammer and strikes the nail, and then he or she raises the hammer again and strikes the nail again. After five strokes, the nail is usually fully included in the board. The worker lifts up the hammer, strikes the nail under visual control, evaluates the result, and repeats the cycle five times. Each cycle includes two motions: The hammer is up in an approximate position and the hammer is down in the exact position under his visual control. The goal of such an action (do not mix with the goal of the task) is *to hit a nail*. According to Russian common language and according to activity terminology, this cycle is a motor action. It includes two motions: lifting the hammer and striking the nail. In order for the nail to be flush, an average worker needed to execute five motor actions. Thus, for hammering 10 nails, 50 motor actions are required. Hence, if for hammering only 10 nails, 50 motor actions are required, then the question arises: How many motor and cognitive actions are required when it is necessary to repair the roof? It is obvious that in answering this question, we need to conduct special analyses of the labor process. We can only say that in fixing a roof, a lot of cognitive and motor actions must be performed.

Let us also consider Leont'ev's (1977, p. 107) example when he attempted to explain the difference between action and psychological operation, where operations are considered as elements of actions. To demonstrate the meaning of the term *psychological operation*, he uses the following example. "It is possible physically divide material object by means of different instruments. Each of which defines how to perform an action. In some conditions will be adequate operation of cutting, and other sawing operation. It is assumed that person should be able to work with corresponding tools, a knife, a saw, and so on" (Leont'ev, 1978/1977, p. 107).

Ambiguity of this example is obvious because one can use a knife to cut bread or cheese, for example, and various types of saws can be used for cutting wood or metal. In the former case, we are talking about everyday home tasks, while in the latter case, we are considering production operations or work tasks. If we want to cut the bread, we need to take it and put it on a table in a required position, take the knife and move it to start position for cutting, and cut the bread. All these actions can be decomposed into motion or motor operations (the action *taking the bread* can be further divided into two motions: *reach for the bread* and *grasp the bread*). Similarly, one can describe the technological operation of cutting metal. Leont'ev, as he tried to show the difference between action and psychological operation, mistook a technical operation for a psychological operation. Dividing a material object, for example, by means of a saw is a technical operation but not a psychological operation, as Leont'ev wrote. Such technical operation can include a number of motor actions and motions. The latter are psychological operations.

The term *operation* proposed by Leont'ev can be confused with such terms as *technical operations, surgery operations*, etc. Therefore, Platonov (1982) offered to replace it with the term *mental act*. From this, activity during task performance could be presented as three-level hierarchies: activity, action, and psychological act (psychological operation). Decomposition of activity on actions and psychological operations or *psychological acts* is generally recognized in AT by different scientists. In case of motor actions, instead of the term *psychological operation* or *psychological act*, the term *motions* or *movements* are used. The terms *motor actions* and *motions* are basic concepts in engineering in studying human work (Gal'cev, 1973). The concept of motion also is important in time and motion study in the United States (Barness, 1980).

In a similar way, we can consider cognitive actions. For example, recognition of a familiar object may be seen as a simultaneous perceptual action, and the rotation of a mental visual image into the required position according to a specific goal is an imaginative action. Usually, in applied study, cognitive actions are not divided into cognitive operations because they have a very short duration. Operations as elements of an action do not have conscious goals and subjects are usually not aware of such operations. In applied AT and particularly in SSAT, there are methods of extracting cognitive and behavioral actions and operations (Bedny and Karwowski, 2004b). It should be noted that several simple actions can be integrated into a more complicated one by a high-order goal. However, the ability to do so is restricted by the capacity of short-term memory. The ability of such integration of actions is important during algorithmic description of task performance.

The next important aspects of our discussion are analysis of the subject–object relationship and units of analysis of activity. We start our discussion with such terms as object of study, subject of study, and then units of analysis. The subject–object relationship is critical in Rubinshtein's (1957) concept of activity. Nature becomes an object only during interaction with a subject. Activity according to Rubinshtein and later to Leont'ev is an object-oriented system. It is important to distinguish such terms as object of study (*ob'ekt izucheniya*) and subject of study (*predmet izucheniya*) in AT where there is a single object of study and diverse subjects of study. The same activity can be described from various perspectives, highlighting its various aspects. Thus, activity serves as an object of study, whereas analyzed aspects of activity are the subjects of such study.

Object of study also should be distinguished from units of analysis. In the book edited by Nardi, she wrote: "In activity theory the unit of analysis is an activity" (Nardi, 1997, p. 73). Similarly, Kaptelinin and Nardi (2006, p. 32) have defined activity as *the basic unit of analysis*. However, activity is an object of study and units of analysis are unified components into which we divide the whole for the purposes of studying the components and their integration into a dynamic whole (Vygotsky, 1956).

The choice of proper units of analysis is of key importance for studying a complex whole. A complex whole is an object of study and the units of analysis serve as a means in the study of this whole.

A production process is a collection of tasks that are organized in a particular order. The same task can be described from a technological or activity perspective. From an activity perspective, the task can be considered as a separate activity restricted by the task's goal and the time frame of performance. The term *separate activity* was utilized by Leont'ev (1997, p. 102; 1981). Task is regarded as a situation-bounded activity directed to achieve a conscious goal of a task. Actions and operations are major units of activity (Leont'ev, 1981, p. 65). According to Leont'ev (1997/1978) and Rubinshtein (1935), cognitive and behavioral actions and operations (the constituent elements of the actions) should be used as units of analysis of activity. SSAT also considers such units of analysis as a member of the algorithm and function block (Bedny, 2006).

Activity as an object of study should also be distinguished from an object of activity. In a subject–object interaction process, an object is incorrectly understood as an objective or a goal. Let us consider Engeström's scheme, which is very popular in the West. We will not discuss all aspects of this schema but just its interpretation of the subject–object relationship. Engeström (1999) depicted activity as a triangular system (see Figure 4.1).

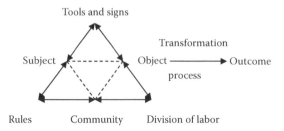

FIGURE 4.1
Triadic scheme of an activity system according to Engeström.

The presented schema requires some explanation. Vygotsky (1971) introduced the concept of semiotic mediation according to which human activity
is mediated by a material and mental tool. Development and functioning of
the mind is understood as a process that involves using appropriate means
such as either external tools (material objects) or internal tools (mental or
ideal objects such as signs, symbols, words, mental images). Development
of a mental tool is not just an individual but also a socially historical
phenomenon.

Development of the human mind is treated as a process of internalization
when activity that involves external tools is a source for formation of internal
mental tools.

Tools mediate the relationship between a subject and an object. The subject interacts with the object not directly but only by using a tool that can
sometimes be only mental. Hence, Engeström's triadic schema implies that
a subject interacts with an object only by using mediated tools and signs
(top of triadic schema). However, his schema does not include a goal, resulting in interpreting an object in this scheme as an objective (Bellamy, 1997,
pp. 124–125; Kaptelinin and Nardi, 2006, p. 66; Nardi, 1997, p. 73). For example,
on page 66 of their book, Kaptelinin and Nardi wrote: "A way to understand
objects of activities is to think of them as objectives that give meaning to
what people do. Concrete actions can be assessed as to whether or not they
help (or otherwise) accomplish the objectives."

Bellamy (1997, p. 124) stated: "...tools and symbol systems mediate between
the individual (the subject of the activity) and the individual's purpose (the
object of the activity)." These authors mix concepts of the object of activity
and its objective or purpose. In AT, when we consider subject and object
interaction, an object that is either material or mental cannot be considered as
an objective. An object of activity is something that can be modified by a subject according to the activity goal (Bedny and Karwowski, 2007; Bedny and
Meister, 1997; Leont'ev, 1981; Rubinshtein, 1957). In our opinion, Engeström's
schema should include a goal because a subject transforms an object according to a goal of activity. Let us depict a scenario when a subject may select
goal 1 or 2. Such modified schema is presented in Figure 4.2.

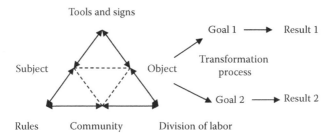

FIGURE 4.2
Modified Engeström's triadic schema of an activity system with two goals.

In order to clarify the relationship between an object and a goal, we will consider a hypothetical example. The rudder performs the task of keeping the ship on a specific course. The helmsperson looks at the compass and turns the wheel if necessary. If the compass needle moves to the left, he or she turns the wheel to the right. If the needle moves to the right, he or she turns the wheel to the left. As per AT, the requirements of a task that are accepted by a subject become a goal of the task (Leont'ev, 1981; Rubinshtein, 1957). Suppose the needle moves to the right. In such a situation, the helmsperson formulates the goal to turn the wheel to the left. In another situation, he or she formulates the goal to turn the wheel to the right. Such activity produces different results: position 1 or 2 of the ship. In the considered task, a compass and a wheel are physical or material tools and the ship is an object of activity. The result can match or not match the activity goal. In the latter, the corrective actions are required. Figure 4.2 makes it clear that the object of activity and the goal of activity are not the same thing. An object is modified (transformed) by a subject in accordance with the goal of activity. According to AT terminology, instead of the term *outcome* in Engeström's triadic schema, we utilized the term *result*. The relationship between activity elements can be presented as follows:

Subject → mediated tool → object → goal → result

Sometimes scientists in general AT confuse actions with tasks.

Let us consider Engeström's (2000) article where he discussed children's medical care. This paper has been analyzed in Bedny and Karwowski (2004b). Engestrom studied work performed by physicians. Our analysis revealed that what Engestrom describes as an action is really a task from an AT point of view. Examination and diagnosis of patients are not an action but rather a diagnostic task that includes not only subject–object but also subject–subject interactions. Engestrom considers a physician as a subject and a patient and his or her father as an object. However, when a physician evaluates a patient's health, it is a subject–object interaction, but when a physician speaks with a patient and his father, it is a subject–subject interaction. Here a subject–object interaction is transformed into a subject–subject interaction, and vice versa. There is a certain inconsistency in terminology in general AT. For example,

Leont'ev (1978) interpreted a motive as a material or an ideal object of need and stated that needs are *objectified*. However, an object cannot be a motive, but it is rather a source of a motive or motivation. The needs can turn into motives when the goal of activity is to satisfy these needs. This corresponds to Rubinshtein's understanding of motives. Rubinshtein (1957) interpreted a motive as an experienced need. Such inconsistency in terminology creates difficulty in interpreting AT terminology in the West.

Without a clear understanding of such terms as goal, activity, action, task, object, and subject, it is impossible to utilize efficiently previously considered triadic schema.

The next concept that should be considered is task and its relation to design. The task concept is very important in AT. A task is understood as an activity, restricted by some time frame and directed to achieve a goal of the task. It consists of a logical system of cognitive and behavioral actions. The general hierarchical scheme of activity includes four levels: *work activity*, *task*, *action*, and *operation* (Bedny and Karwowski, 2003). Work activity can include a variety of tasks that are arranged in time and space in a particular way.

Through changes in motivation and the goal of the task and conditions of its execution, deletion, or addition of various components of the task, it becomes possible to identify essential properties of mental processes and their dependence on the structure and specifics of activity. Such techniques allow for studying cognitive processes in the context of a specific task. The concept of task allows us to study psychological processes in unity with emotions, motives, and goals. In the framework of general, applied AT and SSAT, mental processes should be examined utilizing such structural components of activity as goals, motives, actions, operations, and strategy.

For example, in Zinchenko's (1961) classical experiment, it was demonstrated that memorization is dependent not so much on the nature of the memorized material but on how it is utilized. The subjects had to perform the task that requires organizing presented cards in a particular order. There were pictures and numbers on the cards. One group of subjects was instructed to organize cards according to the pictures and another group had to organize cards according to the numbers. It was discovered that those instructed to organize cards by the pictures were unable to recall numbers; in fact, some subjects insisted that there were no numbers on the cards, while those instructed to organize cards by their numbers could not recall the pictures. This experiment demonstrated that memorization depends not only on the features of the stimulus but also on the way the material was presented. In other words, memorization of material is stipulated by motives, goals, and methods of task performance. Similarly, in SSAT, properties of attention depend on the specificity of individual conscious goal. Subjects attend to the features of the object that are relevant to their goals. From an SSAT perspective, cognitive processes should be studied within the structure of the whole task.

In contrast, Nardi (1997) and then Kaptelinin and Nardi (2006) stated that AT rejects the concept of a task. Nardi (1997, p. 241) wrote, "The concept of a 'task' is suggestive of something atomic, neat, pure," and later on page 242, "By itself the notion of task does not suggest motive or direct force, as the activity theory concept of object clearly does." Unlike cognitive psychology, in AT, such concepts as goals and motives are central when studying task performance. Kaptelinin and Nardi are trying to replace the concept of task by introducing the term *engagement*. The term *engagement* has been criticized by a number of authors in commentary papers under Diaper and Lingaard (2008). Therefore, it would be a mistake to eliminate the concept of task from AT (Bedny and Harris, 2008; Diaper and Lindgaard, 2008).

Kaptelinin and Nardi (2006, p. 119) wrote: "There is an evident difference between higher-level actions and lower-level actions (or tasks)." They state that higher-level actions consist of tasks. However, the hierarchical relationship between tasks and actions is different in AT. Tasks consist of a logically organized system of actions.

Without a goal, there is no task. There is a different understanding of goal in cognitive psychology and in AT. Human goal and goal of the system are not the same. According to Preece et al. (1994, p. 411), the goal may be defined as a state of the system that the human wishes to achieve. Similarly, in GOMS, the goal is something that the user tries to accomplish (Kieras, 1993). The goal is considered as a final state of the system. So the goal is interpreted as something that is externally given to the subject in a ready form and in an externally given standard to which a subject approaches. In AT, a goal is always associated with some stage of activity and includes such stages as goal recognition, goal interpretation, goal reformulation, goal specification, goal formation, and goal acceptance.

You can give three computer professionals the same assignment, and they will come up with three different products due to motivational factors and subjective interpretation and acceptance of the goal. In cognitive psychology, goal includes motivation, while in applied AT and in SSAT, goal is a cognitive mechanism connected with motives. In AT, the same as in any other psychological theory, one can distinguish between such terms as motive and motivation. Motivation is a hierarchically organized system of motives that can be conscious or unconscious. In order to understand these two terms, we compare them with the concept of *vector*. In physics, vectors can be integrated into one resulting vector. Similarly, motivation is the result of integrating various motives. In any particular situation, a person has a variety of motives but only one motivation. Although it is conventional to utilize the term *motive → goal*, one should understand this term as a vector *motivation → goal*. Motives and motivation are energetic and goal is a cognitive component of activity. A motive cannot be shifted to a goal as it was claimed by Leont'ev (1978). With such shifting, the vector *motive → goal* is transformed into a point and a goal directness of activity disappears. According to some professionals who work in the field of personality and social psychology, goal includes motivational mechanisms (Pervin, 1989), but our activity is polymotivated and therefore includes a number of motives. If we follow Pervin's logic that a goal of a task includes various motives, then a person might pursue multiple goals at the same time during task performance. Nonetheless, it is impossible to achieve different goals at the same time. One strives to achieve a goal of a task, the result of which would be used to achieve the next goal, and so on.

Let us consider the concept of design that is at the core of engineering and ergonomics. It facilitates the creation of new products, software, and work processes. Design is also one of the most important areas of AT application. It may be noticed that the term *design* is often misinterpreted. For example, this is how Kaptelinin and Nardi (2006, p. 5) define design: "To us interaction design comprises all efforts to understand human engagement with digital technology and all efforts to use that knowledge to design more useful and pleasing artifacts." So for them, design is just any effort to design useful artifacts. However, in science and engineering, the definition of design is clearly standardized. Let us consider how the term design was defined in the former Soviet Union where AT was developed. For example, Neumin (1984, p. 145), a philosopher with an engineering background, defined design as "creation and description of ideal images or models of an artificial object in accordance with the previously set properties and characteristics with the ultimate purpose of materializing this object." An excellent theoretical analysis of design was also done by Polish scientist Ditrikh (1981), who gave a similar definition of design.

According to American scientist Suh (1990, p. 40), "The output of the design process is information in the form of drawings, specifications, tolerances, and other relevant knowledge required to create the physical entity." This author outlines two main design stages: a creative stage that includes task definition and development of ideas for the possible solutions and a formalized method or analytical process. At this stage, the development of designed models and the evaluation of the obtained data are critically important in the design process. Observation and questioning are used at this preliminary stage of gathering required information. After that, a practitioner should go through formalized steps and develop activity models. Without the second stage, design is simply an art. All definitions of design are similar and include as their main step the creation of models of the designed object. Therefore, during ergonomic design, one has to develop models of activity that describe the activity structure and models of designed objects (equipment, interface). Design solutions can be optimized based on a comparative analysis of such models (Bedny and Karwowski, 2003, 2008a).

In order to understand the relationship between the uniqueness of activity and the possibility of its design, one should be familiar with such concepts as strategy and tolerance. The models of activity reflect preferable strategies of task performance, which are the templates or standards of the designed activity. There are some acceptable deviations of real activity from the normatively described standard of activity performance, and if these deviations are within the acceptable level of tolerance for a particular strategy, then the designer can ignore these variations. It is similar to the manufacturing of a product in mass production where engineers utilize the concept of tolerance. Each part has its unique shape and size. However, if the deviation of its parameters is within a range of tolerance, this part is considered to be similar to others in spite of its uniqueness. Each version of the utilized strategy of activity is unique. However, if each of the implemented versions of strategy of activity falls within an acceptable range of tolerance, then this particular version of strategy should be considered similar to any other versions of strategy that is within the tolerance range. Activity design is an analytical process of creating activity models. According to Suh (1990), the preliminary stage of design includes problem definition and creative process. One also should distinguish a design stage from a production stage. The first stage involves creation of the project of the designed object.

The last terms that we consider in our analysis are problem solving, learning, and training. Problem solving and learning are interdependent areas of psychology. Interrelationship between these areas of study is particularly important in the development of problem-solving skills. However, problem solving can be considered also as independent area of study in psychology. For example, problem solving is important in the study of human thought and in the study of artificial intelligence. From practical perspectives, this area of research is important for discovering solutions of unknown complex problem-solving tasks by workers. Such tasks can be considered as unique only at the first time of their solution. After solutions of tasks are discovered, such tasks can be transformed into routine tasks that do not have problem-solving aspects. After the solution of the problem is found, its execution may not require special training.

The aim of the training is development of new knowledge and skills that can be used in practice. Training in work organizations is used for improvement of performance in a work situation. This means that in practice, the goal of learning or training and the solutions of an unknown problem are not the same. These differences are not clearly articulated by Engeström (1999) in his expansive learning concept.

According to Engeström, learning is considered as the innovation and construction of new forms of collaborative practice. Individuals through collective questioning, criticizing, rejecting, etc., develop some new aspects of the accepted practice and some new form of solving practical problems. In cognitive psychology, this method is described sometimes as a team problem solving. The purpose of such team performance is to facilitate a solution of a new practical problem or discovering a new method of solution to an already known problem. Individual and group activities are facilitated by a researcher-supervisor who also records data for further analysis of team members' performance. Of course, this situation contains some elements of self-learning and acquisition of new experiences. However, the main purpose of this situation lies not in acquiring professional knowledge and skills but in finding a collective solution of a real practical problem. In contrast, the major purpose of learning and training is acquisition of knowledge and skills and achievement of a new level of intellectual development by students.

In fact, what Engeström considered as expansive learning in AT is considered as the development of new methods of solving practical problems by a professional group or a work team. Team problem solving from an activity perspective was described by Shchedrovitsky (1995). A practical solution of real problems by a work team has a major purpose in improving team performance and in developing more innovative strategies for achieving company objectives. Skill and knowledge acquisition and mental development are not major purposes of such situation. Problem-solving skill acquisition is important in the learning and training area. This distinction has not only a theoretical but also a practical meaning. The teaching method that utilizes collective solution of problems by students includes very different goals, objectives, and procedures in comparison to solving really new problems in a production situation. Formation of skills, knowledge, and abilities requires special organization of the educational process.

Students can work together to solve problems that are unknown to them under a teacher's supervision. However, these problems are well known to teachers and professionals. These problems are unknown only to students. Students are taught to acquire effective strategies for solving a certain class of problems. The formation of the required strategies of thinking is of particular importance. Here we are talking about developmental education. In mental development, not only knowledge and skills are important but a person's ability to acquire thinking strategies and to apply them in new situations. Mowrer (1960) designated this as *solution learning*. The solution of learning problems is performed under the guidance of a teacher.

The main purpose of problem solving is discovering and developing new methods of solving the unfamiliar problems in production environment. A discovered method of problem solution can be, for example, utilized by other workers in practice. At this stage, problem solving can be transferred into learning and training. The major purpose of learning and training is development of knowledge and skill. The learning tasks gradually become more complex, they vary in type, and all stages of the solution may be well thought out in advance by the teacher. Areas of problem solving and of human learning studies are interdependent but are not the same. The term *innovative learning* utilized by Engeström does not clearly demonstrate the relationship between problem solving, learning, and training.

The expansive learning cycle according to Engeström (1999) includes the following *seven actions*: questioning → historical analysis (A) and actual-empirical analysis (B) → modeling of new solution → examining the new model → implementing the new model → reflecting on the new process → consolidating the new practice. However, those stages are not actions but simply stages of solution. Engeström mixed seven stages of collective problem solving with seven actions in this cycle. Of course, distinguishing these stages in collective problem solving is useful. However, the stages of activity during problem solving and the cognitive or behavioral actions in problem solving are not the same. Each stage of problem solving includes a number of different cognitive and behavioral actions. As an example, we present the first and the second actions according to Engeström's (1999, p. 235) terminology: "The first action is that of questioning, criticizing, or rejecting some aspects of the accepted practice and existing wisdom. For the sake of simplicity," Engeström called this action *questioning*.

> The second action is that of *analyzing* the situation. Analysis involves mental, discursive, practical transformation of the situation in order to find out causes or explanatory mechanisms...

Described earlier, two incorrectly labeled *actions* are complex stages of activity that can include a number of cognitive and behavioral actions. These stages are involved in the solution of a new unknown for organization problem. Thinking actions are a major type of actions in problem solution. The content of innovative learning can be different depending on its application to a problem-solving situation or learning and training.

Landa's (1984, 1976) algoheuristic theory of learning sees learning as problem-solving centered. Landa distinguished between two classes of problems: one can be solved by algorithms and the other can be solved by heuristics. Based on algorithmic prescriptions, the learner can solve any problem that belongs to a particular class of problems. Heuristic prescriptions do not guarantee success of solution of a particular class of problems. However, they can significantly increase the efficiency of solving such problems. What is called *mental operations* in Landa's theory is called *cognitive actions* in SSAT. These actions are classified according to developed in SSAT principles (Bedny and Karwowski, 2011b).

There are a number of steps for the development of algorithms or heuristics to be used in instructions. One must first define the problem-solving tasks (essentially perform a task analysis). Then it is necessary to uncover the algorithmic process (largely unconscious) used by experts who perform the task. If we consider problem-solving tasks when a problem is not well defined or a solution of a problem is unknown, then a heuristic for their solution should be utilized. Algorithmic or heuristic description of a task-problem enables the trainee to find solutions to a great variety of a particular type of task-problems. Such description of task-problem also helps the trainee to acquire a general algorithmic or heuristic method of thinking (Landa, 1976; Bedny and Meister, 1997). For this purpose, practice is required for the actual execution of cognitive and behavioral actions prescribed in algorithm or heuristic.

To conclude this chapter, we present Table 4.1, which provides a comparative analysis of the basic terminology and integrates the key information selected from the previously presented material.

A comparative analysis of the terminology in different fields of psychology and the analysis of the terminology used in general, applied, and systemic-structural AT demonstrates that some psychologists operate with concepts,

TABLE 4.1

Comparative Analysis of the Basic Terms in Action Theories, Cognitive and Social Psychology, and Activity Theory

Term	Basic Terminology Outside of Activity Theory (AT) and Translation and Interpretation of AT Terms	Original Meaning of Basic AT Terminology
Action	Action is a goal-oriented behavior. It is organized by goals, information integration, plans, and feedback (Frese and Zapf, 1994)—action theory, *German action theory*. In motivational action theory, action is a goal-oriented behavior that is described in four stages: motivational predecisional, volitional predecisional, volitional actional, and motivational postactional (Heckhausen, 1991). Action during task performance is defined in seven approximate stages of user activity that have one loop organization: establishing a goal, forming intention, specifying action sequence, executing action, perceiving system state, interpreting system state, and evaluating system state with respect to the goal and intention—*Norman's approximate theory of action (the United States)*. Action is defined as an intentional, goal-directed behavior that is specific to human beings (Kotarbinski, 1965)—*praxiology*. Action is a task that involves no problem-solving components (Preece et al., 1994)—*HCI field*.	Action is the most important component of human activity. Activity is carried out by some aggregation of actions subordinated to partial goals (goal of actions) that should be distinguished from the overall goal of activity (goal of activity during task performance) (Leont'ev, 1978). Actions have their components or means by which they are carried out. These components of actions are called operations. Actions can be cognitive and behavioral (Leont'ev, 1978, 1981; Rubinshtein, 1957). The main elements and units of activity analysis are mental or cognitive and behavioral actions (Bedny and Karwowski, 2007, 2011a; Leont'ev, 1978; Rubinshtein, 1957; Zinchenko, 1995).
Activity	The term *activity* means an active state of performer. It is a term applicable as a synonym for action, movement, behavior, mental process, and physiological functions (see *Dictionary of Psychology*, Reber, 1985).	Activity is understood as a purposeful interaction of a subject with the world, a process in which mutual transformation between the pools of *subject–object* are accomplished (Leont'ev, 1978). Activity is a self-regulated system that integrates cognitive, behavioral, and motivational components and is directed toward achieving a conscious goal of activity (Bedny and Karwowski, 2011a).

(Continued)

TABLE 4.1 (*Continued*)

Comparative Analysis of the Basic Terms in Action Theories, Cognitive and Social Psychology, and Activity Theory

Term	Basic Terminology Outside of Activity Theory (AT) and Translation and Interpretation of AT Terms	Original Meaning of Basic AT Terminology
Object, subject–object relationship	The object of activity is considered as a synonym of activity objectives (Bellamy, 1997; Kaptelinin and Nardi, 2006; Nardi, 1997; etc.). An object (in the sense of *objective*) is held by the subject and motivates activity, giving it a *specific direction* (Nardi, 1997, p. 73).	An object of activity that can be material or mental (symbols, images, etc.) is something that can be modified by a subject according to the activity goal (Bedny and Karwowski, 2007; Leont'ev, 1981; Rubinshtein, 1957; Zinchenko, 1995). There is not just a subject → object interaction but also a subject ↔ subject interaction (Bedny and Harris, 2005). In general AT, the object of study is activity and the main units of analysis are cognitive and behavioral actions. Smaller units of analysis are psychological operations (Leont'ev, 1978, 1981; Rubinshtein, 1957).
Units of activity analysis	In AT, the unit of analysis is an activity (Kaptelinin and Nardi, 2006; Nardi, 1997).	SSAT utilizes the following units of analysis classified according to the developed principles: cognitive and behavioral actions and operations, functional micro- and macroblocks, and operators and logical conditions. Actions and operations are utilized in morphological analysis. Functional blocks are utilized in analysis of activity self-regulation. Operators and logical conditions are utilized in algorithmic description of activity (Bedny and Karwowski, 2007).
Task	The concept of task is not used in AT and the notion of task does not suggest motive or direct force, as the AT concept of object clearly does (Kaptelinin and Nardi, 2006). Higher- and lower-level actions consist of tasks (Kaptelinin and Nardi, 2006).	Task is a goal given under certain conditions (Leont'ev, 1978). The work process consists of a number of task the performance of which is restricted by time constraints (Bedny and Meister, 1997). Task is a type of activity that requires achievement of a certain goal in specific conditions (Leont'ev, 1978, 1981; Rubinshtein, 1957). A general hierarchical scheme of activity includes four levels: *work activity–task–action–operation* (Bedny and Karwowski, 2007; Bedny and Meister, 1997). Task consists of actions and can be described as a logically organized system of cognitive and behavioral actions directed to achieve the goal of a task (Bedny and Karwowski, 2007; Bedny and Meister, 1997).

(*Continued*)

TABLE 4.1 (*Continued*)

Comparative Analysis of the Basic Terms in Action Theories, Cognitive and Social
Psychology, and Activity Theory

Term	Basic Terminology Outside of Activity Theory (AT) and Translation and Interpretation of AT Terms	Original Meaning of Basic AT Terminology
Goal, goal–motives relationship	In the goal setting theory, goal integrates two primary attributes, content and intensity (Lee et al., 1989). Goal is a state of a system that the human wishes to achieve (Preece et al., 1994, p. 411).	Goal is a conscious desired result of the subject's own activity. Goal cannot be considered as an externally given standard. It should be interpreted, accepted, and formulated by a subject (Bedny and Karwowski, 2007, 2011a). *Motive(s) → goal* creates a vector that lends activity a goal-directed character. Motives are energetic, whereas goals are cognitive components of activity. Activity during task performance can have several motives and only one overall goal of task. The more intensive the motives are, the more effort a subject spends to reach the goal of task (Bedny and Karwowski, 2007, 2011a).
Design	Interaction design comprises all efforts to understand human engagement with digital technology and all efforts to use that knowledge to design more useful and pleasing artifacts (Kaptelinin and Nardi, 2006, p. 5).	Design is the creation and description of ideal images or models of an artificial object in accordance with the previously set properties and characteristics with the ultimate purpose of materializing this object (Neumin, 1984, p. 145). A similar understanding of design is expressed by Suh (1990) in the United States.

describe their characteristics and features, and at the same time never give
them clear-cut definitions. This is why psychological terminology is often
difficult to grasp outside of the context of a given discussion. Moreover, the
same term might be defined differently by different professionals depending
on their interpretation and background.

General AT is a psychological theory that derives from Marxist philosophy,
which played some positive role in the development of Soviet psychology
and at the same time had a negative impact on the development of psychology manifested primarily in the dogmatic application of this philosophy and
the excessive politicization of psychology.

For example, based on political reasons, Vygotsky's cultural-historical psychology has been criticized by many psychologists, including Leont'ev, for
a long time. Leont'ev's book was partly a result of the political climate at
that time. His book *Activity, Consciousness and Personality* was presented as
a bible in the field of AT in the West. Any theoretical discussions of AT that
are reduced only to Leont'ev's concept of activity are meaningless because

they ignore Rubinshtein's school of psychology and the fundamental works in applied and systemic-structural AT.

Leont'ev's book should not be perceived dogmatically by accepting all theoretically philosophical discussions it presents. It should be understood that the book does not cover a lot of key aspects of AT and ignores the fact that development of AT took place within the various schools of Soviet psychology that had some significant differences.

Analysis of the considered material demonstrates that many of the Russian–English translations incorrectly interpret the original meaning of the basic terminology. This may be explained by difficulties in translation and mainly by the wrong interpretation of the words' meaning, as well as by misunderstanding of the context of translated works and by the failure to realize that the meaning of a specific paper can only be understood in the context of other publications on the topic. Moreover, another complication takes place. English is the international language of science; however, many scientific terms have different meanings in different languages.

In this work, we consider how different authors utilize the AT terminology. When touching the fundamental concepts of AT, some authors misinterpret them and often fail to capture their original meaning. Such basic concepts of AT as design, goal, task, actions, operations, and motivation have been presented inaccurately. For example, some scientists reject the concept of task; they cannot explain how one should understand the term action and how actions can be extracted from activity and be classified. In applied AT and SSAT, such concepts as goal-oriented task, task-solving process, motivation in task performance, task analysis, and cognitive and behavioral actions have always been central.

Some scientists in the West have expressed their skepticism about the benefits of applying AT (see collection of articles under Diaper and Lindgaard, 2008). This is not very surprising because general AT could not be applied to the study of human work and to the study of HCI in particular. This fact was very well recognized in the former Soviet Union by applied psychologists and human factors specialists. We believe that the debate about basic terminology in AT will help Western scientists to understand the advantages and weaknesses of general AT in studying human work and will shift their attention to applied and systemic-structural AT.

4.2 Activity and Embodied Cognition Theories: Comparative Analysis

Analysis of the relationship between different fields and schools of psychology and their influence on theoretical and applied studies becomes particularly important in psychology where we can observe the rapid development of new and sometimes conflicting psychological theories.

Such theoretical concepts as situated actions, action theory, situated cognition, distributed cognition, and embodied cognition tried to borrow some ideas in activity and became popular recently in psychology. In this work, we will discuss the embodied cognition that attempts to assimilate some basic ideas of AT.

The major philosophical assumption of embodied cognition is the statement according to which cognition cannot be considered as an intrapsychic phenomenon and should be studied as a process of interaction with the environment (Clark, 1999a,b; Thelen, 1995; Thelen and Smith, 1994; etc.). It emphasizes the formative role of the environment in the development of cognitive processes. Embodied cognition conceives cognitive development as a process of goal-directed, real-time interaction between the organism and its environment. The nature of this interaction, which is always situated and has a sociocultural context, determines the content of human thoughts and of cognition in general. This helps to overcome individualistic and dualistic orientation according to which a thought exists inside of the head and is separated from the ones constituted outside the head, for example, in the real world. One basic condition of embodied cognition is the organism's ability for sensory–motor interaction with the environment. When cognition is embodied, it means that it arises from bodily interactions with the world. In this interaction, perceptual and motor capacities of an organism play a leading role. Thelen (1995, p. 69), who is one of the most influential authorities in this field, wrote: "Thought grows from action and the activity is the engine of change." Action here is understood as a human motor action.

As we will show further, the main ideas of the embodied cognition theory were borrowed from AT where the basic principle is the unity of cognition and behavior (Bedny and Bedny, 2001). However, in this case, the theoretical and philosophical positions of AT that were developed in great detail were destroyed. Moreover, the embodied cognition theory diminished and even ignored the important achievements of cognitive psychology, which has made a significant contribution to the development of psychology in general.

In our opinion, a basic requirement for new theoretical concepts and theories is that they have to demonstrate that existing psychology theories, not just in cognitive psychology but also outside of it, did not cover certain aspects of cognition and mental development, or that they have been mistaken, or have proved inadequate. The embodied cognition theory fails in this respect. We have to state clearly that all basic ideas discussed in embodied cognition were carefully described in AT (Bedny and Bedny, 2001; Bedny and Karwowski, 2007; Bedny and Meister, 1997; Cole and Maltzman, 1969; Danials, 2008; Leont'ev, 1978, 1981; Vygotsky, 1960, 1971, 1978; Wertsch, 1981). Moreover, some major concepts of embodied cognition are simply flawed.

It is necessary to remind that the idea of behavioral bases of cognition has been promoted by some cognitive psychologists in the mid-1980s under the influence of AT (see, e.g., Weimer, 1977). These ideas also received some development in ecological psychology (Turvey, 1996). Unfortunately, these studies have not received further development in classic cognitive psychology. In this chapter, we will consider the relationship between embodied cognition and AT because the basic ideas of embodied cognition were formulated in AT where they were profoundly developed. Despite this, scientists working in embodied cognition have limited themselves to the comparative analysis of the traditional cognitive psychology and their theory. This misses the data obtained in AT.

One of the more controversial issues in the embodied cognition theory is debates regarding subject–object interaction versus body–environment interaction. Thus, we start our discussion with an analysis of the main concept of embodied cognition *body–environment interaction*. In AT *body–environment interaction*, the human as a biological entity and how it adapts to the demands of the physical environment is considered. This aspect of human behavior can be better understood in physiological terms. Professionals study how the environment affects human performance, and the terminology that derives from the study of body–environment interaction is adequate to the one used for estimation of cost-effectiveness of environmental improvement during a physical job performance (Bedny and Seglin, 1997; Bedny et al., 2001). By applying ergonomic principles to the design of the work environment, the negative influence of the environmental factors on performance can be removed (see, e.g., Rodgers and Eggleton, 1986). Environmental conditions can also influence the psychological state, and then instead of the *body–environment interaction* terminology, AT utilizes the *subject–environment interaction*. For example, a certain air temperature can increase the monotony of the job and decrease attention. In such situation, we pay attention to changes that can happen in the psychophysiological subsystem of activity (Bedny and Meister, 1997; Zarakovsky, 2004).

The central issue in AT is the study of goal-directed conscious activity when special attention is paid to the operational subsystem (cognitive processes, cognitive and behavioral actions, etc.), intentional subsystem (emotionally motivational regulation), and personality subsystem (provide regulation of activity, taking into consideration the personal and social significance of the situation). A totally different terminology should be utilized when we study activity and behavior from these perspectives. Analysis of such activity subsystems explains the process of mental development, learning, relationship between external behavior and internal cognition, etc. The cultural-historical theory of mental development (Vygotsky, 1971, 1978), general AT, and systemic-structural AT (Bedny and Karwowski, 2008b, 2011; Bedny and Meister, 1997; Bedny et al., 2011; Bedny and Karwowski, 2011a) study these subsystems of activity that, of course, do not exclude, when required, the analysis of the previously discussed supplemental physiological subsystem. Dividing activity into such interacting subsystems is an example of systemic analysis of activity (Zarakovsky, 2004).

The embodied cognition theory concentrates its efforts on the study of activity from a psychological perspective. It is unclear why the embodied cognition theory replaces the psychological term *subject–object interaction* with the physiological term *body–environment interaction*. What new information can we obtain by doing it? To answer this question, we will examine the basic terminology of AT such as subject, object, conditions, and environment. Activity determines the specificity of interaction of a conscious subject with the external world. Activity is an object-oriented, tool-mediated, and socially formed system. During activity, humans create artificial objects that are necessary preconditions for the development of internal cognitive processes. The inner mental world of human beings is not naturally given but is mediated by artificial objects produced by human activity (Leont'ev, 1978; Rubinshtein, 1957; Vygotsky, 1978). Activity is culturally and historically shaped even when a subject privately and individually interacts with different objects. Object-related activity is embedded in socially determined procedures for the manipulation of objects, which is especially true for artificial objects. People live in a world of stable things grounded in particular schemes of action with particular meanings and purpose. Their internal activity utilizes a historically developed system of symbols and signs such as words, numbers, and icons so that objects are not only confronted physically but also are encountered in defined intersubjective contexts.

Sociohistorical analysis reveals two closely related types of activity: *object-oriented* and *subject-oriented*. Object-oriented activity is performed by a subject using tools to transform a material object. The simplest scheme of such activity can be presented as *subject* → *tools* → *object*. Through the use of tools, the object is modified in accordance with the required goal. A physical environment that is not transformed by the subject but influences the strategies of the subject's activity is referred to as conditions of activity performance. Subject-oriented activity describes what is commonly called social interaction, which can be presented as *subject* ↔ *tools* ↔ *subject*. Social interaction or subject-oriented interactions involve two or more subjects. Just like object-oriented interaction, social interaction begins with a subject's goal setting, orientation in the situation, etc.

According to AT, a subject is a person who has consciousness; acquires human language, social norms, and standards; and can evaluate his or her own actions and their consequences not only through a person's biological needs but also based on the social norms. Rubinshtein (1957) has stated that an individual is as a subject who reflects transformed reality in his or her consciousness and, based on this reflection, regulates his or her activity according to his or her own needs and in relation to others.

An object can be physical or mental (mental sign, images) and is something that a subject transforms or changes according to the goal of activity. Objects and phenomena become an object only when interacting with a subject. Nature becomes an object only during interaction with a subject (Rubinshtein, 1957). All of this terminology has a deep philosophical and theoretical meaning in the theory of activity where the substitution of terms such as subject and object with such terms as body and the environment is unacceptable. Such substitution would reduce the psychological phenomena to the physiological one.

The embodied cognition theory contends (Thelen, 1995; Wilson, 2002) that cognitive processes are developed in real-time, goal-directed interactions between organisms and their environment. Goal-directed sensory-motor actions specify the nature of the developing cognitive capacities. Thought and language would not occur without the initial performance of these actions. Hence, the embodied cognition theory underlines the primacy of behavioral goal-directed actions unfolding in real time. Goal is considered as an end state toward directed human externally performed behavior. Such understanding of goal corresponds to the embodiment principle. At the same time, this theory does not consider the goal of activity, which is a result primarily of cognitive actions.

As we discussed before, *motive → goals* create a vector that lends goal-directed activity its directness. Thus, goal is considered as a conscious mental representation of a future result of activity connected with motives (motivation).

Goal should be distinguished from the end state toward which human behavior is directed as it was formulated. Reducing the goal to a formal description of the final situation that can be achieved during activity overlooks important aspects of the goal-developmental process. Goal cannot be given to a subject in ready form. When a goal is objectively presented to a subject, he or she has to interpret and accept the goal. Hence, an objectively presented goal and subjectively accepted goal are not exactly the same. Goal is formulated as a mental representation of a future desired result of the subject's own activity.

In the embodied cognition theory, there is no clear understanding of action as units of activity analysis and the action mainly implies an element of motor behavior. The concept of action is viewed from the common sense standpoint; the principle of action extraction from the general flow of activity and the method of the actions' classification do not exist and are never discussed. Without a clearly developed concept of cognitive and behavior actions, the relationship between cognition and behavior cannot be considered properly in the embodied cognition theory.

The external and internal aspects of cognition are understood in a totally different manner in AT, embodied cognition theory, and cognitive psychology. According to embodied cognition representatives in the traditional cognitive approach, cognition can be understood by focusing primarily on the brain's internal cognitive processes. Cognition is defined as a rule-based information-processing system that assumes the existence of symbolic, encoded representations that enable the system to devise a solution by means of computation. According to the embodied cognition theory, the cognitive approach considers the mind as a device to manipulate symbols and is thus concerned with the formal rules and processes by which the symbols appropriately represent the world. This, according to embodied cognition, de-emphasizes the importance of interaction of cognition with the external environment (Thelen, 1995; Thelen and Smith, 1994). AT criticizes cognitive psychology from a similar position. The concept of cognitive actions is important in this critical analysis. In AT, cognitive actions manipulate by symbols, images, verbal statements, etc., according to the goal of cognitive actions. This is done in a similar way as external or motor actions manipulate by external material objects. At the same time, formal rules are also important in the mental manipulation of the sign system. Formal logic is an important aspect of the study of logical thinking. Formal logic studies such forms of thinking as a concept, deductions, and reasoning. Logical forms of thinking are closely linked to practical activity. In AT, mental manipulation by symbols according to the required goals of actions and specific rules is considered as a system of thinking actions, which interconnects with practical or material actions.

In embodied cognition, the statement that cognition is embodied means that it arises from bodily interactions with the world. This statement contradicts with the prevailing cognitive view. According to embodied theorists, the major source of cognitive development is the sensory-motor experience. However, this idea is not new because it has already been studied in AT in much greater detail. This idea also was discussed by Piaget (1952) in his study of sensory-motor learning and development. Naturally, one would think that embodied cognition theorists would conduct a comparative analysis of the relationship between sensory–motor activity and cognition not only as it is considered in traditional cognitive psychology but also in AT and in the works of Piaget where it is discussed in great detail. It has not been done so far. Embodied cognition theorists cannot state that their theory introduced something new in understanding the relationship between cognition and performed body actions without analyzing what is done in this area in AT in particular.

The relationship between external, behavioral, and internal or cognitive aspects of the activity has always been one of the central issues of AT and has been interpreted differently by various schools within AT. Activity connects the subject with real world and leads to cognitive development of each individual and human kind in general. External practical activity is internalized and becomes internal cognitive activity. External, motor activity involving material world does not only depend on cognition, but cognition also depends on behavior. In which the relationship between external and internal activities in AT was described. Moreover, this problem has been discussed not only in theory but also from the practical aspects.

Here we only remind some previously considered theoretical principles.

The genesis of mental cognitive activity from external behavioral activity and primarily from human labor promotes a very important principle of AT called *unity of consciousness and behavior* (Rubinshtein, 1957) or more generally the *unity of cognition and behavior*. Internal activity is shaped with the help of external behavior and then can be performed independently (Bedny and Bedny, 2001).

This principle is critically important in the study of human work. Interdependence of cognition and behavior is also central in the works of Vygotsky (1971, 1978) and Leont'ev (1978). However, each author considers the relationship between these components of work in a different light. For example, Vygotsky and Leont'ev paid great attention to the concept of internalization but had two different interpretations of the internalization.

For example, Vygotsky introduced a new determinant in the process of internalization called a mental tool-sign, which performs a mediation function. If the external physical tool is changing physical objects, the mental tool modifies the internal, mental activity. Mental tools define socially developed mental operations. Such mediation tools do not possess individually psychological features but rather are of culturally historical nature.

Vygotsky views higher mental functions as having social-historical rather than biological origins. His idea of using a sign as a psychological tool is an example of successful application of semiotic ideas to demonstrate interdependence of internal cognitive functions and external social activity. It also explains the transformation of interpersonal experience into intrapersonal thought processes.

A different version of study of the relationship of external and internal activities was developed by Rubinshtein (1957). He insisted that a subject does not internalize ready-made standards but rather utilizes exploration and interaction with the objective world as the source of its reflection. During this process of dynamic reflection, human consciousness is developed. Rubinshtein introduced the unity of consciousness and practical activity as a vital principle of mental development. According to Rubinshtein, every human act changes not only the object and situation but the subject as well. Through activity, the subject not only changes the situation but also develops the self. In this process of dual interaction, the instruments and the products of actions are changed and in turn change the subject.

Rubinshtein's idea about dual interaction can be considered as a predecessor to the concept of self-regulation. The sensory-motor activity of children with material objects makes it possible to assess the results of their action and, through this self-evaluation process, develop mechanisms of thinking.

According to Thelen (1995), *cognition is embodied* means that it arises from bodily interaction with the world. In this interaction, sensory-motor activity plays a central role. This basic statement embodied cognition theory simply uses discussed before ideas. Moreover, the statement the *body interacts with the environment* is not correct, because in psychology, the process of interaction of the subject with the environment is important. The term *body–environment interaction* is used in work physiology. Body and subject are not the same. The term *cognition is embodied* cannot replace the basic term *internalization*. From SSAT, internal and external activities are interdependent and are shaped based on the self-regulation mechanisms.

Embodied cognition ignores all aspects of interrelationship between external and internal activities obtained in AT, including the concept of internalization. The statements that embodied cognition conceived cognitive development as a goal-directed process and the nature of interaction between organism and environment is situated and happened in real time are simply repetitions well known in AT ideas. Moreover, the relationship between conscious and subconscious in this process is totally ignored. Further, embodied cognition utilizes contradicted terminology. This theory describes the interaction between the organism and environment on the one hand, and Vygotsky insists that this interaction has a sociocultural context. An organism or body interacting with the environment cannot have a sociocultural context. The term *sociocultural interaction* is applicable for *subject–object interaction* or *subject–environment interaction*. All these terminologies were carefully developed in AT.

It is important also to compare such ideas as self-regulation activity and situated cognition.

According to the embodied cognition theory, interaction of the organism and the environment is always situated (Thelen and Smith, 1994). The term *situated* is also important in situated cognition theory (Kirshner and Whitson, 1997) and situated action theory (Shuchman, 1987). Situated cognition means that cognitive functions are time specific and have to be adequate to real-time interaction process. The term *situated* overemphasizes improvisatory features of cognition (Nardi, 1997). For example, according to Shuchman, action is situated and cannot be planned in advance. However, human activity and even animals' behavior include anticipatory mechanisms. This clearly demonstrates that human activity depends not only on the situation.

The concepts *situated cognition*, *situated action*, and *interaction between organism and environment is situated* are currently popular in some fields of psychology. The term *situated* means that our cognition, action, or interaction depends on the situation. However, our cognition, and therefore interaction with the environment, functions based on what was in the past, exists at present, and will happen in the future. Cognition is a combination of pre-planned, situated, and anticipatory or forecasting mechanisms. Our cognition can move us from the present to the past or to the future. This becomes particularly clear when we study our activity or cognition as a self-regulated system. Without understanding activity self-regulation, there is no possibility to understand what the term *situated* means.

The idea according to which interaction between organism and environment is situated becomes meaningful only when the theory of self-regulation and models of activity regulation are developed. Embodied cognition does not discuss any ideas about self-regulation. A description of activity self-regulation can be found in the works of Bedny and Karwowski (2004a, 2007), Bedny et al. (2014) and Bedny et al. (2004). Activity and human cognition are a combination of situated and preplanned components. Without self-regulation, situated cognition or behavior cannot be interpreted properly. The development of the theory of activity self-regulation has a long history in AT.

Based on the previously presented material, we can briefly summarize the main points of the comparative analysis of AT and embodied cognition. Embodied cognition formulated the following basic statements: The environment is a part of the cognitive system; cognition is for action; off-line cognition is body based; cognition is situated and time pressured; we off-load cognitive work onto the environment (see Wilson, 2002). Analysis of data presented earlier demonstrates that all these ideas were described more correctly and in great detail in AT and in its applied branches.

The embodied cognition theory utilizes the *body–environment relationship* terminology that contradicts with the basic concepts of AT. This terminology excludes the social determinant of mental development, which is of fundamental importance in AT. The body–environment relationship terminology that is used in embodied cognition is known as *subject–object–conditions* system in AT. At the activity level, it is not just a body that interacts with the environment but a subject who has consciousness and exists in particular social and physical conditions (environment) and interacts with objects or other subjects. As Rubinshtein (1959) wrote, nature becomes an object only during interaction with the subject. A subject acquires a particular culture. He or she can formulate conscious goals, be aware of what is going on, be motivated and/or emotionally related to the situation at hand, etc. A body cannot perform such functions. Such type of functions can only be performed by a subject.

The basic ideas of embodied cognition that *the environment is part of the cognitive system* in AT are known as subject → object interaction during task performance. The statement *the mind alone is not a meaningful unit of analysis* is incorrect. According to AT, the mind is not a unit of analysis but an object of study. Units of analysis are, for example, cognitive actions and operations.

The body–environment relationship is rather a physiological terminology that can be utilized when studying physical conditions of work. Embodied cognition uses distorted AT terminology and ignores a tremendous amount of data collected in AT through its history.

The next two ideas for embodied cognition are *cognition is for action* and *offline cognition is body based*. In AT, this idea is associated with the fundamental principle of unity cognition and behavior. Moreover, the statement *cognition is for action* is incorrect. From these statements, it follows that cognition is not a system of cognitive action in embodied cognition. Only external behavior consists of a variety of actions. In AT, cognition is a process and, at the same time, is a system of cognitive actions and operations. Cognitive actions tightly depend on behavioral actions, and vice versa. Interaction of cognitive and behavioral actions provides formation of internal mental plane of activity—a process of internalization, or formation of mental actions based on mechanisms of self-regulation.

Another important statement of embodied cognition is *cognition is situated and time pressured*. This means that cognitive functions are under pressures of real-time interaction with the environment. This is generally a true statement but it does not have sufficient theoretical justification. In AT, situated features of cognition and external behavior derive from mechanisms of activity self-regulation. Cognition and activity are treated as a goal-directed self-regulative system. Activity is constructive and dynamic and has not only adaptive mechanisms but also mechanisms that provide forecasting and functioning based on these predictions.

Analysis of considered data demonstrates that cognition and activity in general have to be conceptualized as a combination of prespecified and situated components. Separate actions and activity should be treated as a goal-directed self-regulative system and therefore as a situated system. Self-regulation is possible only during interaction with the environment. Self-regulation emphasizes responsiveness to the environment and possible future events. Our activity, which includes not only external behavior but also cognition, is always improvisatory and constructive. At the same time, a self-regulative process does not entirely exclude preplanned elements. Our activity is performed in real-time interaction with the environment. The environment is not a part of our cognition. The environment, which includes objects and conditions, is a source of information extracted by a subject for goal-directed activity.

Emotionally motivational mechanisms play an important role in the interpretation of information and in the regulation of activity in general. An emotionally motivational factor cannot be applied to the *body-environment interaction* concept. It can be used for the analysis of subject-object interaction or for the analysis of person and environment interaction and so on. Not only the physical world but also social interaction has significant impact on development of human activity. Activity can be viewed as continuously formed strategies of performance, which are required for achievement of the goal of activity. Goal should be accepted and formulated by a subject but not the body.

4.3 Task Description/Identification versus Hierarchical Task Analysis

Task analysis is the main approach in the study of human work. The purpose of task analysis is to describe and evaluate what an individual worker or team is doing in the system in terms of cognitive and behavioral actions for achieving a goal of task and finally the system goal. Analysis of the relationship between human actions and hardware configuration allows to solve design issues, taking into account the human factor.

Usually task analysis is viewed as a number of relatively independent techniques that have terminology that sounds similar but has different meanings. Data that are obtained by analyzing basic tasks in the man–machine systems are used for design solutions, training, and selection of personal. SSAT considers task analysis as a unified and standardized approach.

Before conducting task analysis, it is necessary to determine that tasks should be performed in the system. For every system that performs various functions, a decision should be made as to whether these functions should be performed by human or technical components of the system. After defining human functions, the list of the corresponding task and their interconnection should be defined as the next step. Determining a list of such task and their organization in any new system is a critically important and at the same time complex and time-consuming job. Tasks in every human–machine system have usually some logical organization. In this chapter, we will demonstrate that the relationship between tasks cannot be reduced to a hierarchical relationship. At the same time in cognitive psychology, some psychologists reduced the relationship of tasks to a hierarchical organization. This idea is known as hierarchical task analysis (HTA) technique (Annett and Duncan, 1967).

In order to demonstrate the difference between understanding the principles of task organization in SSAT and the HTA technique, we will shortly consider a difference between task description/identification and task analysis.

A task is a critical element in studying any technical system. It is also a main component of human work that determines the basic steps by which a human interacts with a machine (Meister, 1999). Tasks are deliberate manmade elements of work during interaction of a human with a technical system or software. The first step of task analysis is task identification and their following abbreviate description. Under task identification, we understand describing all necessary tasks that should be performed by all workers in an existing system, or discovering all potentially necessary tasks in a newly designed system. Identifying a list of tasks that need to be performed in a newly designed system might require complex procedures for analyzing system functions. At this step, analysis of existing systems that have some similarities with a new one can be useful. Ergonomists might have difficulties defining tasks even in the existing system. Task identification is the first and critical step in task analysis. It is accompanied by abbreviated and general task description. It is meaningless to just simply give a name or a title of the task without specifying its purpose and general description of work that has to be done for each task. A detailed description of the tasks and characteristics of equipment with which workers interact during task performance is the following stage of task analysis. At this stage, specialists select the most critical tasks for further detailed task analysis. Therefore, the purpose of task description/identification is a general description of all sets of tasks in the system (Meister, 1985).

Description/identification of the tasks is closely related to the analysis of their organization in the system. Often, determining the final organization of tasks in the system is only possible after task analysis is completed.

In the simplest cases of analyzing existing systems, a researcher can ask questions that in general form are represented as follows:

What tasks are to be performed? At this step, a specialist defines the task name, its purpose, and a general description of task performance without any specific details.

In what sequence will tasks be performed? Some tasks can be performed in any sequence, others are performed in a certain unchangeable order, and some tasks have changeable logical organization.

How critical are some of the tasks? Answering this question helps to determine which tasks should be considered at the following detailed task analysis stage.

We considered only some possible questions in a concise manner. Depending on the specifics of the study, questions can be modified. We want to point out that such questions should be used primarily at the task identification stage.

The apparent simplicity of such type of questions is deceptive. Due to the fact that ergonomists and/or psychologists are professionals who are not working in the system, formulating the questions and understanding the answers might be quite a difficult stage of task description/identification analysis because it requires interaction of various experts and exchange of information. It is even more complicated if this type of question is used at the system design stage when the possibility to gather data by observation is very limited.

At the stage of design of the complex man–machine or software systems, questioning and observation very often are not sufficient. Complex analytical procedures are required.

Man–machine systems often include a large number of components such as displays, controls, and computers. These components are defined by the tasks performed by the operator in a particular system.

This makes it difficult to determine what instruments and controls should be used and what functions should be performed by computer. The operators' responsibilities in such systems include receiving information from various displays, integrating and evaluating this information, manipulating controls, and interacting with the computer. Evaluating various situations and monitoring the system are nothing more than performing various tasks, some of which should be performed in a limited time. It is particularly important to describe tasks that have a high rate of malfunctioning. The second step is to determine types of displays and controls that are used for these tasks' execution. Sometimes there are hundreds of displays and controls for a considered system, making it a difficult step of design.

Incorrect determining of the list of tasks at the system design stage can result, at the later stage, in the lack of required controls and displays that facilitate necessary monitoring of the system. Such situations are specific for unusual and not well predictable states of the system. Lack of necessary controls and displays makes it impossible for the operator to perform adequate monitoring and corrective actions by intervening in the situation. Such situations often require emergency shutdown of the system, emergency landing of an aircraft, and so on. If on the other hand unnecessary controls and instruments are discovered when the system is already functioning, correction of already existing system becomes complex, time consuming, and very expensive. Qualitative methods based on the use of different questionnaires in some cases are not sufficiently effective for task description/identification stage analysis. In such cases, formalized methods can be useful.

In Bedny and Meister (1997), we in abbreviate manner described Galactionov's (1978) method of task description/identification. He used the graph theory to develop an analytical method for determining what information to display in the control room and what controls are required to operate the system. The graph theory analyzes relationships between objects. Objects can be represented abstractly as a set of points, and the relationship between them is represented by lines that connect them. The graph can be considered as a geometrical figure consisting of points (nodes) and lines (arcs) that connect them. Arks demonstrate the frequency of interconnections between elements of the graph. At the first step, a specialist divides a given system into the smallest number of functionally related subsystems. After that, a specialist would specify tasks associated with a specific subsystem. Identification of possible tasks has a preliminary character. Using a graph theory for further analysis, this list of tasks is specified and becomes more precise.

At the next step, possible malfunctioning of a considered subsystem is described. Each possible malfunctioning is considered as a basic critical event. At the next stage, the graph model that describes the underlying exogenous causes of the malfunctions is presented. Each individual model is called the primary cause graph event. It would depict malfunctioning as a basic event associated with causes. Malfunctioning and causes are depicted by circles and their interrelationships by arcs. Based on analysis of such graph, a specialist can evaluate the probabilistic relationship between events and determine the required displays. To protect an operator from being overloaded by unnecessary information, a specialist decides which nodes (events) should not be examined by the operator by analyzing a specific task that is associated with the considered malfunctioning. At the following step, a primary cause graph event can be transformed into a control graph. These types of graphs are used for determining what type of controls should be used. This method is described in more detail in Bedny and Meister (1997).

Thus, the task description/identification stage requires sometimes complex formalized and quantitative methods of analysis. After that, task analysis can be performed. Based on obtained data, a task performance algorithm can be developed. If required, a time structure of the task followed by a quantitative evaluation of task complexity can be performed. Obtained data can be used for a final equipment design stage (Bedny et al., 2006). After design of equipment, such questions as training and staff selection can be discussed.

Currently there are no unified and standardized methods of task analysis. Even the concept of *task* is interpreted differently by different authors. So task analysis is just a bunch of independent and often contradicting techniques.

In the following, we present as an example critical analysis of one such technique known as HTA. This technique has been developed by Annet and Duncan (1967). In the proposed technique, task description/identification is not distinguished from task analysis. Analyzing this well-known technique helps to better understand existing shortcomings in contemporary task analysis.

HTA is a graphical method that presents all tasks as a hierarchical system of steps that should be used to accomplish the system goal. This technique is based on the decomposition of the system goals into subgoals. A higher-order goal can be redescribed in terms of a number of subgoals (Annett and Cunningham, 2000). According to these authors, all human functions (task elements) can be presented as acts of control or servo. Any servomechanism consists of three elements: goal or set point, a power source that produces a variety of outputs, and a feedback indicating discrepancy. This is a unit of behavior, no matter how long or short it is in time. Such unit of behavior is called operation and work as servomechanism (Annett and Cunningham, 2000). The analysis begins with the description of a general task, which is progressively broken down into a series of subtasks, up to the lower level of decomposition. Hence, tasks are described as operations that are broken down into suboperations. The last can be subdivided further. So all tasks are presented as operations of various sizes that in turn may be further subdivided into suboperations. In HTA, the unit of analysis is an operation. The extent of decomposition is determined by a pragmatic stop rule (Annett, 2000).

Let us consider this technique from a contemporary psychology viewpoint. Basic units of analysis are human operator functions that are considered as servomechanisms (Annett and Cunningham, 2000). An operator's task may be described in terms of (a) the stimulus or input conditions, which are simply representations of the current goal discrepancy, (b) the control action required to correct the goal discrepancy, and (c) the feedback indicating the goal discrepancy has been eliminated. This unit of analysis is an operation. According to Annett and Cunningham, human behavior, which is called operation, works according to homeostatic principles.

However, the homeostatic principle of self-regulation is suitable for the physiological level of self-regulation, which is not a goal-directed process, or for analysis simple technical systems (Bedny et al., 2014). Moreover, a goal accepted by a person is conscious and has some subjective coloring (Bedny et al., 2012). An objectively given requirement (objectively given goal) of task is interpreted by a subject and accepted by him or her adequately (subjectively accepted goal). Very often the goal of task can be formulated by a subject independently. In team performance, the goal formation process also depends on interaction with other members of the team. Goal interpretation, goal acceptance, and goal formation are not discussed in HTA. Emotional—motivational components of activity are important in the process of formation or acceptance of a goal of task. The more complex the situation is, the more important the role of these components of activity is. All these data are ignored by Annett and Cunningham.

The authors argue that operation or servo as a unit of analysis can be interpreted and analyzed in a completely different way. However, no method of analysis of such units is actually presented. We simply provide some citation from the text for understanding such technique (Annet and Cunningham, 2000, p. 403).

The basic *input–action–feedback* (I-A-F) structure of operations is capable of being interpreted at different levels of system function—from individual psychomotor and cognitive abilities to social or team interaction. At the psychomotor level, a typical input might be a simple warning signal, the required action turning a switch and the feedback the removal of the warning signal. At the cognitive level, the input may be a pattern of symptoms, the actions a series of investigative or remedial actions, and the feedback a change in the symptom pattern that indicates the problem has been diagnosed and remedied. At the social level, the input comprises the problem faced by the group or team, and the action comprises the way the team must act together to solve the problem (e.g., by sharing information and coordinating individual actions). Different levels of analysis are appropriate to different problems.

It is hard to imagine that such recommendations may be considered as scientific. They contain common words and no specific data and methods of task analysis. It is best to consider a specific example suggested by these authors. Annett and Cunningham (2000) are describing a task performed by the military personnel but it is not clear what this task's function is. On page 407, they wrote: "The goal of fighting might be decomposed into gathering information, using it to assess the threat and respond appropriately to it, and disseminating it to others who might need it." This process is labeled as a task that is divided into four levels and each level is called *operation*. The general number of operations in the considered example is nine (goal hierarchy for command team contains nine hierarchical boxes). According to the authors, all these operations are components of one more general task. The last level includes four boxes: (1) respond to immediate threat, (2) respond to anticipated threat, (3) respond to continuing threat, and (4) deny fire solution. From the presented description of goal hierarchy, it is very difficult to understand what type of devices, displays, and controls is utilized in these operations and what type of activity or behavior is associated with them.

Annett and Cunningham (2000) select for further analysis the military operation *deny fire solution*. The considered operation is really a task that involves visual analysis of the situation directly or with the aid of special instruments, exchange of information among team members, analysis of information from other sources, analysis of potential enemy ship maneuvering and movement, assessment of obtained data, decision making, and selection of adequate controls and responses. Moreover, such tasks are performed in conditions of time restriction and emotional stress. The goal *deny fire solution* has not been presented to a team in advance. This goal is formulated by team members based on analysis of an existing situation. However, the goal formation process is totally ignored in the presented analysis. This is explained by the fact that in HTA, the goal of a system and human goal are not distinguished. Without a description of the utilized equipment and specificity of the interaction of operators in a particular situation and their goal formation process and motivation description of cognitive and behavioral actions performed by the operators in considered

TABLE 4.2

Summary of Operation *Deny Fire Solution*

Variable	Description
Deny fire solution	…….. …… …..
Goal	To deny the enemy a valid fire control solution as determined by relative location, course, and speed
Team members	…….. …….. …..
Teamwork	Reconciling any conflict of interest between air, surface, and subsurface teams and the weapons directors (missile gun directors [MGDs]) and communication with officer of the watch (OOW)
Performance measures	Ship deployed to best advantage with respect to fire control solutions

situation, understanding the behavior of the operators is not possible. Annett and Cunningham (2000, p. 408, Figure 24.2) presented four possible final goals of the task: respond to immediate threat, respond to track (anticipated) threat, respond to continuing threat, and deny fire solution. However, these are not goals of task but rather possible types of decision making that are formed based on the analysis of the situation. The goal of this task is to determine the possible course of action in a specific situation. This example demonstrates the lack of a clear understanding of concepts of goal of activity and goal of a system.

For the analysis of the operation *deny fire solution*, Annett and Cunningham utilized techniques that are borrowed from other sources in psychology and are not specific to HTA. Let us consider how the authors conducted an analysis of the considered operation. A team operation *deny fire solution* is depicted by previously mentioned authors in Table 4.2 (Annett and Cunningham, 2000, p. 409).

The table includes five *variables* and their short descriptions. As an example, we present three *variables* from the five described in the table. Such description of an operation did not really explain what the team did to achieve the goal. This description of the task can be adequate only in the task identification stage, but not in the task analysis stage.

At the next step (see Annett and Cunningham, 2000, p. 413), the authors utilize a questionnaire for analysis of this operation. Factor analysis has been applied to instructor ratings and questionnaire results. Authors obtained eight factors. As an example, we present several of them (see Table 4.3).

It is clear that such data are irrelevant to solving problems related to the analysis or design of performance methods or equipment design. This information is more relevant for analysis of social and psychological characteristics of the group and has no relation to the real task analysis and the design in ergonomics or engineering psychology. Even the authors believe that their method appears to be relatively coarse-grained. The described eight factors do not present useful information about each team member's performance. Moreover, data that are presented in Table 4.3 demonstrate that the utilized method of study cannot be considered as a method that belongs to HTA. Annett and Cunningham (2000) consider *deny fire solution* as one operation from nine operations that belong to the same task. Considered above authors do not distinguish between the

TABLE 4.3

Factors Measured by Team Process Questionnaire

N	Factor Name	Variance (%)	Factor Description
.....	....	....	....
4	Technical breadth	6.2	Experience with other AI[a] systems and equipment
5	Team regard	5.5	Team spirit and cohesion and perceived team competence
.....	....	....	....
8	Compliance	3.8	Perceived willingness of other team members to accept advice and familiarity with others

[a] AI, action information.

concept of goal and operation. For example, *deny fire solution* on Figure 24.2 is considered as both a goal and an operation. In fact, *deny fire solution* is decision-making during performance of the task in a specific situation.

The HTA terminology is used quite differently by various authors because it has no well-defined terminology. Kirwan (1994) in his comprehensive guide to human reliability assessment describes HTA using another example. His interpretation of the basic terms of HTA is often different from the terms used by Annett and his colleagues.

Kirwan (1994, p. 53) gives the following explanation of the difference between the terms goals and tasks:

> ... whereas a goal is a system objective which can be achieved by a varying range of tasks, a task is composed of a set pattern of operations. Sometimes, the distinction between a goal and a task is difficult to determine.

However, according to AT, system objectives and human goal are not the same. System objective at the initial stage appears for a worker as a requirement that has arisen in a particular situation. This requirement should be interpreted by a worker and subjectively accepted by him or her, and it is only after that that the system's objective becomes a personal goal. For example, the system objective cannot be correctly interpreted by the worker. After interpreting, objectively given requirements arise in the next stage, which involve acceptance of the interpreted goal. For example, a worker may modify the goal and reduce the accuracy requirements. Further, goal and task are not the same in AT. According to Leont'ev (1978), a task is a goal that is presented in specific conditions. There is no task without a combination of goal and conditions. A subject is to accept or formulate the goal in specific conditions and then to perform specific actions for goal attainment.

In his book, Kirwan (1994, p. 53) described "Filling road tanker with chlorine" as an example of HTA. The author presents the main components of HTA in Figure 4.3.

According to Kirwan, the top-level system goal "Fill tanker with CL2" is decomposed into five tasks. Task 2 is further decomposed into the next level

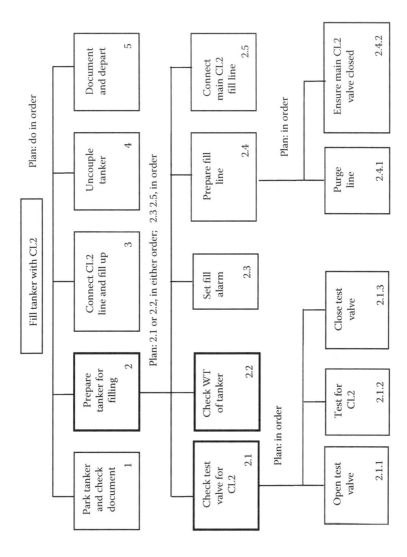

FIGURE 4.3
HTA—CL2 example.

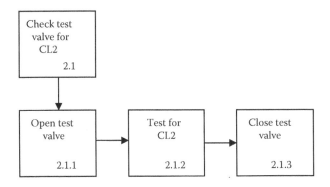

FIGURE 4.4
HTA—CL2 example (fragment).

of subtasks (subtasks 2.1–2.5). As an example, we present some explanations of the decomposition process: "…the task located at the second level down in the HTA box numbered '2' and called 'Prepare tanker for filling' is achieved via five tasks, which are in five boxes found at the next level of the HTA. The task 'Check test valve for chlorine' (2.1) has three operations below it."

Let us examine data related to the task "Check test valve for chlorine" (2.1). It includes three operations: (1) Open test valve (2.1.1.), (2) test for CL2 (2.1.2), and (3) close test valve (2.1.3).

We extract the corresponding boxes for a more detailed analysis (see Figure 4.4).

This description does not define what really should be done by a worker.

Box 2.1 is simply a task title and boxes (2.1.1), (2.1.2), and (2.1.3) are titles of each subtask that are performed in sequence. There is no hierarchy. There are a title of the task and three titles of subtasks that should be performed in sequence. There is nothing about real task performance. We do not know how a worker interacts with equipment during performance of a considered task and what cognitive and behavioral actions should be performed by him or her. For example, the title of subtask 2.1.1 "open test valve" does not give information about constructive feature of displays and controls that are used by a worker. It also does not provide any information about cognitive and behavioral actions that are utilized by a worker in performing this task. We do not know the relationship between the design solution of considered elements of equipment and the structure of activity of a worker.

Simon (1999, p. 187) used an example where he demonstrated differences between a linear chain organization and hierarchy. He wrote: "Similarly a polymer is simply a chain, which can be very long, of identical subparts, the monomers." Simon (1999, p. 186) wrote: "we tend to reserve the word 'hierarchy' for a system that is divided into a small or moderate number of subsystems, each of which may be farther subdivided." Further, this author wrote that some systems can be very long and can be considered simply as a chain but not as a hierarchical system. A critical feature in a hierarchical system is span of control. It is a feature of a system that reflects a "number of subordinates who report directly to a single boss" (Simon, 1999, p. 187). In contrast, for Annett and his colleagues, anything that is related to a complex system is organized hierarchically. Thus, Annett's (2000) hierarchical decomposition contradicts Simon's understanding of hierarchy.

These authors do not take into account the fact that tasks in the system can be organized not hierarchically but logically. They can be performed in a strict order as in the production process on the conveyer. They can be organized according to some logical rules. For example, if an operator completes a task, he or she can obtain different results. One result can be designated as *A* and the other as *B*. In the first case, the next task that should follow can be number 1. In the second case, the next task to be performed will be number 2. Relatively seldom tasks can be performed in any order. Therefore, not every type of decomposition is hierarchical decomposition. Subtasks that simply are listed in the order of their execution cannot be always considered hierarchically organized elements. Enumerating task names does not give any idea about the structure of the tasks. A simple listing of subtask titles does not give an idea of what a worker is really doing.

Let us consider subtasks numbered 3–5 on Figure 4.3. These tasks are performed in a linear sequence. These tasks also can be divided into smaller subelements. However, such further dividing is not presented in the figure. We can ask the question of why these tasks are not divided into their elements. Where are the rules of such decomposition? Used by Annet and Cunningham (2000), the homeostatic regulation principle does not give an answer to this question. Feedback can be used after completion of the whole task, after performance of subtasks, and after performance of its smaller elements. Feedback can be also conscious and not conscious. Moreover, the homeostatic regulation principle contradicts the principles of conscious regulation of activity.

We would like to draw attention to the fact that the description of tasks in the boxes (see Figure 4.3) does not give a clear understanding of what a worker does when performing each task. A task description should give a clear understanding of what a worker has to do in order to complete the task, even for specialists who were not involved directly in task observation. Such specialist should be provided with drawings of utilized equipment or tools and description of cognitive and behavioral actions in each subtask. Behavioral actions should be further decomposed into motions. Cognitive action might be decomposed into smaller cognitive operations. Actions and their organization depend on equipment configuration and utilized tools and goal of task. This relationship is the basis for ergonomic design and training. The relationship between cognitive and behavioral actions and equipment configuration is not described.

The term hierarchical structure of human activity is applicable to a combination of several actions that subordinates directly to one goal. Such combination is restricted by span of control, introduced by Simon (1999). Span of control is determined by the number of actions and their elements (psychological operations) and the capacity of short-term memory. Cognitive and behavioral actions and sometimes a combination of several actions have more often logical but not hierarchical organization.

Terminologies used by Annett, Cunningham, and Kirwan do not coincide with each other. According to Annett and his colleagues, a common goal is decomposed into different levels of operation. According to Kirwan, a common goal is divided into subtasks. The term *operation* refers to the last level of decomposition. The task is the basic term of HTA but there is no clear definition of the term task in these studies. The difference between the goal of task and the task is not clear. They are considered as being synonymous. The same goal can be given in different conditions that will create different tasks.

Any already existing system has a formally defined list of tasks and professionals have to read manuals of the system. A system cannot function correctly if there are unknown tasks that should be performed by personnel. However, all HTA methods start with identification of tasks in the existing system by using redundant decomposition procedures when such list of tasks can be found in already existing system documentation. Such artificial decomposition can lead to selecting tasks that do not exist in the real system. In HTA, selection of tasks stems from the method of analysis and does not correspond to really existing tasks, and some tasks can be overlooked.

As considered by Annett and Cunningham's (2000) example, the system already exists and the tasks that are to be performed in this system should be already known.

In such situation, the research can simply be based on analysis of documentation and discussion with experts that work with the system to identify all the tasks running in the system. Instead of consulting the experts to identify all tasks performed in the system. Annett and Cunningham utilize instead their artificial decomposition method to determine a list of tasks and their organization. Decomposition of the work process into progressively smaller units–tasks is also incorrect. Tasks can be organized in a linear sequence. They may have a logical organization. Tasks can be coordinated in time and space. They cannot be seen as smaller pieces of the same task. HTA is an artificial technique that depends heavily on skills and professional intuition of an analyst, and different analysts may produce different lists of tasks for the same already existing system. This drawback of HTA was also pointed by Patrick (1992, p. 1983). Ergonomist, psychologist, or any other specialist cannot define a list of tasks in the already existing system based on the HTA method. In such system, a list of tasks is given in advance, defined at the stage of design of the system. Only for the system that is under design and does not exist yet, a list of tasks can change depending on design solutions. However, HTA is not suitable for task description/identification of a new system.

Human activities in a system may be divided into job operations, jobs (positions), duties, tasks, subtasks, and behavioral elements (Huchingson, 1981, p. 15). For example, job operation includes a combination of duties and tasks essential to accomplish a certain system function. Job duties are a set of tasks within a given position. A task is defined as a composite of related activities performed by an individual and directed toward accomplishing a specific amount of work within a work context. A task element is the smallest logically definable set of perception, decisions, and responses required to complete a task or a subtask (some part of a task). In SSAT terminology to some extent, it is different. However, we fully agree with the described hierarchy of the component of a work process.

We can make the following conclusion. HTA ignores the fact that before an operator is involved in interaction with any type of machine, he or she should know what type of tasks he or she has to perform. When human factors/ergonomists attempt to improve functioning of an existing man–machine system, he or she should describe already existing tasks associated with this system, describe their logical organization, and select the most important of them for further analysis. When human factors professionals are involved in the design of a new man–machine or software system, discovering all necessary tasks for the system functioning can become a very difficult step of analysis.

A critical review of HTA shows that hierarchical description of tasks is not a task analysis. This method with some modification can be used at the task description/identification stage that precedes task analysis. Moreover, for an already existing system, a list of tasks already is known. The hierarchical description of the tasks may not reflect the actual list of tasks in an already existing system.

In a new not yet existing system, HTA cannot be used at all.

HTA shows that there is no clear terminology and even no notion of a task. There are no units of analysis of behavior and internal mental processes. Describing a task in the absence of a clear and standardized terminology leads to an ambiguous interpretation of data by specialists with no direct involvement in the observation of the system. This contradicts with the principles of system design. It is only after the task description/identification stage the specialist can perform task analysis.

Tasks should be described in a written form and accompanied by a layout of a workplace; drawings of equipment, tools, and software; photos of a workplace; position of a worker during a task performance; and so on. Such methods of description facilitate better understanding of what a worker does. It also allows getting some general information on the possible cognitive components of work. When it comes to the design and a real task does not exist, the role of a correct combination of verbal and graphic descriptions of a task increases. A combination of such methods provides a general idea about a task at hand, even in cases when cognitive components play an important role in task performance. In all cases, a task description should be conducted in such a way that it is clear not only to the one who describes it but also to those who will later work with such documentation. This requires development of a standardized language of activity description. Without temporal analysis of task performance, we cannot understand their organization. All of these suggest the need to develop a single, unified set of methods that can be used by experts who study human work. Applied and systemic-structural AT can be useful for this purpose.

Section II

Training and Design

5

Knowledge and Skill Acquisition as a Self-Regulative Process

5.1 Self-Regulative Concept of Learning and Training

In this section, we briefly present some new ideas on the self-regulation theory of learning and training that have been developed in the systemic-structural activity theory (SSAT) framework (Bedny and Karwowski, 2007; Bedny and Meister, 1997) and consider examples of application of this theory to the study of human performance. Learning and training are important areas of work psychology and ergonomics (Patrick, 1992; Seel, 2012). Learning can be defined as relatively permanent changes in external behavior and cognitive processes due to experience. In the area that studies human work, learning and training are two interdependent concepts. Training is a process of acquiring skills, concepts, and attitudes that are necessary for job performance. Training is more specific than learning. It is needed to improve the performance of specific tasks. Learning sometimes is not sufficient for acquiring a required level of task performance. For example, a subject can perform a task with a satisfactory quality but he or she might need much more time than professionally acceptable. Therefore, training is needed to achieve a required pace of performance. Transition to a higher level of performance pace cannot be reduced to increasing the speed of execution. Our research discovered (Bedny, 1987) that transition to a higher pace of performance requires reconstructions of skills that are involved in task performance. During such training, a worker moves from one strategy of performance to another. A strategy is a dynamic and adaptive plan made by a subject for goal achievement that is enabling changes in the approach to a goal attainment. Such dynamic plan is a function of external conditions and internal state of a subject. It includes not only behavioral but also cognitive components such as strategies of attention, etc. In some cases, a specially organized training process can accelerate transition from a less effective strategy of execution to a more efficient one. Hence, learning is a necessary but not the only condition for professional skill acquisition. Learning has a broader purpose than training because it is a prerequisite to training. When we consider declarative knowledge, learning is the basis for the teaching process, and when we consider procedural knowledge, hands-on training is necessary. Usually declarative knowledge in vocational schools is given at a higher level than hands-on training, which gives trainees an opportunity for future growth. Learning principles are bases for training processes. According to SSAT, some basic principles of learning are unity of cognition, behavior, and motivation in learner activity; consideration of learning as a complex goal-directed self-regulative process that integrates conscious and unconscious levels of self-regulation; description of the learning process as a transition from a less efficient to a more efficient strategy of activity performance; and the ability of a learner to achieve the same goal by various strategies. According to SSAT, the basic units of analysis in the learning process are cognitive and behavioral actions and functional mechanisms or function blocks. SSAT utilizes the principle of systemic analysis of knowledge and skill acquisition processes.

In SSAT, the main terms that are used in the learning and training process are task, actions, conscious goal, motives, self-control, self-regulation, feedback, human algorithm, heuristic, strategies, etc. Such notions as stimulus, response, reinforcement, reward, and punishment are practically unused. According to Norman (1976), the structure of our long-term memory is continually modified during the learning process and our knowledge of the same data continually changes. This evolution of stored knowledge in the memory influences the process of acquiring new information and skills. Knowledge is information encoded into long-term memory. For example, declarative knowledge includes images, concepts, and propositions (Anderson, 1985; Landa, 1984). There is also procedural knowledge that is particularly important in the vocational training process. This type of knowledge includes information about mental operations that transform images, concepts, and propositions into long-term memory. If a subject has declarative knowledge, it does not mean that he or she has knowledge about procedures. Let us examine the relationship between these two types of knowledge. The training process is particularly important for the development of an adequate relationship between these two types of knowledge. When presented to students, material should be organized in order for information in memory to be structured. The organizing principles must be logical and reflect a causal–consequence relationship.

Data obtained in cognitive psychology and activity theory testify for the fact that the acquisition of knowledge may be in part performed in internal mental plane without interaction with the external situation. This is accomplished through internal mental actions utilized by the subject. Thanks to this, the memory structure is reconstructed. Imaginative and thinking actions play a particular role in this process. Thus, not only memory but also thinking is essential in such way of knowledge acquisition.

One of the main purposes of teaching is to develop a student's capability to think adequately in various work situations. Professional thinking is not knowledge but what one does with that knowledge. It can be defined as a set of structurally organized mental operations or thinking actions carried out in order to solve a variety of professional problems and make decisions. Thinking plays an important role in the generalization and transfer of knowledge. The process of generalization and transfer requires the development of conscious and unconscious strategies of extracting critical or essential features of an object or phenomena and neglecting those that are not essential. It is important to find out the regularities of the variations of these features from situation to situation.

One can have a wide repertoire of images, concepts, and propositions (declarative knowledge) in one's mind, but only a limited repertoire of mental actions and therefore cannot apply adequate procedures to utilize this knowledge.

When theoretical classes provide a higher level of general declarative knowledge than procedural knowledge obtained at practical training, such discrepancy provides a better opportunity for further professional growth. However, this discrepancy between declarative and procedural knowledge often results in a situation when a subject applies his or her knowledge intuitively.

Novikov (1986) introduced the following classification of declarative and procedural knowledge in vocational education:

The first class of declarative knowledge includes the following:

1. Knowledge received by directly sensing the objects during a task performance
2. Phenomenological (descriptive) knowledge about technical objects (devices, equipment, computer, etc.)
3. Knowledge about functioning technical objects and their construction
4. Knowledge about the scientific basis and principles of organization and operation of technical objects

The second class of procedural knowledge includes the following:

1. Knowledge received by directly sensed performance of actions by perceiving and imagining actions with objects
2. Phenomenological (descriptive) knowledge about actions with technical objects
3. Knowledge about general rules (algorithms) of actions with specific objects under various conditions
4. Knowledge about the scientific bases and general principles of performing technological processes and organizing labor

Spontaneous knowledge that is obtained directly from sensory-perceptual experience is often implicit and cannot be verbalized. In contrast, other categories of knowledge are explicit and can be verbalized.

The other important units of professional experience are habits and skills. In cognitive psychology, there is no clear distinction between procedural knowledge and skills. Procedural knowledge is associated with mental operations that transform information (images, concepts, and propositions) into memory. This knowledge is also utilized when performing physical actions and operations. Procedural knowledge is not sufficient for a trainee when he or she performs an assigned procedure. Such knowledge has to be transformed into a skill through practice. One can see this in gymnastics, when a gymnast knows the technique but cannot perform an element without extensive practice. Practice transforms knowledge into a skill.

Let us examine the relationship between habits and skills from the applied activity theory (AAT) and SSAT viewpoint.

In applied activity theory (AAT) and SSAT, habits are considered to be cognitive and motor actions that are consciously automated via learning and training. Sometimes the high level of automaticity causes transfer of cognitive and motor actions into cognitive or motor operations. Such operations are components of more complex cognitive or behavioral actions. Habits facilitate easy, quick, and economic performance that requires a low level of attention and a high level of efficiency. Habits can be integrated into motor or mental patterns. Such patterns or structures create the first level of cognitive or motor skills. These skills can be efficiently used in a practical situation. For example, an experienced driver can drive a car efficiently with low level of awareness. This driver has car driving *naviks*. A novice can drive a car at a high concentration and emotional tension. This driver has the first basic skill level or the first level of *umeniya*. These are different types of cognitive and behavioral actions that have not yet developed up to a certain level of automaticity. After special training, they can be automatized and turned into *naviks* (habits). The first level of skills (*umeniya*) and habits (*naviks*) are task specific.

There are more complicated or second level of skills or second level of *umeniya* that constitute an individual's ability to organize knowledge and first-level skills that transform into a system efficiently used to perform a particular class of tasks or to solve a particular class of problems. Such second-level skills can be roughly labeled as know-how. They can be applied to a broad class of tasks. This type of skill is the foundation of knowledge-based behavior according to Rassmussen's (1983) terminology. However, such behavior includes lower-level skills and habits as their components.

When these habits are simple, they can be used as building blocks or operations to form more complicate actions. However, there are cognitive and motor operations that almost never become conscious. The other group of operations at the first step of training is unconscious, but at the following step of special training, these operations become conscious, and therefore after their structural organization, they are transferred from operations into conscious cognitive or motor actions. For example, a gymnast's unconscious motor operations can be transferred into consciously regulated actions that make the gymnast's skills more flexible and consciously regulated. In some instances, the transformation of unconscious operations into consciously regulated actions can be undesirable because it causes deautomaticity of skills (Bedny and Meister, 1997). During a skill acquisition process, a student can transfer his or her attention from controlling his or her actions to the goal and results of his or her actions.

AAT classifies skills as sensory, perceptual, mnemonic, thinking, imaginative, or motor depending on the content of actions they are involved in Gil'bukh (1979). The other system of classification distinguishes between sensory (visual, auditory, etc.), mental (calculation, reading, problem solving, etc.), and motor (connected with performance of motor actions) skills (Platonov, 1970).

It is useful to distinguish between instrumental and noninstrumental skills. An operator can regulate his or her cognitive and behavioral actions based on information that is presented on various displays or instruments. He or she can also regulate his or her actions based on informally revealed information such as noise, vibration, resistance of controls, and smell. Instrumental and noninstrumental information are usually presented in combination. So it is possible to categorize skills into instrumental and noninstrumental ones. For instance, noninstrumental skills play an important role in a pilot's activity and can change strategies of task performance in emergency conditions (Bedny and Meister, 1997). There are also such terms as officially required and actually used skills. Operators, while performing their tasks, can discover new methods of task performance. Noninstrumental skills are considered to be actually used skills that are often acquired by an operator independently. Actually used methods should be evaluated and might become an officially accepted method of task performance. Such methods often require acquisition of new skills. The training program should provide an opportunity to actively practice skills required on the job. Reparative performance of the same activity with evaluation of the obtained result leads to changes in the activity structure.

These changes can be described as follows (Bedny and Karwowski, 2008c):

1. *Changes in methods of action performance.* The individual operations are integrated. Unnecessary action components (unnecessary because enhanced skills render them unnecessary) are progressively eliminated. Performance speed is increased and actions that have been performed sequentially before can now be performed simultaneously.

2. *Changes in methods of actions' regulation.* Reference points for actions' regulation are extracted very quickly. For example, during acquisition of sensory-perceptual skills, students develop an ability to extract required acoustic features from their environment. An external contour of regulation is replaced by an internal one. For instance, the visual evaluation of motor actions is replaced by kinesthetic evaluation. Levels of action regulation are changed or a leading level of regulation of actions associated with our consciousness and meaningful interpretation of actions is gradually replaced by a lower level of actions' regulation. As a result, individual attention is transferred from the perception of student's own actions to the action's output. Actions are evaluated not only consciously but also intuitively.

3. *Changes in strategies of activity performance.* Separate actions are integrated into a holistic structure of activity. An individual prepares for (anticipates) subsequent actions while performing preliminary ones. Transfer from one action to another is performed automatically without having to make this transition consciously. As a result, cognitive and motor actions are performed with little or no conscious attention or effort, reducing utilization of short-term memory and thinking for the actions' regulation process. The probability of simultaneous execution of actions increases. Less efficient actions and operations are replaced by more efficient ones.

4. *Changes in emotionally motivational regulation of activity.* The energetic components of activity associated with the emotionally motivational state of a subject are changing during skill acquisition. Subjectively, this state is experienced as changes in efforts required for activity performance, reduction of emotional tension, changes in evaluation of significance of different components of activity, and changes in motivation in general. Emotionally motivational regulation of activity influences cognitive strategies of activity performance.

Changes in the skill structure during practice can be explained by mechanisms of activity self-regulation (Bedny, 1987). Activity acquisition can be considered as a self-regulative process when a subject switches from less efficient strategies of performance to the more efficient ones. Strategies of performance include conscious and unconscious components of activity. The more complicated the skills are, the more intermittent strategies are required during the skill acquisition process. SSAT demonstrates that learning can be viewed as an active regulative process, and strategies of performance can be described based on the analysis of self-regulation mechanisms.

Increasing efficiency of the training process can reduce the number of intermittent strategies of performance. The more efficient the training process is, the more important is the role of the second level of skills that is involved in performing a particular class of tasks or solving a particular class of problems. Acquisition of these kinds of skills enables a student to utilize different strategies of performance within the same activity. The skills become flexible and adaptive to the ever-changing situation. According to Bernshtein (1966), skill acquisition is a complicate constructive process. During each new trial, a student develops a new structure of the skill adapted to the particular trial. Therefore, repetition of trials should not be considered as the same activity performance. Activity is reconstructed during each trial and this process is referred to as *repetition without repetition* (Bernshtein, 1966).

The self-regulative approach of learning (Bedny, 1987), which derives from the model of activity self-regulation, contradicts with studies of the stimulus–response paradigm as a basic unit of analysis for human learning. At the same time, this approach does not consider a learner as an information-processing system but rather as a subject who actively interacts with the situation according to the goal of the activity.

A subject develops his or her knowledge and skills by acting or in other ways performing cognitive and motor actions. Hence, basic units of analysis in the learning process are cognitive and motor actions. Sometimes the learner can develop knowledge and skills by using only cognitive actions with idealized objects. The other important concepts in learning and training are feedforward and feedback connections. In SSAT, these two notions are always considered in the context of specific models of self-regulation. In contrast to understanding feedback as information about the external modification of a situation, in SSAT, feedback can be external or internal (mental). A learner can mentally manipulate ideal objects and evaluate the result of such manipulations. In other words, a learner can act utilizing mental actions. Therefore, learning and training are self-regulative processes that facilitate the formation of external or motor and internal or cognitive actions. Interaction of internal or cognitive and motor or external actions in such self-regulation processes is a critical factor in knowledge and skill acquisition processes. The self-regulative process is the basis for any kind of learning including that of stimulus–response associations. Analysis of the acquisition process demonstrates that even the salivary condition response of Pavlov's dog is a result of the self-regulation process. According to the self-regulation concept of learning, there are conscious and unconscious levels of self-regulation. Learning that derives from association without conscious knowledge of the logical interrelationship between phenomena is based on reinforcement that has informational and motivational functions. Informational functions (cognitive) of reinforcement are limited because there is no understanding of the stimulus–response relationship. In order to understand why one or another association has been formed during the learning process, it is necessary to study a self-regulation process that creates those associations.

In the training process, feedback can be internal (e.g., proprioceptive or mental) and external (e.g., visual, auditory). There are also intrinsic and extrinsic feedback in the learning and training process. Intrinsic feedback is normally available in the training process, while extrinsic feedback is artificially introduced (Annet, 1961). It is important to know that extrinsic feedback is not always useful in training and it is important to develop correct procedures for utilizing it.

Sometimes extrinsic feedback can be redundant and can contradict the natural mechanisms of activity regulation (Bedny and Meister, 1997). In one experiment (Novikov, 1986), a trainee learned how to apply the exact effort to a particular control when using additional orienting points in an oscilloscope. The trainee very quickly performed the required motor actions without errors. However, when the information from the oscilloscope was removed, the trainee could not perform the required motor actions correctly. When the instructor started to use discreet feedback, which included turning on a bulb only when a trainee performed erroneous actions, it resulted in significant improvement in real performance. From an activity self-regulation perspective, it can be explained as follows: In performance that utilized an oscilloscope, the trainee uses only visual external feedback (external contour of self-regulation) and the internal counter of self-regulation associated with muscle feedback is not activated. However, the internal contour of self-regulation in this task is critically important, and in the second scenario when discreet feedback with a bulb was used, the internal contour of self-regulation was activated. This caused increased efficiency of performance in a real situation.

In another study, a special simulator for acquiring filing skills was developed.

This simulator was presenting information about the required trajectory of the file's movement on the oscilloscope's screen during benchwork. Two training methods were used: (1) Trainees started using the simulator immediately and (2) trainees worked without the simulator first and then started using it after. It was discovered that the second method was more efficient. Trainees performed incorrect actions first and then could compare erroneous and correct actions that helped them to form a subjective standard of successful results and a subjective standard of admissible deviations. Activation of explorative activity allowed successful development of the required mechanisms of self-regulation.

The function of feedback in training can be understood better if we know its role in activity self-regulation. Bedny developed the self-regulation theory of learning (Bedny and Meister, 1997), which asserts that self-regulation is the basis for all types or levels of learning. Learning in the self-regulation framework is conceptualized as the development of performance strategies. The self-regulation theory of learning distinguishes between associative and cognitive levels of learning and training. The associative level is a lower, more primitive type or level of learning or training. It is unconscious and does not involve a conscious goal but rather an unconscious set. The cognitive level on the other hand includes conscious components and a conscious goal. Both levels of learning or training are interdependent. It is often necessary to use both types of learning or training. This is particularly obvious in motor learning. For example, in gymnastics, trainers apply methods that require conscious and unconscious levels of movement regulation. According to the self-regulation concept of learning and training, a learner acquires various strategies of activity during performance. The more complex tasks are, the more intermittent strategies are utilized by a learner. At the final stage of the training process, a learner has to acquire flexible cognitive strategies of task performance. During the learning and training process, a learner acquires knowledge and turns them into skills. Changes in task performance during learning are viewed as the reconstruction of performance strategies. Increasing the efficiency of the training process can reduce the number of intermittent strategies of task performance. Without understanding the principles of activity self-regulation, it is difficult to develop effective principles of training (Bedny et al., 2014). Self-regulation is the real *machinery* of the learning process. In our model of self-regulation, the result of the self-regulative process is described by a function block *new experience* that inter-acts with past experience (function block *experience*), reflecting how a new experience reconstructs the past experience (Bedny and Karwowski, 2007). The term strategy plays a certain role in cognitive psychology (Patrick, 1992), making it easier to integrate the self-regulative theory of learning with the data obtained in cognitive psychology.

According to the concept of self-regulation proposed, it is necessary to distinguish between external feedback and internal mental feedback. In the latter case, the person can perform cognitive actions, evaluate their results mentally, and only then perform the actual motor action. Moreover, motor actions can also be assessed internally on the basis of evaluation of motor action programs. In other words, there are internal anticipatory mechanisms for the regulation of motor actions. Such results were obtained by various authors (see, e.g., Angel, 1976; Anokhin, 1962, 1969).

From the general activity theory perspective, the relationship between external and internal components of activity in the learning process should be considered as a process of internalization. In AT, this process is under-stood as the transformation of external activity into internal activity. The most popular concept of internalization has been suggested by Gal'perin (1969), stating that the formation of mental actions passes through a series of stages. The first stage includes familiarization with the task associated with action performance; the second stage involves performance of actions with material objects or its materialized presentation accompanied by audible speech; the next stage includes actions that are performed based just on ver-balization without direct support from an object; the fourth stage involves a learner that performs actions by using external speech but in a whisper; the last stage includes performance of actions based on internal speech. Actions during such stages are abbreviated, become unconscious to a learner, and seem to appear simultaneously. Through these stages, actions are internal-ized. According to Vygotsky, learning is produced by internalization of activity during social interaction (Vygotsky, 1978). He formulated the semi-otic concept of internalization according to which a learner internalizes vari-ous sign systems, particularly a verbal system, during social interaction. The concept of internalization plays a critical role in the explanation of human mental development in general.

From the SSAT perspective, internalization is not the transformation of external components into internal ones but rather a continuous process of mutual influence of internal and external activity through feedforward and feedback loops and formation of internal mental actions (Bedny and Karwowski, 2006). External, material actions include cognitive mechanisms and serve as the basis for the formation of internal mental actions. Internal mental activity is constructed by the mechanisms of self-regulation. As such, internalization cannot be reduced to separate psychological processes and memorization in particular. Internal and external components of activity regulate each other based on the mechanisms of self-regulation. Internal activity is shaped through external activity. At the final stage, mental actions can be performed independently. Self-regulation is a goal-directed process where emotionally motivational processes play a critical role. Therefore, cognitive and emotionally motivational processes in mental development work in unity.

A learner uses a variety of strategies during the process of self-regulation. The final structure of internalized mental actions does not coincide with that of the external material actions. A learner uses intermediate auxiliary operations of various modalities that are temporal in nature. They are not included in the final structure of internalized mental actions. Utilized strategies are individualized and also depend on a learner's past experience. In the process of internalization, a learner compares actions of various modality and operations that can be performed in a material form with the actions that involve manipulation of various sign systems. Such a comparison significantly increases the efficiency of internalization. Therefore, internalization is not just a transformation of external components into internal ones. Internal activity is constructed based on the mechanisms of self-regulation and interaction of external and internal components of activity.

The contemporary studies in the multimodal interface design demonstrate the applicability of these theoretical data (Oviat, 2012). Multimodal systems represent a new direction in computing moving away from conventional windows, icons, mouse, and pointer interfaces, which combine user input modes such as speech, pen, touch, manual gesture, and body movement. Multimodal interfaces allow utilizing flexible input modes and therefore a choice of various types of cognitive and behavioral actions. A user can utilize and compare actions of various modalities during task performance to achieve the same goal, which makes it easier to perform various computer-based tasks. Multimodal interfaces can be useful in training because they facilitate comparison of material, verbal, and other symbolic actions. Such interfaces are also instrumental in the development of individual strategies of task performance.

The self-regulative theory of learning includes several basic steps. The first step involves defining learning tasks based on task analysis. Then it is necessary to uncover the basic strategies of task performance at the different stages of learning and training processes. The final stage includes the development of algorithmic description of basic strategies of task performance. Algorithmic description of various strategies is used as a prescription of what a student should do during learning and training. Some intermittent strategies have only temporal character and can be utilized only for acquisition of more complex final strategies. Some complex task-problems cannot be described algorithmically. In such situations, the instructor utilizes heuristic descriptions, which can increase the efficiency of solving learners' task-problems. But they do not guarantee solutions. The degree of uncertainty of utilized instructions can be significantly decreased during the learning and training process. It is important to know that human algorithm in the learning process usually describes general principles of task performance or solving specific problems. Because of that, such algorithms are applicable to all problems that belong to a specific class.

Human algorithms describe the logical organization of human cognitive and behavioral actions and operations. Performing tasks or solving various problems according to algorithmic and heuristic prescriptions helps students acquire required professional knowledge and skills. Therefore, written instructions play an important role in the self-regulation theory of learning. Landa (1984) developed the algoheuristic theory of learning, which is derived from the general activity theory. Bedny and Meister (1997) and Bedny and Karwowski (2007) developed formalized principles of algorithmic description of human performance, and these principles are utilized in the self-regulation theory of learning. We will describe the basic principles of algorithmic description of human activity during task performance in Section 7.1 and 7.2.

5.2 Introduction to the Learning Curve Analysis

The learning phenomenon has received increasing attention in studies of productivity in various organizations. The basic ideas of the learning curve approach derive from psychological studies of skill development, and the main one is that performance gradually improves through repetitive trials that lead to the reduction in performance time and a number of errors. Analysis of the relationship between practice and performance improvement originated in psychology as a study of the acquisition process. In Pavlov's (1927) study, salivation in response to the conditioned stimulus is plotted on the vertical axis, and the number of trials is depicted on the horizontal axis. The study demonstrates that salivation (conditioned response) gradually increases over trials and approaches an asymptotic level. This is an acquisition process in classical conditioning that is depicted by a learning curve.

Utilizing a similar method, one can develop learning curves when studying the acquisition of any professional skills.

Studies have shown that the learning curves can be divided into two general groups: (a) curves with negative acceleration, when in the beginning, the basic skills are formed fast, and then the process of final development slows down more and more, approaching some stable level, and (b) curves with a positive acceleration when development of skills is slow at the first stage and then the acquisition process speeds up. The first way is typical when a blind trial and error strategy dominates in mastering the skills. The second way is specific to development of skills when understanding of the process is important. As soon as the method of task performance is understood or figured out, the acquisition process becomes quick and errors are quickly eliminated.

The shape of the learning curve is affected by the nature of the task, the idiosyncratic features of the trainee, and the training method. By combining individual learning curves, one can discern a general type of curves during performance of a specific task. Averaging curves permits one to discover general features of the skill acquisition process. At the same time, this can result in loss of some important information about the specifics of skill acquisition. Therefore, during the training process, the researchers should use both individual and average learning curves.

Learning curves can be described by mathematical formulas that permit one to predict the process of skill acquisition and define the time required to achieve a stable level of performance. The relationship between skill acquisition and practice can be often described by the *power law*. For example, the performance time of repetitive tasks is reduced exponentially to a stable level when further improvement is practically impossible. This law can be used for predicting the task performance time (Gal'sev, 1973; Wright, 1936). The main equation for the power law is described by the following formula:

$$T = C_1 X^b$$

where

T is the average task performance time when a task was executed X times
C_1 is the performance time of the first task

Parameter b has a value from 1 to 0 and describes the worker's learning rate and corresponds to the slope of the curve. If b is close to 1, this demonstrates the high learning rate that leads to a fast approach with the minimum task execution time.

Similarly, the cost of the produced part can be estimated using a number of practices for the skill acquisition instead of the performance time. The learning curve is also used to study the relationship between the training process utilized by a company and the quality of the product produced by the company (Khan et al., 2011). In general, learning curves can be used to analyze the process of improvement. Therefore, learning curve analysis is applicable not only for studying the process of improving separate individual skills but also for analysis of improvement of team performance, or to improve performance of the entire organization. This is specifically relevant in economics and management. However, when professionals attempt to utilize the power law, they overlook the fact that very often the acquisition process is accompanied by changes in strategies of task execution. This means that sometimes it is required to use not one but several equations with different parameters, or use other mathematical methods to describe the process of acquisition in organizations.

Learning curves are a useful tool for analyzing skill retention, where it is necessary to compare learning curves when workers just obtained a stable level of skill making the first batch of parts, and then after a defined period of break when a worker has to make the second batch of parts, a new curve is developed (Bohlen and Barany, 1966). Analysis of tasks that involve perceptual-motor skills demonstrates that continuous tasks have better retention in skill performance in comparison to discrete tasks (Fleishman and Parker, 1962). In our study, subjects perform a complex manufacturing task and then performed it after a 2-month break. It was found that workers who made the batch of parts for the second time after a break reached the same level of stability much faster than with the first batch, although their performance time was the same in the beginning. This demonstrates that the task performance time after a break can be approximately the same, but the learning rate can be faster after a brake (Bedny, 1979).

Plateau of acquisition curves is used as a criterion that demonstrates the stabilization of skills. Stabilization is not a unitary process. Some parameters can be stabilized earlier and the others later. For instance, one should utilize criteria that demonstrate the stabilization of skills based on temporal, spatial, and force (physical effort) parameters in studying the stabilization of motor skills. The coefficient of dispersion of these parameters of motor action execution for repetitive performance is an index of stability of a skill in percent (Novikov, 1986).

Such coefficient for temporal parameters is calculated according to the following formula:

$$V_t = \left[\frac{\Sigma(t_i - T)}{NT} \right] \times 100$$

where
   t_i is the performance time of an individual motor action
   T is the average performance time of an action
   N is a number of cycles

Such coefficients can be calculated the same way based on other parameters. For example, to calculate a coefficient based on the physical effort, one can use the following formula:

$$V_f = \left[\frac{\Sigma(f_i - F)}{NF} \right] \times 100$$

where
   f_i is the effort involved in the performance of a particular action
   F is the average effort

The coefficient for the spatial parameters can be calculated similarly. Comparison of these coefficients permits one to analyze the skill stabilization process according to considered parameters when workers perform not just the same but also different tasks. Measurement of these parameters can be performed in various ways. For instance, spatial parameters can be measured by marking a work surface with squares and then an arm position can be videotaped registering the position of an arm in each square.

Learning curves are important tools for the analysis of human strategies during task performance. Changes in strategies of performance during the acquisition process can be discovered by utilizing a variety of skill acquisition curves. The shape of the learning curve is affected by the nature of the task, the idiosyncratic features of the trainee, and the methods of training. Learning curves have irregular features such as peaks, troughs, and plateaus. These fluctuations can be explained not only by accidental factors but also by changes in the performance strategy. The plateaus demonstrate that using a specific strategy is no longer effective when improving performance. The troughs indicate a worsening of performance as the trainee attempts to transfer to a new strategy. A new strategy is tested by a trainee, and if it is evaluated as a more efficient one, the new strategy is used and inefficient ones are discarded. The formation of various strategies and transition to a new strategy are not completely conscious processes. The more complex the task is, the more strategies are utilized by a trainee. Final strategies of specific tasks' performance often cannot be acquired without intermittent ones. Sometimes it is possible to reduce the quantity of such strategies. Using a new strategy often requires more time initially and produces more errors than the previous less efficient strategy. Any strategy has a series of stages: formation, implementation, evaluation, correction, and perfection. Hence, skill acquisition is a self-regulative process. Undesired strategies are eliminated and even forgotten by subjects. This process can be performed at the conscious and unconscious levels. At the conscious level, the trainee deliberately selects a strategy and tests it. Very often a subject can acquire a number of strategies during performance of the same task. Depending on the specific situation, he or she utilizes the more preferable strategy. The ability of the subject to use a number of strategies for the performance of the same task increases the reliability and precision of task performance.

Some strategies are temporary. They are used for the acquisition of other required strategies. In other words, they are not included in the final structure of developed skills and in most cases even not remembered by trainees. This explains the fact that the more complex the task is, the more time is required for the acquisition process. That is why the duration of the acquisition process is a more sensitive indicator of the complexity of the task than the time of its execution. All described data demonstrate that learning can be considered as a self-regulative process that involves in formation of various strategies of activity (Bedny, 1979). When the strategy selection process is conscious, a student develops a set of possible strategies and deliberately selects the best one based on conscious or unconscious criteria. However, in most cases, an instructor should know what strategies should be used by a learner and develop a training process accordingly.

One important principle of the activity theory is the genetic study of psychological functions. The essence of this principle is to study various psychological phenomena during their development. Scientists study the psychic phenomenon based on the analysis of the sequential stages of its formation, reconstruction, and dialectic genesis. This principle of study has been introduced to AT by Vygotsky (1960), Rubinshtein (1959), Leont'ev (1972), and others. In this and the following sections, we apply the microgenetic method of study that derives from this principle. In SSAT, the microgenetic method of study is understood as an analysis of the process of activity structure formation during a relatively short time period (Bedny and Meister, 1997; Bedny et al., 2000). This method is based on analyzing skill acquisition during an experimental study. If the regularities of the activity structure formation and the specificity of the acquired activity structure are known, we can evaluate efficiency and reliability of human performance more accurately and uncover the relationship between various equipment or software configurations and the efficiency of human performance.

When an activity is already acquired and stabilized, users become less sensitive to changes in equipment or software design. The qualitative differences can be discovered only under unusual circumstances or stressful conditions. A complex task that requires a long and complicated process of skill acquisition can be performed as efficiently as a much simpler one once a skill has been fully acquired. At the same time, activity that has a more complicated acquisition process is more sensitive to disturbances and stressful work conditions and proves to be less reliable. Therefore, the analysis of the acquisition process or microgenesis of activity can predict the reliability and efficiency of task performance in adverse work conditions. Thus, studying acquisition processes is very useful. The relationship between practice and improvement in performance is of fundamental concern not only in the study of learning and training but also in the analysis of task complexity, efficiency of equipment, usability of software, and so on.

A formative (teaching) type of experiment is a version of the microgenetic method of study. The essence of such experiments is that a researcher actively interferes in a subject's activity during an experiment and guides a task acquisition process. Usually, this method is utilized by pedagogical psychologists, but it can also be applied in ergonomic psychology and work psychology. In order to use a genetic principle of study in an experiment, we need to develop methods of registering and analyzing a subject's activity. One such method is development of a time structure of formatting and already formed or acquired work activity (Bedny, 1979). Another method of study is the development of learning curves that depict activity acquisition. Using obtained data, we can analyze various strategies for activity acquisition. Certain features of tasks, equipment, and software configuration are considered as independent variables, whereas specific characteristics of activity acquisition process are considered as dependent variables.

In contrast to analyzing the training process, in which attention is concentrated on the correlation between teaching methods and learning curves, in the genetic method, attention is paid to the specificity of task performance, equipment, and software design that is used in the work activity and the acquisition process. For instance, subjects in the experiment perform the same task using different versions of equipment or software for further comparison of the acquisition processes and efficiency of the design.

The genetic principle assumes that the training process is not closely supervised by the trainer and is rather a process of independent skill acquisition. The experimenter interferes in the acquisition process sporadically, only when it is important or is necessary according to the carefully developed plan. There is supervised learning and self-learning or independent learning when a subject uses explorative strategies. Self-learning is particularly important for computer-based tasks. Learnability is an important feature of a user interface. Analysis of task learnability gives important insights into the difficulty of task performance and hence usability of an interface. An ability to support learning or self-learning is an important aspect of usability of human–computer interaction systems. Co-relation between practicing and performance improvement is of fundamental concern not only in the study of learning and training but also in the analysis of task complexity, efficiency of equipment, usability of software, and so on (Bedny and Sengupta, 2005). Tasks with various levels of complexity can be performed with the same efficiency at the final stage of acquisition and their performance time can be identical, but more complex tasks require longer acquisition processes because a worker uses more intermediate strategies of performance during the learning process (Bedny, 1979). Complex tasks are less reliable in extreme and stressful conditions. We have concentrated our efforts not on training methods but rather on the acquisition of activity in various working conditions. Thus, we use the term *acquisition curve* instead of the term *learning curve*. There are various methods for developing activity acquisition curves. In some cases, an acquisition curve can depict the process of activity acquisition for a group of subjects. Acquisition curves are also developed for individual subjects. At the next step, curves developed for the individual subjects can be compared with the one developed for a group of subjects. It is also very useful to develop an activity acquisition curve that depicts not just acquisition for a complete task but also for separate elements of the task (Bedny and Meister, 1997). Such method is useful for the analysis of subskill dynamics in task performance. This method also allows us to analyze skill reconstruction and skill flexibility. The experimenter can include or exclude some task elements, change their sequence, and develop acquisition curves for individual elements of activity and for activity as a whole.

The learning curve is an important tool for the analysis of basic principles of formation of different skills. The most general regularities of the individual skill formation process are not dependent on the specific tasks. Hence, in our research, we focus on the more general aspects of the individual skill formation process that can be used in any type of training. In the following three chapters, we will consider the acquisition process in studying blue-collar workers and the analysis of the acquisition process in learning by observation, and finally, we describe the acquisition process of computer-based tasks.

5.3 Analysis of Changes in the Structure of Skills during the Acquisition Process

In this section, we will demonstrate how this method of study is applied to the analysis of the skill acquisition process of bench-repair workers. This study has a general theoretical value, as it allows analyzing the general regularities of the skill acquisition process. These data are also important for the time study during the training process. Our subjects were vocational school students. To graduate from such a school in considered specialty, one had to put in 2 years of study. Vocational school students had to attend lectures several days a week and also develop their skills in vocational school workshops (apprentice training). At the final stage of their education, students underwent on-the-job training at the factory. On-the-job training (3–6 months) was used only to train workers for jobs requiring the lowest qualification, and the vocational system prepares a qualified workforce for industry. A bench-repair worker not only has to acquire skills for bench-assembly work but should also have some experience working on a variety of metal-cutting equipment. Students involved in our study (trainees) had some experience in bench-assembly work. However, they did not have experience in performing lathe work. Therefore, in our experiment, a trainee performed this kind of work for the first time. In addition, we involved in this experiment instructors who had a great deal of experience in doing lathe work. We selected for this experiment two trainees with high academic performance (trainees number 4 and 5), one with average performance (trainee 2), and two with poor academic performance (trainees 1 and 3). Students were selected for experiment by an instructor who worked with these trainees for almost a semester. The selection of students with a variety of academic performance can better identify individual strategies of performance and detect possible types of errors when trainees perform a specific task, which is an important technique for the analysis of the skill acquisition process.

Individual properties of personality impact the skill acquisition process because individual differences dramatically affect adaptation to objective requirement. Activity strategies that are derived from idiosyncratic features of personality are called *individual style of performance*. This style is a

critically important notion that links an individual to his or her job performance (Bedny and Seglin, 1999a,b). Past experience is determined by two
main factors: the individual features of personality and the specifics of education and training.

So when experts analyze the learning and training processes, it is useful
to select students with different individual properties of personality and/
or varying academic performance. Moreover, the analysis of individual
academic achievement is a useful indicator of individual features of personality. This indicator is specifically useful for practitioners who do not
have a psychological background. One important technique in the study of
the individual style or performance strategies is the analysis of individual
learning curves. It should be noted that different people can perform the
same job with equal efficiency by using their own method or individual
style of work that can be shaped both consciously and unconsciously.
Frequently, it takes a significant period of time to develop a viable style
of performance. It is an important goal of practitioners and psychologists
to identify an individual style of performance that is helpful in the skill
acquisition process. Knowledge of individual style of trainees' performance
facilitates the development of an adequate method of training enabling an
instructor to devise instructions that help use trainees' strengths and com-
pensate for their weakness. Selection of students with varying academic
performance can better identify individual strategies of performance and
detect the possible type of errors when trainees perform a specific task.
Selection of such students and an analysis of their learning curves are useful methodological techniques for analysis of the skill acquisition process.

In our experimental study, we also select three instructors. They had good
professional skills to perform lathe work. Trainees and instructors had to
turn parts as shown in Figure 5.1. The first version of the part the subjects
had turn is shown in Figure 5.1a. When the subjects reached stabilization of
skills in turning the first part, they started turning the second version of the
part as shown in Figure 5.1b. During a task performance, they use only one
lathe cutting tool.

Figure 5.1a shows that a part consists of three elements. Two elements that
are located on the right side of the part have identical shape and width of
protuberance and grooves.

The third element is located toward the end of the part on the left-hand
side and has protuberance and groove that are two times wider than the
protuberance and grooves of the first two elements. Figure 5.1b demonstrates
that the second version of the part consists of the same elements. However,
the position of elements is changed. The element with wide grooves and pro-
tuberances is now in the middle position.

Parts' dimensions are shown in Figure 5.1. Thus, the trainees cut the
chip to 0.5 mm during longitudinal feed and cut the chip to 2 mm during
cross-feed. Three instructors also turned a part with the same configuration. However, the depth of cross-feed was not 2 but 3 mm. Quality

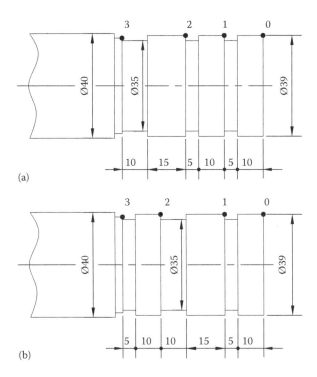

(a)

(b)

FIGURE 5.1
The first and the second version of the part: (a) shaft version 1 and (b) shaft version 2.

requirements for trainees' work were lower than for instructors' work. Therefore, instructors performed a similar but more complex task. It was acceptable for this experiment because we did not compare the performance time of these two groups but compared only the dynamics of the skill formation process for each group.

For a novice, such a task was sufficiently complex. In order to participate in the experiment, trainees had to go through preliminary training that involved performing longitudinal feed and cross-feed (without real turning of details). They learned to work with the lathe's dial and also twice tried turning a cylindrical detail and making a groove. It was only after that preliminary training that they participated in the experiment. The trainees were instructed on how to perform a required task. Instructions were given to each subject separately and the experiment itself was conducted separately with each subject. The experimenter observed and timed (chronometric) subjects with the assistance of one instructor. Before each trial, an instructor-assistant set a cylindrical blank into a three-jaw chuck. So trainees and instructors who were the subjects in the study were only involved in performing the turning operation by using a lathe cutting tool. The trainees worked for 4 days during which they turned the first version of the parts (see Figure 5.1a). Every day they had to turn 10 parts. They worked for 4 days because on the fourth day, they achieved stabilization of the skill level. On the fifth day, students had to turn the second version of the part (see Figure 5.1b). They had to turn 10 parts also. Conditions of task performance were similar to those that were described earlier. Three instructors also turned a part with a similar configuration. Every day they had to turn 10 parts. Instructors worked for only 3 days.

Instructors worked on the first version of the part for 2 days because on the first and second day, they showed a stable level of skills. Instructors worked with the second version of the part on the third day and we did not discover any changes in their task performance. When building learning curves and performing statistical analysis, the experimental results were approximated with precision of up to 1 s, which was sufficient for analyzing the skill acquisition process. For performance of chronometrical analysis, the part was broken down into defined and measurable elements. Each element had beginning and end points depicted by bold dots. The end point for the first element is the beginning point for the next element. A stopwatch was turned on when the cutting tool touched point 0 and switched off when the tool reached point 3.

Let us consider the experimental study in more detail.

Figure 5.2a shows the learning curves that depict the dynamics of the formation of students' skills on the first day where they worked on the first version of the part. It can be seen that with increasing part number, the task performance time sharply decreases.

A significant variation in the performance time of each part is also clearly visible. Some students demonstrate periodical deterioration of task performance. The dynamics of the skill formation process of well-performing trainees is very different from the dynamics of skill formation of low- and moderate-performing trainees. For example, trainee number 2, who according to his or her preliminary evaluation has background of an average performance, demonstrated slower performance than trainee number 3, who was evaluated as a weak student. However, his quality of performance was much better than that of weaker trainees. An experimenter had to make a remark to weak trainees about their poor performance and the need to improve their quality of work. After that, weak students decreased their speed and increased quality of performance.

Figure 5.2b depicts the dynamics of the skill formation process on the fourth day of task performance. It can be seen that the curves become smoother, their inclination is decreased, and variation in performance time between trainees is sharply reduced. Performance time between good and weak trainees differs slightly. One can reasonably conclude that when the skill is relatively stable, the difference between good and weak students virtually disappears. Our statistical analysis started with considering the statistical differences in task performance in the first and fourth day (see Figure 5.2a and b). Figure 5.3 depicts the dynamics of the skill formation process on the first and the fourth days based on the comparison of means of students' performance time.

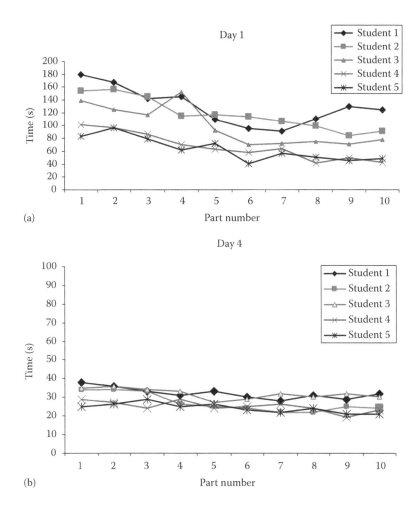

FIGURE 5.2
Learning curves that depict students' skill formation process: (a) on the first day and (b) on the fourth day (the first version of the part).

The average time of task performance on the first day was 95.6 s, and on the fourth day, it was 28.0 s, that is, on the first day, students spent 3.4 times more time than in the fourth day. The mean difference in task performance was estimated for correlated groups. This difference in time performance in the first and the fourth days was statistically significant (paired t-test $t(4) = 5.62, p < 0.005$).

Learning curves in Figure 5.4 depict the skill formation process when trainees were working with the second version of the part on the fifth day.

The difference in task performance time on the fourth and fifth day was statistically significant (paired t-test $t(4)$ 2.87, $p < 0.05$). The F-test of unequal variance revealed that student performance time varied more on day 1 than

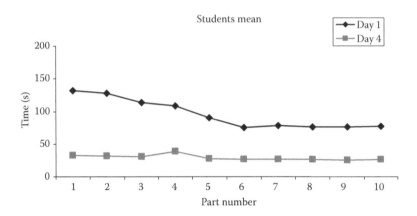

FIGURE 5.3
Learning curves that depict students' skill formation process on the first and fourth days based on comparison of the mean performance time.

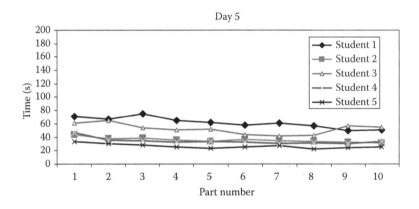

FIGURE 5.4
Learning curves that depict students' skill formation process on the fifth day (the second version of the part).

day 4 ($F_{(4,4)}$ = 62.78, p = 0.002) and day 5 more than day 4 ($F_{(4,4)}$ = 15.62, p = 0.02). It is interesting that rearrangement of parts' elements led to disruption of skill and poor performance mostly for weak trainees. This is clearly evident when comparing the performance of the weak and good or average performing students. This can be explained by the fact that good students acquired a more general strategy of task performance, while weak students acquired a rigid pattern of cognitive and motor actions.

Let us now analyze the results produced by experienced craftsmen or instructors. Figure 5.5a and b shows the dynamics of the three instructors' skill formation on the first and the second days when they worked with the first version of the part.

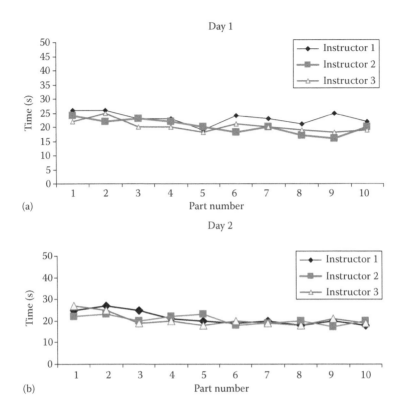

FIGURE 5.5
Learning curves that depict instructors' skill formation process: (a) on the first day and (b) on the second day (the first version of the part).

The curves in the considered figures demonstrate that the instructors completed the task at about the same time, and their performance quality is much higher than that of the students and stays at about the same level. The relationship between qualitative and quantitative indexes remains approximately the same during all trials. These data demonstrate that instructors have a stable level of skills for the task at hand.

On the third day, the instructors performed the second version of the part.

Figure 5.6 depicts the curves of the skill formation process for instructors working on the second version of the part on the third day.

These curves clearly demonstrate that changes in the order of the part elements do not influence the qualitative or quantitative characteristics of instructors' skills. According to analysis of variance (ANOVA), there is no difference in performance time for instructors in the first, second, and third days ($p > 0.3$).

Thus, the task performance time does not change with restructuring of the skill because in contrast with the trainees' skills, instructors' skills are resistant to changing conditions and can be quickly restructured and adapt to new requirements.

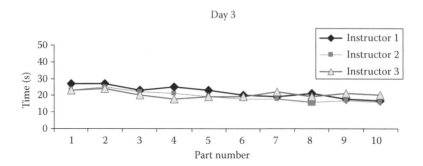

FIGURE 5.6
Learning curves that depict instructors' skill formation process on the third day (the second version of the part).

Comparison of the curves in Figure 5.2a and b demonstrates that relative stabilization of skills among students was reached on the fourth day after 30 repeated executions of the task, and the instructors showed a stable level of skill immediately after they start to perform this task (see Figure 5.5a and b).

Analysis of each learning curve position is also of some interest. Students' curves 1 and 2 (see Figure 5.2a) begin to have a similar angle starting from part six. We do not present results obtained on the second and third day. But it is useful to note that on the second day, all five curves start to take approximately the same position starting from part six. This means that all three elements have similar dynamics of acquisition in time.

These data demonstrate the development of the temporal structure of activity. At the beginning, performance of two identical elements is integrated into a holistic pattern of movements. Integration of the third element into a coherent pattern of activity follows. This was also clearly seen by observing trainees' work. At the beginning, we could see how a holistic rhythmic pattern of activity was developed for the first two elements. Only then could we detect how a rhythmic pattern of activity was developed for the performance of all three elements. It was discovered that during the skill acquisition process and the formation of rhythmic pattern of activity, not only vision and kinesthetic senses but also acoustic information plays an important role in the manufacturing of the part.

Let us consider the students' time performance of separate elements of manufacturing parts.

The learning curves (see Figure 5.7a and b) show that the first element requires less execution time than the same second element of the part.

Trainees spent more time on the third element that was two times wider than the first and second elements. Performance times were different across elements (computed over trials) ($F(2,18) = 100.88$, $p < 0.0001$; all t-tests Tukey HSD $p < 0.05$). A similar result was obtained when we compare differences

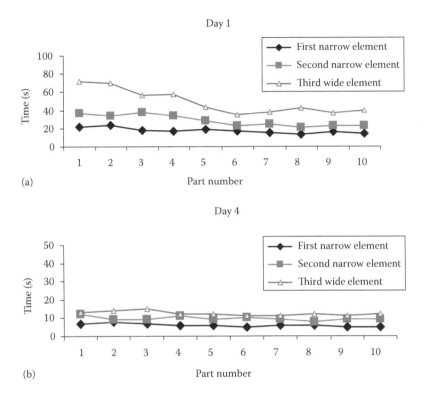

(a)

(b)

FIGURE 5.7
Learning curves that depict students' skill formation process of performance of separate elements of the parts: (a) on the first day and (b) on the fourth day (the first version of the part).

in time performance of elements 1, 2, and 3 in days 1, 4, and 5. According to ANOVA, in day 1, $F(2,18) = 68.91$, $p < 0.0001$; in day 4, $F(2,18) = 100.88$, $p < 0.0001$; and in day 5, $F(2,18) = 31.48$, $p < 0.001$.

This means that the position of the element affects the time performance of this element.

Considered data also demonstrate that the dynamics of skill acquisition for three elements of manufacturing parts is not the same. For instance, execution time of the first element of cutting a part was stabilized earlier than the performance time of the second and third elements. On the fourth day, students' performance time of separate elements was considerably reduced. The difference in the execution time of each element was also reduced. If we compare the dynamics of skill formation in performing the individual elements, we can see that the rate of skill formation for the first two elements is higher than for the third element.

Let us consider the dynamics of the students' skill formation for individual elements of task for the second version of the part on the fifth day (see Figure 5.8).

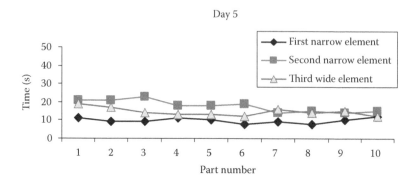

FIGURE 5.8
Learning curves that depict students' skill formation process of performance of separate elements of the parts on the fifth day (the second version of the part).

Comparison of the learning curves shows that the change in the order of the part's elements leads not only to increase in the performance time but also to changes of the learning curves' position relative to each other (each curve depicts the acquisition process for each element). So curve 3, which was previously located at the top, took the middle position, and curve 2 took the top position. This means that the position of the element affects the performance time of this element.

It should be noted that the third element of the part (curve 3) was the widest and therefore its execution time should be longer than the time spent on the narrow elements. Thus, the element's position is the main factor that affects the duration of the element execution.

If we analyze not only the motor actions but also the internal cognitive regulation of activity, it becomes clear that change in the order of elements also results in changes of the nature of the cognitive activity regulation. Indeed the permutation of the part's elements leads to complication of counting on the limb, to changes of the specifics of comparison of the previous result on the limb with the following one, to changes in the strategy of attention, and so on. It is interesting to note that the first elements of the first and second versions of parts were the same. However, switching the positions of the second and third elements also led to change in performance time of the first narrow element, even though the first narrow element did not change position. The difference in the time performance of the first narrow element in the fourth day and the performance time of exactly the same element in exactly the same position in the fifth day was statistically significant (unpaired t-test $t(17) = 16.62$, $p < 0.0001$). These data demonstrate that students form a task execution program not only for separate elements of the task but also for the whole task, and changes in positions of the second and third elements require the formation of a new program of performance for the whole task. This is the cause of longer performance time of the first element of the parts that remained unchangeable.

Weak students were particularly sensitive to the restructuring of the task, demonstrating a similar result as good students in the final stages of performance of the first version of the part. However, the permutation of the elements led to breach of their skill. Observations and discussions with students showed that students use a variety of strategies in performing the same task. For example, good students form general principles of using a limb, while weaker students try to remember intermediate data on a limb. The earlier material shows that past experience and individual properties of personality impact the process of skill acquisition. Individual strategies are easier to identify and study during skill formation and not at the stage when skills reach their stabilization level.

When skills are already developed, the difference in the task performance sharply reduces in its external manifestation, but if we make the task more complex or change the conditions of its performance, individual differences between trainees become apparent again. It is also necessary to pay attention to the fact that the trainees with high ability for a specific job reach a stable level of skill acquisition much faster than low-ability trainees. Another important aspect in analyzing skill development is the fact that the more complex the task is, the longer is the skills acquisition process. It was discovered that the more complex the task was, the more intermediate strategies were used by trainees, and in contrast, the simpler the task was, the less such strategies were used by trainees (Bedny and Meister, 1997). Therefore, the duration of the skill acquisition process depends on task complexity.

Bedny and Zelenin (1989) also demonstrate that the acquisition process for different elements of a skill is not the same. They present two curves for skill acquisition of a production operation. One curve reflects the skill acquisition of only one component of production operation and the second curve depicts the skill acquisition of a production operation as a whole. The speed of skill acquisition of a separate element was significantly higher than the speed of skill acquisition of the whole operation. Hence, the skill acquisition process is heterogeneous and useful for the analysis of skill acquisition to utilize not one but multiple learning curves. This is also beneficial in understanding the structure of the skills.

Let us now consider the results of studies with the instructors for acquisition process during the performance of separate elements of a task (see Figure 5.9a and b).

The curves' analysis before and after the parts' element rearrangement shows that their performance time is practically unchanged. All curves occupy a similar horizontal position. The performance time of the first narrow element on the fifth day after changes in positions of the second and third elements was similar to the performance time of the same element on the fourth day. The difference was not statistically significant. As we have demonstrated, when students performed the second version of the part, the first narrow element required more time on the fifth day than on the fourth day.

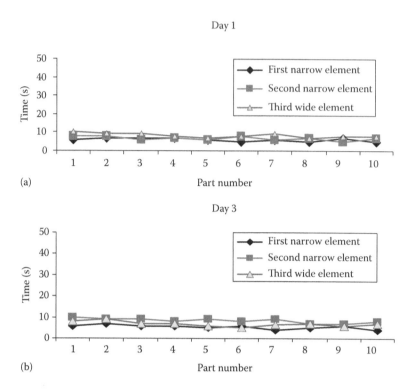

FIGURE 5.9
Learning curves that depict instructors' skill formation process of performance of separate elements of the parts: (a) on the first day (the first version of the part) and (b) on the third day (the second version of the part).

It should be noted that curve 3, which in the beginning had the top position, has taken a position that was held by curve 2 before. In other words, curves 2 and 3 reversed their positions. Increase in the performance time depicted by curve 2 is accompanied by approximately the same reduction in execution time depicted on curve 3. Such changes do not alter the total task performance time. It should also be noted that the curves are much closer to each other than trainees' curves after rearrangement of part's elements. Experimental data show that the main factor affecting the duration of performance of individual elements is not the width of the element but its position in the sequence. When the position of an element changed, so did performance time. The position of elements was changed from day 1 to day 3. The effect of the element on the performance time also changed from day 1 to day 3, as a result of the position change (element by day interaction, $F(2,54) = 12.7$, $p < 0.0001$). Performance was typically slower for wide than narrow elements (day 1: wide element = 8.3 s, narrow element = 6.8 s, $t(18) = 17.9$, $p < 0.005$). However, the same narrow element took more time to complete than the wide element, when the order of elements was changed such that the wide element preceded the narrow one as shown on Figure 5.1.b (day 3: wide element = 6.9 s, narrow element = 8.4 s, $t(18) = -3.3$, $p < 0.005$).

Skills have no unchanged and ridged features. Developed skills are characterized by mobility in changing conditions. Skills should be considered as flexible strategies of performance that can be used for a specific type of task.

Learning curves have irregular features such as peaks, troughs, and plateaus not only because there are some accidental factors but also because learners change their performance strategies. Plateaus often demonstrate that used strategies are no longer effective in improving performance. The troughs can demonstrate that attempts to transfer to a new strategy are accompanied by a decline in the efficiency of execution. New strategies are tested by a trainee, and if evaluated as more efficient, they are used further. Some intermittent strategies are used only for developing new, more efficient ones. They are never used as possible strategies of task performance when training is completed.

Since increase in task complexity is accompanied by increase in the number of utilized intermediate strategies, lengthening the duration of skill acquisition is an important indicator of task complexity. This criterion is often more effective when comparing task complexity than such indicators as task execution time or number of errors in the final stage of training. Learning curves clearly reflect the duration of the skill acquisition process and can be utilized in the experimental evaluation of task complexity.

The basis for the formation of performance strategies is activity self-regulation. In the process of skill development, people constantly assess their results and learn to reconstruct activity strategies and to correct errors. Optimization strategies of activity can be based on subjectively selected criteria of success. These criteria are dynamic and can change over time resulting in the changing relationship between qualitative and quantitative indicators, desire to achieve stable performance over time, or desire to simplify or complicate a task at hand based on motivational factor. Therefore, such criteria are not sufficiently precise and are based on *subjective feelings* developed as a result of personal experience.

In contrast to an experienced performer, trainees do not have general skills and the ability to flexibly regulate their work activity. Therefore, even small changes in a task lead to deterioration of performance. Trainees or novice are error-prone, and any changes in task performance can lead to a sharp increase in their execution time.

Changes in strategies of performance during an acquisition process can be discovered using learning curves, and averaging curves are useful in such analysis. However, averaging learning curves can result in missing some important information. Therefore, in this study, we demonstrate the possibility of using a learning curve in the framework of developmental principles of task analysis, when a task can be analyzed in the process of its acquisition. This method is consistent with the principles of genetic analysis of activity.

In conclusion, we would like underline some aspects of skill automaticity. For example, Schneider and Shiffrin (1977) have distinguished between control processing and automatic processing. Control processing requires consciousness, involvement of memory, and concentration of attention. On the other hand, automatic processing involves little or no consciousness and occurs with little mental effort in general. At the beginning of practice, motor and cognitive skills are regulated consciously, and at this stage, control processing dominates. After extensive practice, skills can be performed faster, without conscious control and with less effort.

It is important to know that sometimes deautomaticity can happen when complicated skills can be lost. Losing a skill can be either permanent or temporary. If a person does not perform an activity for a long time, the skills can temporarily disappear (extinction). Pavlov (1927) shows that extinction is not disappearance. Under certain circumstances, a response that disappeared can be recovered. In the case of deautomation of skills, we often need to avoid purposeful attempts to restore them. In such a situation, it can be useful to postpone acquiring such skills for a long period of time. Often a skill can be recovered spontaneously or through specifically organized training. There is also another aspect of analyzing skills deautomation. After a complex skill has been automated, any attempt to regain conscious control of such skill may result in deautomaticity of the skill. Loss of such skill happens more often when the skill is acquired by rote repetition. If, from the beginning, the trainees try to consciously acquire various components of the skill and vary its performance, deautomaticity happens very rarely.

Analysis of the skill development process is important not only for the creation of training methods but for discovering a more efficient method of task performance. Considering the skill acquisition process as a process of formation of flexible strategies and utilizing a learning curve allow studying the activity structure where a worker performs a specific task during activity formation. This method significantly extends the possibility of using learning curves in task analysis.

5.4 Analysis of the Acquisition Process in Learning by Observation

Learning by observation is the major point of the social learning theory (Bandura, 1977). In activity theory, learning by observation refers to the area that is known as social interaction (obschenie). Theoretical analysis of learning and training through observation has a certain practical value especially in considering the relationship between group and individual experience in professional learning and training when people are observing others and utilize their own past experience to acquire new knowledge and skills. This aspect of learning and training is well known in gymnastics where motor skills are extremely complex. For example, it is a well-known fact that if a trainer transfers a promising young gymnast from a low skilled group into a highly skilled group, it speeds up his or her rate of developing complex gymnastic skills. Such transfer has two aspects: On the one hand, the younger gymnasts were able to repeatedly watch how the more experienced gymnasts performed complex gymnastic elements that contributed to the rapid development of gymnastic skills, and on the other hand, it has contributed to a reevaluation of the complexity of considered skills by a novice member. As a result, the new elements were now seen as more accessible and less dangerous. This can be verbalized as follows: If everyone can perform a given element so can I. In other words, observation led not only to formation of more adequate cognitive mechanisms of skill acquisition but also changes in emotionally motivational mechanisms of skill acquisition.

The method of training that derives from behavioral traditions requires trainees to perform some elements of a task and evaluate and correct his or her behavior. There were no such traditions in activity theory and observation was considered as a component of the training process. Depending on the specificity of acquired skills, the demonstration of a new method of performance can be presented in various ways at various stages of the training process. Therefore, learning and training by observation are not considered theoretical concepts of learning or training but simply useful methods that can be utilized in learning and training processes. There is evidence that learning by observation can be successfully utilized in computer training. Some aspects of the skill acquisition process in performing computer-based tasks will be considered in the following chapter. Learning and training by observing others have specific meaning in the production environment. Usually this way of learning or training is combined with the real performance of tasks. We consider training by observation in experimental conditions that replicate a real task performance in the production environment. Already mastered tasks and tasks that are in the process of being acquired can be identified in any production environment. The workers should understand new methods and techniques. This would enable them to perform their tasks in a prescribed manner. Workers may be expected to acquire new knowledge and skills through oral or written instructions. When new tasks become more complex, special training may be required. However, in most cases, workers should be ready to perform a new type of work without special training based on oral or written instructions or perform a task after a short specific on-the-job training. Time standards for a new task (execution time for considered task) should be changed according to the dynamics of skill acquisition for this task. If some workers begin to take on new tasks earlier than others, we need to know how workers who start to perform new tasks later than others take into account experience of their coworkers. Learning by observation absorbs characteristics of informal practices of social interaction. In these conditions, learning by observation becomes essential. We will consider learning and training by observation from the self-regulation perspectives. We describe this method in terms of its applicability in the production environment. Observational learning is the main component of the social learning theory suggested by Bandura (1997). One example of such training is modeling. According to Bandura an expert acts as a model and trainees observe his/her behavior. Learning by observation includes observation of actual performance or videos of task performance.

Learning by observation cannot be reduced to perception and to reproduction of observed events. This activity includes various cognitive processes. Observation is mental activity that depends on a subject's conscious goal, motives, and methods of activity performance. For example, subjects who are observing the same events may interpret them differently. In the well-known experiment that has been conducted by Zinchenko (1961), subjects were instructed to organize the cards either by pictures or by numbers. Those instructed to organize cards by pictures were unable to recall the numbers. Some subjects even insisted that there were no numbers on the cards. A similar result has been obtained when subjects attempted to classify cards according to numbers. They had difficulty recalling the pictures. This example demonstrates that learning by observation cannot be reduced to the sequence of separate cognitive stages such as paying attention, remembering, reproducing actions, and becoming motivated. A learner selects information and interprets the same events in various ways depending on his or her goal and on the specificity of activity in general. Observation is a voluntarily regulated activity due to the presence of conscious goals and motives where functions of attention are important and attention is considered as a goal-directed self-regulated system. Strategies of attention depend not only on cognitive but also on emotionally motivational mechanisms. From this follows that learning by observation should be understood as an independent task that has its own goal and motives. People use different strategies in achieving a tasks' goal. So observation of the same data can produce completely different results. Attention includes involuntary mechanisms for which the physical characteristics of the observed situation are crucial. In all these cases, we are talking about involuntary attention that either dies down or is transformed into voluntary attention. These aspects of the observations should be considered especially when it is carried out without external supervision, which is specific to learning by observation in the production process.

Thinking plays a special role in the task of observation. It acts as an important component of observation, which organizes the cognitive processes in line with the goal of the task. An observer has to interpret the meaning of observed events. The meaning of environmental events provides understanding of their role in specific situations. Thus, observation of the situation depends on a subject's past experience and understanding of the meaning of environmental events. Involved in this process are mechanisms that also evaluate the significance of the situation elements. Such phenomenon as inattentional blindness, which is described as *looked but failed to see* (Herslund and Jorgensen, 2003), cannot be reduced to perceptual observation. Perceptual and meaningful blindness are interdependent. Thus, observation should be considered from systemic self-regulation positions and described utilizing such functional blocks as goal, motivation, significance (sense), meaning, and so on. Thanks to the analysis of feedforward and feedback connections between these blocks or mechanisms, it is possible to discover strategies of observation that include such processes as searching, comparison, recognition, and selection of essential features, judgment and reasoning, decision making, selection of observational tools, validation of findings, reformulation of observational goals, and remembering of obtained data. In some cases, observation is so complex that it may be separated from execution and becomes an independent task that requires special training. This task can be specified or even reformulated. Sequential formulation of new, more accurate, and specific observational tasks allows obtaining necessary information about the observed phenomena and behavior of other people.

Formation of adequate observational strategies requires special training. Vocational training in this area is performed under the supervision of an instructor or more experienced colleague. Such training often requires not only mental but also motor activity. A trainee imitates performance of considered tasks or performs some elements of a task in order to identify its most important features. Sometimes without practical testing of an observed performance, many important components of activity remain unknown.

In order to develop better observation skills, an instructor can use a list of questions that allow trainees to focus their attention on various aspects of the situation. Selection of some specific aspects of the observed phenomena is often associated with the process of verbalization. For example, an ability to distinguish certain perceptual properties of a situation can be associated with an ability to verbally describe these properties. Students' observation strategies can be monitored by specially prepared instructions.

In our further analysis, we consider group learning of motor skills where observation is an important component of the skill acquisition process. Observation of motor skills can be so complex that cognitive components of mastering motor skills play a leading role. Knowledge of the group skill acquisition process through observation for new types of tasks that are introduced into the production process has an important practical value. The relationship of individual and group experience is an important factor in work time studies (Bedny, 1981). The purpose of time studies for the acquisition of new types of tasks is to determine changes in standard time required to do given tasks in accordance with the dynamics of the acquisition process until this process is completed and the time standard is the same as for an already acquired task. The acquisition process should be considered when workers perform a variety of complex tasks that are periodically changed.

In our analysis of the acquisition process, we encountered situations that suggest that workers can acquire new types of work mostly without special training directly on the job due to their past experience. Workers normally perform a variety of tasks in a specific period of time, and periodically, new equipment, tools, or software are introduced. In the production environment, process engineers, specialists in the time study, and safety engineers are also involved in the analysis of the acquisition process. They usually record acquisition process data for a new task, evaluate execution time for a required level of precision, rate errors, and analyze whether the utilized methods contradict with the identified technological requirements and safety regulations. However, these aspects of analysis of the skill acquisition process are rather technical ones and we do not consider them.

The main factor in the acquisition process is repetitive performance of the same task.

The learning curve demonstrates performance improvement with increase in the task number, until the curve reaches its stable level. When the same task was performed by various workers, it was discovered that improvement depends not only on individual repetition of task performance but also on the number of repetitive performances of the same task by all workers. This means that workers learn by watching their colleagues perform the same task. Therefore, observation is an important component in the development of new work methods. Moreover, workers adopt in their style of performance only those components of work that best fit their subjectively acceptable work style. It has been discovered that the acquisition process often correlates with a general number of repetitive performances of the same task by all workers. So some specialists in the time study suggest taking into account the number of parts produced by all workers when analyzing the acquisition process and introducing a new time standard for task performance because rational methods of work performance are developed based on experience accumulated by all workers (Gal'tsev, 1973). This means that learning through observation is one of the most important factors in the acquisition process. Such viewpoint assumes that if the first worker completes 30 parts (30 tasks, respectively), and then the same task will be performed by the second worker, then he or she starts to perform at the level corresponding to the 31st part performed by the first worker. In our opinion, this statement exaggerates the importance of such factors as learning or training by observation and ignores the role of exercise or practice in learning and training.

It is natural to assume that observational learning is important for acquisition of new skills, but the role of practice in improvement of skills should not be reduced. Thus, our goal was to determine the relationship between practice and observation during acquisition of new skills by a group of workers.

In our studies, we have chosen a model of production operation for bench-assembly work to be an object of study. A laboratory experiment setting allowed us to control various versions of a task without violating safety requirements. In order to conduct this experiment, a special physical model of production operation was developed (see Figure 5.10).

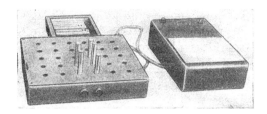

FIGURE 5.10
Physical model of production operation for pin installation.

This model consisted of a pinboard that contained 30 holes for metal pins. There were two push buttons on the front of the board and a box containing pins behind the pinboard. At the left-hand side, there was a panel with 30 cells that lit up when each space was filled. The panel also contained a stopwatch that was turning on when a subject pressed two buttons and turning off when the last hole was filled in and a subject pressed two buttons again. The results were measured with 0.1 s precision and rounded to the next whole second. The right-hand side panel not only allowed measurement of performance time but also registered the sequence of performance. In this experiment, regular (without a flute) pins were used. After completion of each task, an experimenter would file a brief record of his or her observation, if necessary.

Participants were senior students of the University of Civil Engineering of South Ukraine. All of them have taken courses in industrial engineering and in ergonomics. They perceived this experimental study as a regular laboratory class. All of them were motivated to complete this experimental study and receive a good grade. Moreover, laboratory classes always aroused keen interest among students as it allowed them to obtain new information about principles of organizing a production process. The participants' task was to fill a pinboard with thirty pins. The instructions indicated that students had to work with optimum pace and at the same time find out the most effective sequence of installing pins into the holes while working with both hands. When trainees attempt to choose the best method of work, they had to discover a sequence of installation of pins and the principle of the symmetrical movement of the hand, select an adequate pace of performance, discover a rational way of attention distribution and the best way of grasping pins, and so on. For the students it was a task that required evaluating their work. Thus, it was not a simple, routine task because it required self-assessment of their performance.

The experimental procedure was as follows: 20 male sophomore students were selected for the experiment. Their age was 19–20 years old. Out of these 20 subjects, 5 subjects were selected by chance for the second group in order to eliminate the marked effect on the time of task performance caused by individual characteristics of the subjects in each group. At the same time, the differences in speed characteristics between these groups were not critical, since we were not interested so much in the absolute value of the task performance time but rather in the dynamics of change in the time of the task performance. As a result, there were 15 subjects in the first group and 5 subjects in the second one who had to fill a pinboard with 30 pins in the more efficient manner. The general quantity of performed trials in both groups was the same and was equal to 150 repetitive executions of the task.

The first group of subjects combined task execution with learning by observation. In this group, some subjects could observe the task execution of their colleagues. The second group with five students performed the tasks independently without being able to observe the task performance of others.

The experiment with the first group, where we used learning by observation, was organized as follows: Fifteen subjects were divided into five subgroups. Each subgroup included three subjects who worked independently from other subgroups and observation of performance in the independent subgroups was eliminated. The scheme of the experiment with the first group is shown in Table 5.1.

From this table we can see that each subject performs only 10 tasks. The first subject performing the tasks could not observe the task performance of others. However, the second subject had an opportunity to observe the first one, and the third subject observed performance of the first and second subjects. Subjects could not hint each other.

The scheme of the experiment with the second group where observation of other subjects was not possible is presented in Table 5.2.

From this table we can see that in this group, there were five subjects who worked independently and each subject performed 30 tasks. Both tables also

TABLE 5.1

Scheme of the Experiment with the First Group of Subjects

Number of Subgroup	Subjects' Number in Each Subgroup			Number of Tasks Performed in Each Subgroup
	The First	The Second	The Third	
	Number of Tasks Performed by Each Subject			
1	10	10	10	30
2	10	10	10	30
3	10	10	10	30
4	10	10	10	30
5	10	10	10	30

TABLE 5.2

Scheme of the Experiment with the Second Group of Subjects

Number of Subjects				
1	2	3	4	5
Number of Trials for Each Subject				
30	30	30	30	30

demonstrate that the general number of tasks in both groups was 150. During the research, the experimenter could give hints in a form of questions to help subjects in critical situations. However, this was done very rarely and the instructor never informed the subjects what should be done. Hints were used in critical situations to point out a more appropriate search of adequate methods of a task performance.

Based on obtained data, the learning curves demonstrate the dynamics of skill formation. This experimental method allowed us to analyze the dynamics of skill acquisition in group performance with the presence of the observation component of learning and compare it with individual performance of the same task when observation was not possible. It is important for time study during acquisition of new tasks to take into account how experience of predecessors in group performance affects the rate of skill formation.

Let us consider data of the first group of subjects where learning by observation was a component of the skill acquisition process. Learning curves based on the average task execution time depict the acquisition process for the first, second, and third subgroup of subjects. These curves reflect the dynamics of skill formation for subjects who performed the same task in sequence and observed the work of predecessors (Figure 5.11).

The learning curves show that the average time for task performance in the first group is higher than in the second group, and in the second group, the average time for task performance is higher than in the third group. The average performance time for the first group was 35.1 s; for the second group, it was 31.7 s; and for the third group, it was 30 s. The second group reduced the task performance time by 9.6% (3.4 s) compared with the first group, and the third group reduced execution time when compared with the first group by 14.6% (5.1 s).

Paired t-test revealed a significant difference between the first 10 trials and second 10 trials ($t(4) = 20.30$, $p < 0.0001$), the first 10 trials and last 10 trials ($t(4) = 27.66$, $p < 0.0001$), and the second 10 trials and last 10 trials ($t(4) = 8.00$, $p < 0.001$). Due to the fact that the difference in time of the task

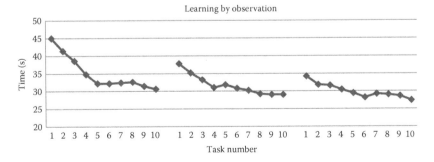

FIGURE 5.11
Dynamics of skill formation in learning by observation for three groups of subjects who performed the same task (each subject performed 10 tasks).

performance in the first, second, and third groups is statistically signifi-cant, and all groups had the possibility to perform the task with the same number of times, we can conclude that the observation of others improves task performance.

We want to stress the fact that the second subgroup of subjects is spending significantly more time on the first few tasks compared with the execution time of the last tasks performed by the first subgroup of subjects. Only after repeated executions of the task did the second subgroup of subjects achieve the results of the first subgroup and then begin to show results that are better than the results of the first subgroup. A similar pattern is observed when we compare the results of the third subgroup and the second one.

We want to remind that the first subgroup did not have an opportunity to observe the work of predecessors. The third subgroup was in the best position because its subjects had an opportunity to observe the first and sec-ond subgroups while the second subgroup watched only the first one. This explains the fact that the execution time in the third group was even lower than in the second subgroup. One must also take into account that all three subgroups were involved in independent performance of the tasks, that is, observation was only a part of their skill acquisition process. Such conditions are similar to the production environment where one can combine indepen-dent performance and learning by observation.

Let us analyze data presented in Figure 5.12 that depict the dynamics of the skill formation process during sequential task performance of each sub-ject independently without learning by observation.

We utilize the learning curve that demonstrates the average performance time in individual performance of 30 trails and the average performance time in learning by observation (3 subjects together perform 30 trails) for better visual comparison of the obtained result.

The average performance time for the first 10 trials was 36.94 s; for the sec-ond 10 trials, it was 30 s; and for the third 10 trials, it was 25.8 s. So the task

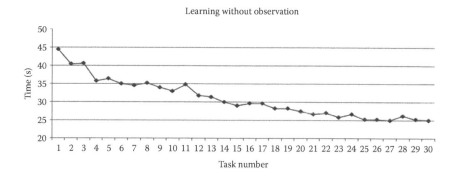

FIGURE 5.12
Dynamics of skill formation in learning when each subject performs the same task without observation (subject performed 30 tasks).

performance time is 19% lower in average between the 11th and 20th trials and 30% lower for the 20th to 30th trials.

Analysis of learning curves shows that when subjects perform 30 trials without observation, there is a more significant reduction in the task performance time. The curves have a smoother character and there is no significant deterioration in performance when subjects start their performance as in the previous study.

Hence, the average execution time of the last 10 tasks for the individual performance is shorter than the average performance time of the last 10 tasks for the group performance that includes learning by observation. In the group performance with learning by observation, the average performance time of the last 10 tasks (the third group) was 30 s, and for the individual performance without observation average time of the last 10 tasks, it was 25.8 (tasks 21–30).

Statistical analysis demonstrates that group 3 (learning by observation) was significantly slower than the group that learned by performing the task themselves on their last 10 tasks (trails 21–30) (between-subject t-test $t(11) = 6.15$, $p < 0.0001$). Similarly, group 2 (learning by observation) was marginally slower than the second 10 tasks of the group that performed the task themselves (between-subject t-test $t(11) = 1.76$, $p = 0.11$).

This suggests that for this kind of task, individual practice is more effective than observation.

We want to remind that we have compared two conditions: In the first case, subjects made only 10 tasks but had an opportunity to observe the work of their predecessors, and in the second case, subjects performed 30 tasks, but were not able to watch the other. Obtained data demonstrate that the observation has a positive effect on the skill development, but direct execution is a leading component of the skill acquisition process that cannot be neglected.

In the earlier analysis, we concentrated on the learning curves. However, in order to detect differences in the skill acquisition process in various conditions, one would need to conduct a more detailed qualitative analysis of data. Hence, analysis of the curves should be combined with data obtained by observing the subjects.

The following are some of those observations: In the first group where there were three subgroups that combined training and observation, many subjects were trying to take 3–5 pins by the left hand and then insert them by the right hand into the hole of the panel. With this method, subjects did not effectively use two hands because the left hand was in a static position most of the time. Keeping pins in the left hand resulted in muscle tension. Insertion of the pins by one hand is faster than with two hands simultaneously. However, working with two hands significantly reduces the task performance time. Other subjects in the first subgroup were trying to take 4–5 pins with two hands, which made it very difficult to install pins into the holes, and there were cases when subjects dropped the pins on the panel. It is interesting to note that the inefficiency of this method was quickly understood by subjects in the first group and subjects of the second and third groups did not repeat such mistakes.

Typical errors were the absence of symmetrical movement of both hands that complicated coordination of hands and distribution of attention. In critical situations, the experimenter reminded that subjects should use two hands and try to reduce the distance of their movements. Gradually, subjects started choosing efficient work methods. The most obvious inefficient methods were gradually eliminated by the first group of subjects. Such rough errors were not repeated by the second and third subgroups of subjects. Gradually, the search for a rational work method narrowed from chaotic search of performance methods to defining the most rational sequence of pin installation. One of the typical errors was when the subjects tried to fill in the first row of holes on the panel, then the second row, and so on. Then the experimenter had to make a remark that for this method to work, they need to raise their hands up. In the production environment, verbal exchange of information between workers usually is utilized sporadically. Similarly, the experimenter's remark can be considered as sporadic verbal exchange of information between workers. Moreover, the experimenter did not give direct instructions on how to correctly perform the task. He just paid attention to the very serious mistakes that were repeated by the subjects. Finally, the subjects discovered the best method of task performance where they used both arms to fill in the center row starting with the space that was closest to them and then working upward and outward.

So it was discovered that when using observation, the roughest errors were detected by the first subgroup of subjects. The second subgroup of subjects took into account errors of the first one, and the third subgroup of subjects took into account the error of subjects in both subgroups. So subjects in the second subgroup frequently performed better than the subjects in the first one, and the subjects in the third subgroup performed better than the first and second subgroup. In some cases, the subjects of the following subgroups repeated the errors of previous groups or made a new type of errors, but the cause of errors was noticed quicker by those who had an opportunity to watch the others. Thanks to this factor, the speed of corrections increased in subsequent subgroups. Analysis of the results shows that the best method of work was found on two occasions by the second subgroup, in three cases by the third subgroup, and none by the first subgroup.

The second group of subjects with independent performance and no training through observations used similar ways to find the best method of performance. We will briefly consider some distinctive features of searching strategies of the best method of execution utilized by the second group (individual performance). This group of subjects had a greater number of reexecutions because they could not take into account the experience of the predecessors. So the number of strategies used in this group was significantly greater, and as a result, this group evaluated the results of their own performance more efficiently.

It should be noted that similar elements of task performance can be found for task execution with and without observation. For example, such elements of task as installation of pins in the holes or taking pins out of the box can be found in each version of the task execution. They require not only a similar method of motor performance but also specific strategies of attention. Therefore, there was a better transfer of skill elements when subjects switched from one method to another for an individual type of performance due to the greater repetition of similar components of a task. Behaviorists define learning as changes in behavior due to experience. For example, a student should not just listen and observe others but also perform observable responses to confirm that learning has taken place. In contrast, Bandura demonstrated and proved that people can learn just by watching other's performance. Such interpretation of learning was a step forward in understanding learning compared to behaviorism. AT views this is an obvious fact that does not require any proof. A person can act with ideal objects the same as with the material ones utilizing mental actions and mental feedback of the results of these actions. In the game of chess, people can move various figures on the board in the mental plane and estimate results of such movements in accordance with the goal of activity. A chess player uses not only perceptual but also imaginative thinking and other actions. Thus the activity approach does not require special evidence that learning is possible not only by doing but also by observing.

Let us consider stages that are involved in learning by observation that were described by Bandura (1977). He has proposed to utilize the following four basic steps that are necessary for the modeling of the behavior of the observer (1) attention or paying attention → (2), retention or remembering → (3), motor reproduction or reproducing overt actions → (4), motivation or being motivated to imitate an observed behavior. Therefore, first students are to pay attention; then they should remember and reproduce what they observed; finally, they should be motivated to perfect their imitation. However, learning by observation cannot be presented as a sequence of these steps. For example, Bandura points out that reproduction of an observed behavior at the time it is observed is very useful because it provides an opportunity for a learner to evaluate results immediately and improve his or her behavior. We have to pay attention to the fact that a learner most often cannot immediately reproduce behavior correctly because a real learning process always requires a number of trials and a complex training process is needed. Learning cannot be reduced to memorization (retention) and later behavioral reproduction when a person is motivated to do so. A learner actively selects information according to the goal of observed task and personal significance of the element of the situation at hand. Motivation is needed not only for motor reproduction of previously perceived behavior but for any other previously listed steps of behavior, including attention.

Learning by observation is a problem-solving task that cannot be considered as a linear sequence of presented steps. It involves complex cognitive activity. From the activity theory perspective, every task includes goals, motives, explorative, and executive cognitive and motor actions. It includes evaluative processes. For instance, attention is impossible without the goal of activity. Activity is a complex self-regulative system. So a person formulates a goal of observation and, based on it, regulates his or her attention and, then based on the formulated or accepted goal, develops a mental model of the situation, evaluates significance of situation, promotes a hypothesis, performs explorative and executive actions, and so on. Therefore, observation involves complex strategies of activity that are derived from self-regulation. Involuntary attention is the one influenced by external stimuli. Due to a motivational factor and the goal of activity, this type of attention can be transformed into voluntary attention. Otherwise, involuntary attention can shift to other elements of the situation due to its instability.

As described by Bandura, sequential steps are just elements of such activity that are presented in an incorrect order. Not only the motor reproduction stage but also the correct observational stage very often requires special training. Observed events should be adequate to a person's past experience. Bandura's four components or steps of behavior should not be considered as isolated cognitive processes but rather as a combination of various cognitive processes. Observation includes interpretation of perceived events making thinking an important part of observation. A person can interpret what he or she saw and, then based on this, demonstrate not only similar or related behavior but also a totally new type of behavior because the same goal can be achieved by using different strategies or due to creativity of an observer and his or her motivational state. In our study, it was discovered that even when the order of pin installation was the same, the strategies of attention, grasping, etc., could vary.

Bandura separated observation and behavior. According to him, thanks to observation, the subject develops new types of monitoring responses. However, the subject not only learns to give correct responses but also learns to observe and interpret the situation. It is interesting that in our example, the final method of motor execution was not very complex but the observational stage of the acquisition process was rather complex. During training, strategies of observation in group performance gradually changed. Observation and response are interdependent elements of activity. They influence each other through feedback. For example, positive or negative assessment of response may change the strategy of observation. Summarizing the earlier analysis, we can make the first general conclusion. Learning by observation cannot be represented as a sequence of such steps as attention, retention, motor reproduction, and motivation. Observation is a human activity that integrates cognitive, executive, motivational, and evaluative components. Observation just as any other activity is based on the self-regulation mechanisms. Thanks to the variety of observation strategies, their reconsideration and reevaluation take place.

Our study revealed that the task execution time depends on the sequence of pin installation, choosing the right techniques to perform activity elements, and the degree of improvement of the chosen methods of work. This allowed identifying two groups of factors affecting the process of skill acquisition in training by observation. The first group of factors is related to searching for the best sequence of task performance that can be identified relatively easily due to observation. Any other aspects of task performance that are accessible by observation belong to this group. Another group of factors is identified by improvement of selected work methods that often cannot be easily discovered by observation. For example, it was discovered that subjects involved in group performance paid attention to changes in sequence of pin installation and the correct way of using two hands simultaneously. Subjects took into account the individual experience of their colleagues that depended on mastering separate components of movement regulation to a lesser extent. For example, effective strategies of attention and the ability to take the pins out of the container and install them correctly could not be identified in learning by observation. It is important also to understand that in learning by observation, the subject may face a situation that can be expressed in the following words: the *subject looks but cannot see*. The subject must learn to notice what is important in a particular situation. Subjects needed direct practice for this purpose. In some cases, the first or second group of factors prevailed. The ability to identify the relationship between these factors is of some importance in the time study of task performance during the skill acquisition process.

In some cases, when the first group of factors dominates, after finding a rational method of performance by the worker that was observed by others, his or her method of performance can be acquired immediately by his colleagues. Then a stage of acquisition lasts only until the workers find the most efficient method of task performance and all subsequent stages of the acquisition process can be ignored. When the first group of factors dominates, there is a clear dependence of the task performance time on the accumulated group experience and on the related observation process. With the predominance of the second group of factors, there is a great dependency on the individual experience. Executive activity performs important evaluative functions at the first stage of skill acquisition. Analysis of factors that are identified during task performance allows not only to correct motor components of performance but also to correct the strategies of observation.

Analysis of dynamics of the acquisition process and comparison of the task execution time at the end of the experiment in particular allowed us to conclude that the main factor influencing the task execution time is discovering the correct order of putting pins in holes. According to our classification, the best sequence of task performance belongs to the first group of factors that depends on the observation process. The second group of factors associated with an individual experience has less impact on reducing the task execution time because taking pins and installing them into holes are not totally new types of motor activity for subjects. They can start performing this type of motor activity effectively enough immediately. A reduction in the task execution time, depending on the second group of factors, may not be significant for such types of tasks.

We have considered learning by observing by using an example that is adequate for the industry. It has been shown that pure observation is rarely used. Usually observation is combined with verbal interaction. The more complex the performed task is, the more important is the verbal interaction. This is particularly adequate in contemporary computerization of work where external motor activity precedes complex cognitive activity. It should be noted that observational learning occurs when employees interact with computers or learn from each other on how to use all kinds of new gadgets. Currently, a typical situation is when software developers learn various coding techniques from members of their teams. Depending on their expertise, members of the software development team can become mentors for other members while being a mentee or on the receiving end for another specialty. Subject matter experts are not professional trainers. They are colleagues and members of the work team.

When trying to transfer their knowledge to the coworkers, the subject matter expert often just shows them how to resolve an issue at hand without a proper explanation of why this resolution would work. They are also often working too fast for a novice to follow their key strokes.

When asked for an explanation, they often cannot verbalize why they took certain steps. The following are some of the reasons:

1. Just learned it mechanically without understanding the cause–outcome relationship
2. High level of automaticity of learned skills that are difficult to verbalize

The ability to be a mentor for other members of the team depends on the type of knowledge possessed by employees. Skills that are based on meaningful learning are more easily verbalized than skills that are based on rote memorization. The cognitive component prevails in this kind of job, and without a good understanding of the content of the skill, its acquisition is not efficient.

We considered learning by observation utilized by experienced workers in production settings, but some tasks can be so complex in nature that even experienced workers may require considerable training to perform required tasks satisfactorily. Such training should be facilitated by a training department.

Regularities of the acquisition process that are adequate also for the development of a training program are identified in our experimental study. However, in such cases, a number of additional requirements for training skilled workers or novice are needed. In such circumstances, training is carried out under the guidance of an instructor. In this situation, the instructor utilizes written and oral instruction. However, when in training, observational learning is also applied where workers can watch someone perform a task or watch a videotape. The instructor has to organize in a specific way written and oral instructions with demonstration. If trainees are experienced workers, verbal interactions are very useful. Usually both an instructor and workers are involved in developing new methods of task performance. Therefore, training procedures are modified during training. Material presented in this chapter allows to us conclude that there are two basic distinct factors in learning by observation. The first group of factors is related to essential characteristics of tasks that can be discovered while observing the task performance by others. Based on such data, a subject can immediately perform adequate cognitive and behavioral actions to achieve more efficient task execution. The second group of factors is related to essential characteristics of a task that cannot be discovered by observation but can be identified only during repetitive performance of the task. Adequate cognitive and behavioral actions can be gradually developed during repetitive task execution.

The relationship between the considered factors that affect learning by observation is important in analyzing individual and group experience and in studying the acquisition process in the production environment.

5.5 Acquisition Process in Computer-Based Task Analysis

5.5.1 Description of Tasks and Performance Measures during Skill Acquisition

The development of a training process based on the analysis of learning or acquisition curves is widely used in the study of traditional kinds of work, but there are no clearly developed methods of creating such curves for computer-based tasks. In order to utilize genetic principles to the analysis of computer-based tasks, it is necessary to create procedures for developing different kinds of learning or acquisition curves. Utilization of such curves and their comparison for the analysis of skill acquisition is an important tool for applying genetic principles to the analysis of computer-based tasks. We remained that the microgenetic method of study is understood as an analysis of the process of activity structure formation during a relatively short period of time (Bedny and Harris, 2005). The purpose of this method is to study task performance at the various stages of skill acquisition. A researcher can sporadically interfere with the trainee activity during the experiment and guide the task acquisition process. In such study, the development of learning curves is an important stage of task analysis. So the main purpose of this work is to demonstrate the method of creation of such curves for the study of computer-based tasks and demonstrate some performance measures that are derived from the analysis of the skill acquisition process. In this work, we utilized modeling a computer-based task (Sengupta et al., 2011). There were four groups of tasks based on their level of complexity. The versions of task varied from trial to trial within each group of task complexity. We conducted an analysis of the activity formation process during task performance for each group of tasks.

The task we have chosen for our study is a model of computerized task that requires a relatively short skill stabilization period to achieve an average skill acquisition level. In spite of this fact, as we show later, the tasks presented to the subjects were sufficiently complex. Variation in the level of complexity was achieved by manipulating different features of tasks and particularly by changing the location of the tools on the screen in relation to the other features of the task and the sequence of actions. All data obtained in this study cannot be presented in one chapter. So we describe the data that are most illustrative for the basic principles of genetic analysis.

Let us consider experimental design and data collection. We chose a laboratory experiment because it gave us an opportunity to carefully control the variables of the task performance. In order to conduct the experiment, a special computer-based model of the task was developed. To correctly design the model of the computer-based task, we needed to know the context of the task performance. Some tasks are designed for the user that would work with a few screens for years. Other tasks are intended for the ever-changing workforce where the workers are hired for 4–6 weeks during peak season. Some tasks can be performed upon request and/or under stress or time limit or accompanied by various interruptions. There are also computerized tasks that are used by consumers, who might perform it once or just a few times. The study of the task acquisition should depend on the conditions in which this task is going to be performed. For example, in the first previously mentioned case, the allowed acquisition time can be much longer than in the next three cases. In the second case, training cannot exceed half a day. In the third case, which is related to the stressful work conditions, the concentration should be on simplifying task performance by reducing the workload on working memory, simplifying decision making, etc. If the software is intended for the consumers, it should be self-explanatory because its usage is not preceded by any training at all.

In this study, we strived to understand how task sequence requirements and display structure impact the usability of a graphic user interface. For this purpose, we have utilized the genetic principle of study. Usability of a graphic user interface refers to the ease of use, efficiency, effectiveness, and satisfaction while interacting with the system through the interactive elements on the interface display. The display structure is the arrangement of the elements (icons, menus, etc.) on the interface screen, which influences the understanding of the system and strategies of interaction with the display. The quality of the graphic user interface can affect the efficiency of performance of the users and thereby the usability of the interface.

The task was to change the position of the letters and impart the features to these letters according to the ones presented on the screen goal, which varied from trial to trial. Any sequence of actions could be possible by the user, but the user (subject) was only instructed to reach the final arrangement. There was a limitation on the sequence of the tool selection. The subject could not complete the tasks unless he or she understands that the task performance sequence depends on the tool arrangement characteristics of the task. A subject could complete the task by using various sequences of actions. The software allowed the subject to use the tools in an order that corresponded to each version of compatibility. The sequence of actions could not contradict the tool arrangement that presented constrains in the task performance. Violation of existing constrains resulted in impossibility to change the features of the objects. This feedback informed the subject about erroneous actions and he or she could correct the sequence of actions. Hence, these tasks in the acquisition process included self-learning components that can be observed during the performance of computer-based tasks. From trial to trial, subjects reduced possible errors until they achieved a strategy that was perceived by the subject as a final one. Once the subject learned the possible strategies of performance, he or she achieved the plateau on the learning curve that signified that the skill acquisition has been completed. Each group of subjects has been involved in performing a particular type of task. After completing one version of the task, they were involved in performing a new version of the task.

The main goal of the study was to observe how subjects developed preferable strategies for achieving the task requirements. The setting was similar to the real situation when users attempt to perform relatively new tasks. Usability of such tasks can be evaluated based on the analysis of the self-learning process. The limitations that have been imposed by the program and the necessity to discover possible strategies made these tasks sufficiently complex. Self-training is organized in accordance with the principles of self-regulation of activity. Right or wrong performed actions and their results were evaluated by the user, and based on these data, subsequent strategies of task performance can be changed.

The features of task that are manipulated by the subjects during trials were as follows:

1. Position: the location of the letters with respect to each other
2. Color: the color of the cell containing the letters
3. The format of the letters

These features resulted in three functional groups (based on the interface guideline of functional grouping) in the interface. It is commonly observed in a variety of software. Almost all interfaces include functional groups for the users to understand the general functions of a tool group. For example, in Microsoft Word®, this is observed in the format group and the alignment group (see Figure 5.13).

The tools designed for manipulating these features and their functional groupings are depicted in Figure 5.14. Their functional grouping and manifestation in the interface are given in Figure 5.15. Therefore, the main

FIGURE 5.13
Format and alignment group in Microsoft Word®.

Position		Color		Format	
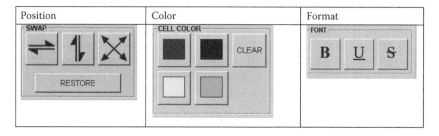 Swap horizontal position			Red	**B**	Bold
			Green		
Swap vertical position			Yellow	U	Underline
Swap diagonal position			Blue	S	Strikethrough

FIGURE 5.14
Tools for task designed as icons on the interface with intended functional grouping.

Position	Color	Format
SWAP ⇌ ↕ ✕ RESTORE	CELL COLOR ◼ ◼ CLEAR ☐ ◻	FONT B U S

FIGURE 5.15
Functionally grouped structure of tools in the interface.

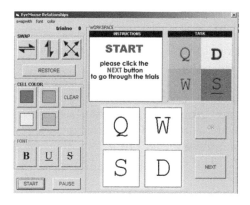

FIGURE 5.16
The experimental interface.

focus of the task is to alter the features of the objects with available tools. Initially, the letters had no special features.

The interface for task performance is given in Figure 5.16. According to existing basic elements of activity such as *subject → task → tools → object → method → result* (Bedny et al., 2006), it is possible to extract three functionally relevant areas on the screen: the tool area, the object area, and the goal area.

The tool area consists of the tools that are used for the transformation of the objects. The object area consists of the letters to be manipulated according to the goal of the task, while the goal area demonstrates the arrangement that must be achieved as a result of the transformation of the object area. Analysis of subjects' activity strategies in previously described areas corresponds to the method of study related to the functional analysis of activity (Bedny and Karwowski, 2007; Bedny et al., 2014). The object and tool areas of the task are generally seen on the interface. The goal area in most cases is not presented. In self-initiated tasks, the goal is formulated by a subject independently. In these cases, any activity stage associated with the various elements of the screen when the purpose of interaction is formulation of the goal of activity can be considered as a goal area of the screen. If it is impossible to define the stage of activity associated with the analysis or formulation of the goal and associated with it areas of the screen, only the object and tool areas are utilized during task analysis. Allocation of areas on the screen, depending on the goal of the activity, is considered in SSAT as one of the methods of functional analysis of activity. The functional purpose of various elements of the screen depends on the goal of the task or goal of actions. The goal of the task is fixed and therefore the object and tool areas on the screen also have fixed meaning. The goal of actions is not the same and depends on the purpose of actions. In such situations, the object and tool areas of the screen are not fixed. Thus, depending on the functional purpose of utilized elements of the screen at different stages of activity performance, these elements can be related to different areas of the screen. In most cases, areas on the screen can be extracted based on the task's goal. In such cases, the allocated area of the screen has a constant meaning and corresponds to the goal of the task. The interface for task performance with three functionally relevant areas on the screen is given in Figure 5.16.

Task performance begins after pressing the "START" button and finishes after pressing "OK" button. Performance of the next task (trial) begins after pressing the button "NEXT." During trials, only one out of four squares inside of the object area can be activated at a time. For this purpose, the subject clicks the corresponding square. As a result, the borderline of the corresponding square is highlighted by a bold line. After that, the features of the square could be changed.

According to the AT principle of unity of cognition and behavior, eye movement and mouse movement registration were utilized. The software for the interface was additionally coded to capture the mouse event and eye movement data. Eye movement was registered by using ISCAN eye tracking system. Although the eye movement point of regard coordinates were recorded, there were inherent difficulties with the analysis of the point of regard coordinates due to equipment restrictions. As a result, the eye movement data were obtained through the analysis of the video of the point of regard.

At the next stage of task analysis, we provide a classification of tasks according to their compatibility, tool arrangements, and design of experiment.

Before explaining the various interfaces that influence task complexity, a description of the utilized terminology is required.

Compatibility: The functional relationship between the display structure and the embedded task sequence requirements. If the display structure or layout (top to bottom or bottom to top) supports a particular sequence, then it is defined as compatible; if not, it is incompatible.

Tool arrangement: The display structure or the arrangement of the interface elements on the screen, which supports a particular operation. It can be top to bottom or left to right or any other spatial sequence. By manipulating the tool arrangement and compatibility, we can change the complexity of the task. In the pilot study, it has been discovered that the most preferred sequence of various features' manipulation was the positions of letters followed by the color and then the format of the letters.

There were only two versions of tool arrangements and four embedded task sequences associated with them. The first version of tool arrangement was position → color → format. The second version of tool arrangement was color → format → position. A combination of tool arrangements and four embedded task sequences is described as follows.

Compatible from the top: It is an interface that uses tool arrangement, which is consistent with the embedded task sequence (Figure 5.17). In this case, the tool arrangement is from top to bottom with the positioning tools being on the top, the color tools being in the middle, and then the format tools on the bottom.

The embedded task sequence is also in the same order as that of the layout of the display, that is, the positioning tools being the first, then the color tool and then the formatting tools.

Compatible from the bottom: In this case, the tool arrangement is from bottom to top with positioning tools being the lowest tool groups, then the format, and then the color tools. The task sequence embedded with the interface is

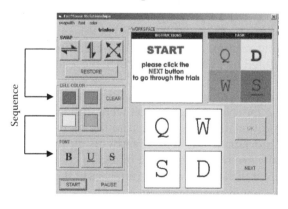

FIGURE 5.17
Compatible from the top. (Embedded sequence: position → color → format; display from the top: position → color → format.)

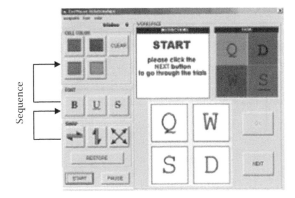

FIGURE 5.18
Compatible from the bottom. (Embedded sequence: position → format → color; display from the bottom: position → format → color.)

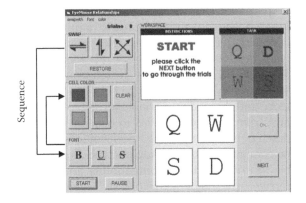

FIGURE 5.19
Incompatible from the top. (Embedded sequence: position → format → color; display from the top: position → color → format.)

also based on the same order, which is from the bottom: positioning first and then the format and then the color (see Figure 5.18 for details).

The incompatible from the top interface used the same display as that of the compatible from the top interface, but with a different embedded sequence in the interface, which was not congruent with the display structure of the tools. In this case, the embedded task sequence was positioning, format, and then color (see Figure 5.19 for details).

Incompatible from the bottom: The display used in this case was the same as that for the compatible form from the bottom, but with a different task sequence utilized, which was not congruent with the display order from the bottom. In this case, the sequence used was positioning, color, and then format, which was not congruent with the display layout, from the bottom (see Figure 5.20 for details).

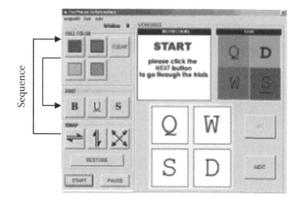

FIGURE 5.20
Incompatible from the bottom. (Embedded sequence: position → color → format; display from the bottom: position → format → color.)

Hence in this experiment, two display layouts had compatible task sequences, whereas the other two were incompatible due to nonmatching of tool arrangement and task sequence. This resulted in four groups with different levels of compatibility of task sequence requirements with the tool display (tool arrangement). It should be noted that the design principles were deliberately violated in order to observe the difference between groups with respect to the task complexity and dependent variables. The basis of using two displays (from the top and from the bottom) was to observe any effect of the visual scanning, top-down or bottom-up, on the performance. In general, the simpler task was the one compatible from the top and the most complex task was the one incompatible from the bottom. Compatibility and tool arrangement and their combination were the major factors of task complexity in this experiment.

The tasks forced the user to explore different strategies of task performance using the trial and error strategy. The users could not complete the tasks unless they understood the rules and limitations in sequence of task performance through self-learning. Once the users learned the rules, they performed the tasks using the stable strategy that they chose as the most preferable.

On the basis of the earlier discussion, the final experimental design is shown in Table 5.3. The four previously described interfaces are analyzed in the following text.

The within-group effect in the learning phase on the subjects' performance will be considered further based on the acquisition curves of the groups, and the model for ANOVA will be formulated on the basis of the final design addressing both the between-group factors of compatibility and tool arrangement and the within-group effect of the learning phase.

A summary of the between- and within-subject analysis can be presented in the following way:

Between-subject analysis: (a) main effect of compatibility (C); (b) main effect of tool arrangement (T); (c) interconnection of compatibility and tool arrangement ($C \times T$).

TABLE 5.3

Factorial Design for the Experiment

	Tool Arrangement 1	**Tool Arrangement 2**
Compatibility 1	*From top*—position, color, format	*From bottom*—position, format, color
	Sequence (*compatible*)—position, color, format	Sequence (*compatible*)—position, format, color
	Subjects—8	Subjects—8
Compatibility 2	*From top*—position, color, format	*From bottom*—position, format, color
	Sequence (*incompatible*)— position, format, and color	Sequence (*incompatible*)— position, color, and format
	Subjects—8	Subjects—8

Within-subject analysis: (a) phase of learning (exploratory and post); (b) interaction of phase and compatibility ($P \times C$); (c) interaction of phase and tool arrangement ($P \times T$); (d) interaction of phase of learning, compatibility, and tool arrangement ($P \times C \times T$).

Sixteen subjects have been involved in the preliminary study (four subjects in each group). Thirty-two subjects took part in the experimental study for the major experiment. They were divided into four groups (eight subjects in each group). All subjects (15 female and 17 male) had at least several years' experience working with computers. In this work, we present new methods of analysis of the acquisition process.

We have employed performance time data, mouse movement data, and eye movement data for the analysis of the acquisition process. New measures of motor performance based on the mouse movement data include three measures. They are described below.

5.5.2 Measures Based on the Mouse Movement Data

Time per Click (TC): This measure was utilized based on measuring time between two clicks. It was suggested that this time becomes shorter during the skill acquisition.

Motor Efficiency (Click Efficiency) (E): For each task, there were a minimum number of actions that could be used to accomplish it. However, the users either used a minimum number of actions or used more depending on the number of errors they made and the strategy they chose. Efficiency is based on the following formula:

$$E = \left[\frac{\text{Number of actions required } (n_{ideal})}{\text{Number of actions used } (a_{actual})} \right]$$

which is expressed as a ratio or percentage.

Real efficiency E_{actual} is less than 1. The better the motor efficiency is, the more E_{actual} approaches 1.

Efficiency in this case is affected by the following factors:

1. Errors due to misapplication of tools
2. Errors due to embedded task sequence
3. Errors due to omission of a feature
4. Excess clicks due to inefficient strategy

The efficiency basically reflects the errors due to the compatibility of the task sequence with the tool arrangement and hence is the indicator efficiency of the subjects' performance. The lower the efficiency, the higher the number of actions used to complete the task, and hence, the lesser the usability of the interface. This is due to the fact that subjects make mistakes because of the inconsistency of the task interface relationship. This results in the lower usability of the interface caused by the incongruence in the task and interface features.

Distance Mouse Traversed per Time (Mouse Traversal Rate [MT]): Here, movement in terms of distance traversed is considered:

$$MT_i = \frac{M_i}{t_i} \text{ pixel/s}$$

where
  M_i is the mouse traversal in task i
  t_i is the time to complete task i

This measure reflects the speed at which the users are performing the task as well as how efficient they are in accessing the tools. The larger traversal indicates more efficient movement and performance. Mouse movements depend upon various factors. Mouse movements mostly corroborate either with the events during the task performance or with the indecision on part of the users when they are not able to find the icon needed to perform the task or the subtask. The lower the mouse traversal per unit time, the more time is required by the users to perform a certain action. Users may also do more clicks in less time. We have described three measures that are based on the mouse movement data. Below we consider four measures that are based on eye movement data.

5.5.3 Measures Based on Eye Movement Data

Search efficiency during task performance, which has been defined as the number of visits required and also the processing time per visit to the various areas of the screen, can be obtained by studying the eye movement data. The total number of fixations along with the fixation duration is taken as the gaze at the particular area of interest. It was necessary to account for shorter fixations in the later stages of the experiment. The shorter duration threshold was taken as 100 ms as has been recommended by Yarbus (1969).

In this study, the eye movement data have been analyzed and the classification for each area on the screen presented in Figure 5.16. Defining areas of interest based on already preidentified tool, object, and goal area is an important aspect in the eye movement study. Tools, goal, and objects were the main elements of the task in the functional analysis of the subject's performance. Using tool, object, and goal areas sets a general paradigm for identifying strategies of task performance utilized by different users. For example, in this case, the total number of visits to the tool area by the eye will be calculated on the basis of the number of transitions the eye made to the tool area during the task performance. Four measures based on eye movement data are the following.

Visual Fixation Time in Different Areas of the Screen during Skill Acquisition (VT): These data represent the amount of confusion that existed at the initial stage of task performance. During skill acquisition, the duration of visual fixations on various areas of the screen should be decreased. Moreover, this duration can decrease unevenly for different areas depending on the specificity of a particular task.

Hence, if VTg is the visual fixation time in the goal area, VTt is the visual fixation time in the tool area, and VTo is the visual fixation time in the object area for the *i* trial, then the total visual fixation time for the *i* trial is given by

$$VTg + VTt + VTo = VT_i$$

Total Number of Eye Visits to Different Areas per Click (VN): These data also represent the amount of confusions that existed at the initial stage of task performance. During skill acquisition, the number of eye visits to different areas of the screen should be decreased. The total number of visits (VN) here is calculated as follows:

[Number of visits to the goal area + number of visits to the tool area +

number of visits to the object area for a particular trial]

Hence, if VNg is a number of visits to the goal area, VNt is a number of visits to the tool area, and VNo is the number of visits to the object area for a particular trial *i*, then the total number of visits for the trial *i* is given by

$$VNg + VNt + VNo = VN_i$$

Average Processing Time per Visit (tv): If the total time required to complete the trial *i* is *T* and the number of visits during the trial *i* is *V*, then the average processing time per visit is given by $tv = V/T$. The reason it is called average is due to the fact that the amount of time spent during each visit within the particular trial may not be the same, and hence, the average time is used in this case.

Ratio of Eye Visits to the Object Area to the Number of Eye Visits to the Tool Area (ETO): It is calculated based on the following formula:

$$\text{ETO} = \frac{\text{Number of eye visits to the object area}}{\text{Number of eye visits to the tool area}}$$

It represents the difficulty of the user in executing actions in the tool area. If ETO is >1, then it represents difficulty for the user to execute actions in the tool area. The minimum value of ETO depends upon the familiarity of the user with the interface. When this ration is equal to 1 it can be considered as acceptable since the user is visiting the area to perform an action utilizing the mouse (click). A higher ratio indicates that the user is facing difficulty in associating the task with the tools and thereby moving back and forth between the tool area and other areas of interest related to the task.

We also utilize such measures as total time of task performance and number of errors during the skill acquisition process. Some other measures can be developed based on the described principles. It is to be noted that obtained measures can be calculated as average data for all groups and as measures related to a particular subject. Therefore, we can develop a learning curve or a curve of acquisition process for a group or for an individual. This helps us to study individual strategies utilized by subjects during skill acquisition. Later we will demonstrate in an abbreviated manner how some of these measures can be used for task analysis by utilizing the genetic principle of study (task analysis during the skill acquisition process).

In this study, an acquainted set of experiments with different subjects were used. At the first acquainted set of experiment, subjects manipulated only one single feature of the task during 12 trials. The second acquainted set required manipulation of just two features of the tasks. Therefore, the subjects were sufficiently familiar with the strategies of performance. This to some degree reduced the effect of learning. Nevertheless, if we can discover the effect of learning even in this situation, then our methods are sufficiently sensitive. In general, an acquainted stage depends on the specifics of the task. For example, in our case, it is sufficient in each two acquainted sets to have 3–5 trials. It requires expert analysis in each case to determine a number of trials for the acquainted set or for the preliminary experimental study. It is also important to consider that not only cognitive factors but also emotional-motivational adaptation factors should be taken into account. These factors influence the skill acquisition process.

In the main set of experiments, subjects performed the whole task. In this set of experiments, all subjects performed 16 trials. The first five tasks belong to the exploratory learning stage and the final set of five tasks is considered to be the postlearning stage (Carrol, 1987).

There were four groups of tasks with various levels of compatibility and tool arrangement. Between–within ANOVA model was used for the statistical analysis of data obtained in these groups. The compatibility of task sequence and tool arrangement were the between-subject factor and the learning phase was the within-subject factor. Only two levels of within-group factors were considered and the variable type for each analysis was the same. Since only two blocks were used in this case (exploratory and postlearning), the sphericity test was not necessary.

5.5.4 Experimental Study Based on Motor Activity Analysis

A general analysis of the users' performance demonstrates that they have utilized explorative strategies. The users do have one fixed goal in mind, which is the final arrangement. However, during trial and error, the users have different subgoals that they can change during trials based on the feedback information about the task performance. Each action gave them a feedback based on which the user understands better and what kind of restrictions is imposed by the interface. In most cases, the users would change their subgoals when they encounter the limitations or the rules of the task. If the output of the actions was not what the users desired, they would change the subgoal and therefore the strategy of their task performance. Once they are accustomed to all the rules, the performance became more stable and the errors and changes in the strategies of performance became infrequent. Thus, we consider how allocation of various areas on the screen and their internal structural organization impact the strategies of users' performance during the skill acquisition process. One of the important methods for the analysis of the self-regulation of activity is the method of study when conditions of the task can vary and the specialist analyzes the possible strategies for achieving the same goal of the task. Based on the previously described methods, we developed learning curves that demonstrated possible strategies of task performance during the skill acquisition process. Let us consider the previously discussed measures in task analysis during the skill acquisition process.

The total task performance time was the simplest measure for the performance comparison. These data are presented in Table 5.4. In this table, N is the number of subjects in each group.

From Table 5.4, it can be seen that the more complex the task is, the more time is required for its performance. However, the difference was not statistically significant and we consider it as a tendency. There were also differences in standard deviation (SD). The more complex the task was, the more the variations were present in the task performance.

Statistical analysis revealed that the compatibility had no major effect on the task performance. There was no effect of compatibly and tool arrangement on the total task performance time. Within the groups, there was a

TABLE 5.4

Mean and Standard Deviation of Total Task Completion Time for the Exploratory and the Postlearning Stages[a]

Level of Compatibility	Level of Tool Arrangement	N	Exploratory		Postlearning	
			Mean	Standard Deviation	Mean	Standard Deviation
1	1	8	125.73	16.51	98.49	18.67
1	2	8	126.37	28.99	108.98	24.16
2	1	8	130.20	37.46	103.70	20.87
2	2	8	139.42	30.84	114.75	28.16

[a] It should be noted that statistical data are selected and presented in such a manner that it permits the reader to understand the basic principles of its utilization. For example, statistical data have been collected for other experimental results in a similar way as presented in Table 5.4. However, we showed such data only one time for demonstration.

significant effect of the stage of learning ($F(1,28) = 11.2$, $p < 0.01$). All the other null hypotheses were accepted.

The total task performance time can be used only for the preliminary data analysis. Sometimes tasks have approximately the same total performance time at the final stage of skill acquisition, but the task learnability can be different.

Moreover, the task complexity very often cannot be evaluated based on the total performance time because they have different amounts of elements to be manipulated during their performance. Hence, the task that requires more time can be easier than the shorter one. Learnability is a more sensitive criterion for the evaluation of task complexity (Bedny and Meister, 1997). Figure 5.21 demonstrates learning curves that were developed based on such measure as Motor Efficiency (Click Efficiency) E.

Acquisition curve was developed based on 16 main trials. Figure 5.21 shows an average group's efficiency across different tasks. An initial observation suggested that for incompatible tasks, the efficiency at the exploratory learning stage of performance was quite low. The dotted area represents the exploratory learning stage and the postlearning stage in the graph (see Figure 5.21).

Qualitative analysis of these curves demonstrated that incompatible groups have lower efficiency than the compatible ones. During the trials, the efficiency of all groups of subjects that perform different types of tasks increased. At the end of the trials, three groups of subjects that performed various types of tasks described earlier excluding the compatible from the top tasks demonstrated similar efficiency.

The difference in motor efficiency between groups that performed compatible tasks and the incompatible ones was more significant in the beginning of the trials than at the end of the trials. Therefore, the more complicated the tasks are, the more significant efficiency improvement can be observed. The variation in average efficiency of the groups can be observed in Figure 5.21. Let us consider ANOVA for the between-group effects. There was a

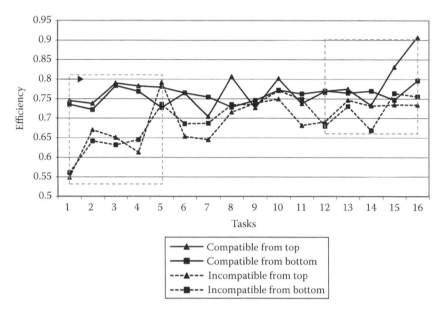

FIGURE 5.21
Average efficiency across different types of tasks for four groups of subjects.

significant main effect for the factor of compatibility ($F(1,28) = 11.1$, $p < 0.01$). There was no significant interaction effect. The tool arrangement did not have any significant effect on the performance efficiency.

The learning stage has a significant effect on the performance efficiency ($F(1,28) = 4.21$, $p < 0.05$). The between–within interaction of stage and compatibility had a major effect on efficiency ($F(1,28) = 18.32$, $p < 0.001$). There was no other between–within interaction observed.

For the compatible from the top task group, the efficiency remained more or less at the same level with a mean of 78% (SD = 5%) for the exploratory stage and a mean of 76% (SD = 7%) for the postlearning stage. Incompatible task groups, however, showed drastic improvement in performance; their efficiency increased for top tool arrangement from a mean of 61% (SD = 9%) to a mean of 77% (SD = 10%) in the postlearning stage and for bottom tool arrangement from a mean of 64% (SD = 8%) to a mean of 72% (SD = 5%).

The between-group hypothesis (H_1: $\mu_{Compatible} \neq \mu_{Incompatible}$) for efficiency was accepted at the 0.05% significance level (statistically significant data). However, due to the interaction effect, this consideration should be given only to the exploratory learning stage. The null hypothesis (H_0: $\mu_{From\ top} = \mu_{From\ bottom}$) for tool arrangement was rejected. In the postlearning stage, the null hypotheses for both compatibility and tool arrangement were rejected.

Groups that performed incompatible tasks showed significant improvement in efficiency. For tasks from top tool arrangement, the mean was equal to 61.5% and SD 9% at the exploratory stage. In the postlearning stage, efficiency increased and became equal to 77.7% with SD 10.3% ($p < 0.005$).

For tasks from bottom tool arrangement in the exploratory stage, efficiency was equal to 64.3% with SD equal to 8.1%. In the postlearning stage, efficiency achieved 72.4% with SD = 5.5% ($p < 0.005$). Efficiency also increased for compatible task groups during the comparison of exploratory and postlearning stages. However, these changes were not statistically significant.

For compatible groups of tasks, the hypotheses for differences in the efficiency in the exploratory and the postlearning stages were rejected. However, for incompatible groups, the hypotheses for increased efficiency in the postlearning stage (H_0: $\mu_{Exploratory} = \mu_{Postlearning}$) versus the exploratory learning stage was accepted at a significance level of $\alpha = 0.05$.

Below we will consider learning curves that were developed based on time per click data.

The time per click data were initially used to estimate when the subjects had reached a steady state of performance. These data were used for defining a preliminary number of trials. Here we will describe how this measure has been used in the main experiment. The mean time per click of the different groups across tasks is given in Figure 5.22.

Let us consider some data as an example. For the group performing the compatible from the top task, the time per click has reduced from a mean of 1.51 s (SD = 0.18 s) in the exploratory stage to a mean of 1.30 s (SD = 0.12 s) in the postlearning stage. The group that performed compatible from the bottom tasks reduced the time per click from a mean of 1.44 s (SD = 0.22 s) in the exploratory stage to a mean of 1.32 s (SD = 0.23 s) in the postlearning stage.

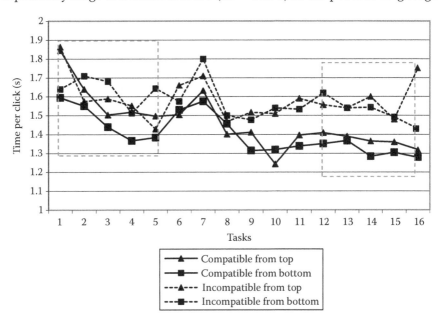

FIGURE 5.22
Time per click across tasks for different groups.

The group that performed the incompatible tasks showed similar drastic improvement in performance.

The ANOVA for the between-group effects demonstrated that there was no significant interaction effect. There was a significant main effect for compatibility ($F(1,28) = 6.3$, $p < 0.01$). The tool arrangement did not have any significant effect on the time per click measure. The learning stage had a significant effect on the time per click ($F(1,28) = 21.6$, $p < 0.0001$). No interaction effects for the between–within factor were observed. We did not consider any other statistics related to this measure.

In general, the results suggest the significant effect of compatibility on subjects' performance as far as time spent in between clicks was concerned, which means that incompatible interfaces are time consuming. However, the time per click reduced, as has been expected due to the increase in the pace of performance as the subjects got a better understanding of the task sequence. However, the effect of compatibility and the significant ($p < 0.005$) difference between compatible and incompatible task performance show that even when the subjects already knew the task sequence, there were difficulties in accepting the task sequence, which resulted in the excess time between the actions.

At the next stage we will demonstrate how mouse/traversal time (mouse traversal rate [MT]) can be used to develop learning curves (see Figure 5.23).

As has been described earlier, this measure evaluates the mouse movement in pixels/s. The major hypothesis when we utilize this measure was that various task sequence requirements and interface relationship can

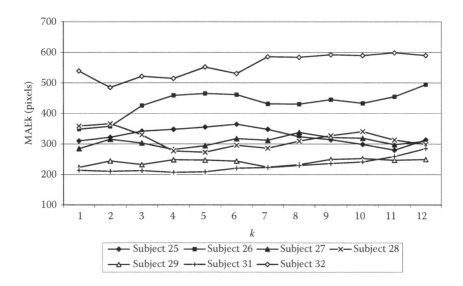

FIGURE 5.23
Average mouse movement distance per second for group 3 (incompatible from the bottom interface).

affect the mouse traversal rate. Analysis of this measure demonstrated the following. The main effect of compatibility or tool arrangement on the mouse distance traversed per unit of time was not observed. Hence, $HB1_0$ and $HB2_0$ were accepted. The interaction effect was also not observed. Within-group tests showed a significant effect of the learning phase ($F(1,28) = 9.08$, $p < 0.01$). Therefore, $HB3_0$ was rejected at 0.05% level of significance.

The mouse traversal rate of the group performing compatible from the top tasks changed from $M = 308.7$ (SD = 80) at the exploratory learning phase to $M = 366.9$ (SD = 67.2) at the postlearning phase. For the group performing the compatible from the top task, there was an increase in the mouse traversal rate from the exploratory phase ($M = 363.7$, SD = 71.7) to the postlearning phase ($M = 381.9$, SD = 60.9). For the group performing the task using incompatible from the top interface, the mouse traversal rate increased ($M = 328.2$, SD = 73.9) to the postlearning phase ($M = 346.7$, SD = 65.8), while for the group that utilized the incompatible from the bottom interface, there was also an increase in the mouse traversal rate from the exploratory learning ($M = 328.2$, SD = 101.6) to the postlearning phase ($M = 381.4$, SD = 129.3). Therefore, this study discovered that subjects required less time in the postlearning phase to traverse the same distance. This measure is particularly sensitive to the effect of learning.

For the study of individual differences in skill acquisition, it is useful to utilize the acquisition curve for individuals. An example of subject's individual curve developed based on the mouse traversal rate is presented in Figure 5.23. This figure demonstrates that the 26th and 32nd subjects demonstrated individual differences in performance in comparison to other subjects according to considered measures. For the correct interpretation of these differences, it is required to compare individual curves in relation to other measures. Moreover, a more accurate interpretation of the data can be obtained during comparative analysis of acquisition curves based on the diverse measures.

The material presented earlier is restricted to the measures that are based on the mouse movement and mouse log data. This is not an exhaustive list of measures and other measures can be used. Below we consider measures that utilize eye movement data.

5.5.5 Experimental Study Based on Eye Movement Analysis

We consider the possibility of using eye movement data to analyze the acquisition process in more details. The objective was to consider the relationship between eye movement and task interface features. At the first stage we utilize percentage of total dwell time in tool area for development of learning curves using the following formula:

$$\frac{\text{Total dwell time in tool area}}{\text{Total dwell time}}\%$$

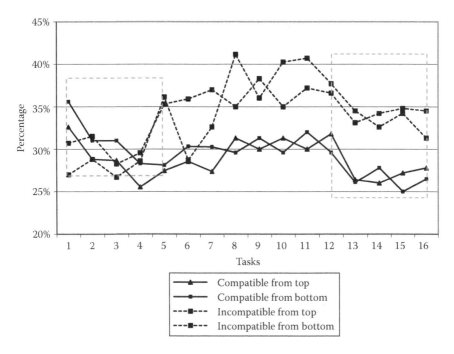

FIGURE 5.24
Percent of the dwell time in the tool area.

A comparison within the groups for the difference in the exploratory and the postlearning stage and the effect on the eye movement due to learning has been analyzed at the first stage.

The percent dwell time in the tool area (PDTT) is presented in Figure 5.24.

The ANOVA for the between-group effect demonstrates that there was no significant interaction effect. There was a significant effect of compatibility ($F(1,28) = 12.1$, $p < 0.001$). The tool arrangement did not have any significant effect on the PDTT. There was no significant between-subject effect observed. However, interaction of the learning stage and compatibility seemed to have an effect ($F(1,28) = 5.91$, $p < 0.05$).

The ANOVA showed the within-subject effect. The learning stage had a significant effect on the percent dwell time in the tool area ($F(1,28) = 68.8$, $p < 0.0001$). There was also an interaction effect of the compatibility and the learning stage ($F(1, 28) = 33.2$, $p < 0.0001$), indicating different dynamics in percent dwell time for groups performing various tasks in terms of their compatibility. No other interaction effects were observed.

Groups that were performing incompatible tasks had significantly higher PDT in the tool area than for the compatible ones. The null hypothesis for PDTT was rejected at 0.05% level of significance. This significance was observed in the postlearning stage.

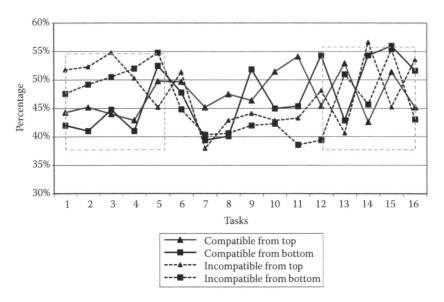

FIGURE 5.25
Percent dwell time in the object area.

Similarly, we can develop learning curves based on the total dwell time in the object area by applying the following formula:

$$\frac{\text{Total dwell time in object area}}{\text{Total dwell time}}\%$$

The percent dwell time in the object area (PDTO) during the trials is presented in Figure 5.25.

The ANOVA for the between-group effects demonstrated that there was no significant interaction, though there was a significant effect on the tool arrangement ($F(1,28) = 16.8$, $p < 0.001$). Compatibility did not have any significant effect on the PDTO. The null hypothesis for the tool arrangement (H_0: $\mu_{From\ top} = \mu_{From\ bottom}$) has been rejected. An interaction effect ($F(1,28) = 6.84$, $p < 0.01$) of the compatibility and the tool arrangement has been also observed. The stage of learning had a significant effect on the PDTO ($F(1,28) = 7.9$, $p < 0.0001$). No interaction effects for the between–within factor was observed. The null hypotheses for the effect of compatibility (H_0: $\mu_{Compatible} = \mu_{Incompatible}$) have been rejected for both stages of learning. Null hypothesis for the tool arrangement has been accepted for both stages of learning. The PDTG in the goal area has also been also analyzed. The within-group null hypothesis for the learning stages has been rejected ($F(1,28) = 36.8$, $p < 0.001$).

We can also aggregate eye and mouse movement measures.

It was possible to determine the ratio of the eye visits to the mouse visits in the area of interest. As an example, the tool area acquisition curve that is

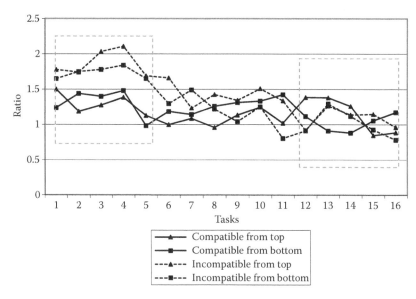

FIGURE 5.26
Ratio of eye to mouse visits in the tool area.

utilizing this measure is presented in Figure 5.26. It can be observed that there were a higher number of eye visits than mouse visits at the exploratory learning stage.

The within-subject ANOVA shows the main effect of the learning stage ($F(1,28) = 36.67$, $p < 0.001$). An interaction effect of the learning stage and compatibility ($F(1,28) = 8.26$, $p < 0.01$) was also observed. As an example, we also present a means plot of the eye to mouse visit ratio (compatibility stage) in the tool area (see Figure 5.27).

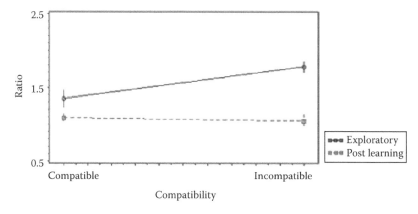

FIGURE 5.27
Means plot of the eye to mouse visit ratio (compatibility stage) in the tool area.

Groups that were using the incompatible interface had higher means in the exploratory learning stage than in the postlearning stage. The groups that were using the compatible interface had less difference in the eyes to mouse visit ratio at the exploratory stage compared with the groups using the incompatible interface.

Previously presented data allow us to proceed to discuss the results. The purpose of this work was to demonstrate the microgenetic method of study being applied for the evaluation of the task usability, complexity, etc. The essence of this principle consists of the study of human activity in the process of its formation and development. This method is particularly relevant for the study of computer-based tasks. This can be explained by the fact that users constantly acquire new tasks and more often than not through the self-learning process in which they utilize explorative strategies. Thus, learnability is an important feature of a user interface. The more complex the task is, the poorer is the learnability of the task. Analysis of learnability demonstrates that the basis for the development of exploratory strategies is the process of self-regulation.

In this study, we developed four versions of task with various levels of complexity. The more complex the task is, the less is the task usability. Analysis of the task learnability utilized by users in activity performance is an important experimental approach to the evaluation of the task complexity and therefore usability of the interface. According to ISO 9241, some usability measures such as evaluation of the effectiveness of performance, efficiency of performance, and satisfaction have been suggested. Some of these measures are not precise and ambiguous. For example, effectiveness can be evaluated based on the calculation of percentage of goals achieved. However, in activity theory, there is an objectively given goal, a subjectively accepted or formulated goal, etc. Therefore, the percentage of goals achieved is not a precise measure of user performance.

Similarly, the time to complete a task as a measure of efficiency is a very general criterion. Satisfaction measures without objective data can be very subjective. In general, these measures do not give us sufficient information about the activity structure during task performance. Therefore, in this work, we described usage of the microgenetic method and derived from it measures of usability evaluation of the computer-based tasks.

In order to utilize the genetic method, it is necessary to develop the methods of registration and analysis of the activity acquisition process during performance of the computer-based tasks. Development of the learning curves that depict the activity acquisition process and their interdependent analysis is an important method of the genetic study. Learning curves are widely used for the analysis of training but not for the analysis of the task complexity and evaluation of the usability of the interface. Usually in psychology, integrative learning curves are utilized. They are not sufficiently informative. Moreover, there is no well-defined method of learning or development acquisition curves for computer-based tasks.

In our work in contrast to the study learning process, we used acquisition curves for the usability evaluation. They have helped us to evaluate task complexity and understand some changes in the structural characteristics of a task.

Activity during performance of a computer-based task can be considered as a complicate structure comprised of logically, hierarchically, and functionally organized elements.

According to the functional principles, depending on its purpose in the activity structure, the same element of the screen can be related to different functional elements of activity.

Therefore, hand and eye movement data in computer-based tasks should be related not only to particular elements of the screen that have some technological characteristics but also to some elements of activity that have a particular functional purpose in the activity structure. The critically important functional elements of activity are goal, object, and tool. The principle of the extraction of such element in activity theory is different in comparison to cognitive psychology. Depending on its functional purpose in the structure of activity, the same elements of the screen can be different at different stages of task performance. Therefore, analysis of eye and mouse movement for different areas of the screen at different phases of learning helped us to understand the functional structure of activity depending on the stage of skill acquisition. In some cases, it is necessary to distinguish between areas of the screen that are allocated based on objectively given goals of task and areas of the screen that are allocated based on users' subjectively formulated goals of subtasks at different stages of skill development. Our study demonstrates a combination of the genetic method of study and the functional analysis of activity, where the last one is considered as a self-regulative system.

The other principle that was utilized in this study was the principle of unity of cognition and behavior. This has been facilitated by combining eye movement registration with mouse movement registration and analysis of the interdependency of obtained data. Yarbus was the first one who introduced direct registration of the eye movement, but he considered it from totally different theoretical perspectives of cognitive psychology by attempting to prove that behavioral activity is tightly connected with cognitive activity and then demonstrated how cognitive processes depend on the goal, motives, and strategies of activity performance. In cognitive psychology, these aspects of his study of eye movement were never introduced. In this work, we paid attention to these factors by utilizing some elements of functional analysis of activity when activity is considered as a self-regulative system. This was performed by dividing the screen into three functional areas. Analysis of total eye visits to different areas of interest at different phases of learning, calculation of percentage of distribution of eye visits to different areas of interest (goal, object, and tool areas), calculation of percent dwell time in different areas of interest at the different stages of learning, etc., demonstrated how subjects changed their strategies during the acquisition of different tasks. These data allowed us to evaluate various versions of the interface.

It should be noted that there is no consensus regarding the level of familiarity of users with a particular task during the study of the acquisition process. In our experiment, subjects received significant training in performance of separate elements of tasks. Even in such conditions, the acquisition process is a sufficiently sensitive method for task analysis of computer-based tasks. In any practical situation, scientists have to specify the level of user's familiarity with particular tasks during the analysis of the acquisition process.

6

Introduction to Ergonomic Design

6.1 Basic Characteristics of the Design Process

The main aspect of the study in the first book of this series by Bedny et al. (2014) is the analysis and design of computer-based tasks. An operator's interaction with a computerized system is not limited to performing computer-based tasks. It also includes the performance of various tasks involving highly automated technological systems when interacting with various displays and controls. Therefore in this book we primarily discuss those aspects of task analysis and design where an operator interacts with displays and controls in man–machine systems. However, some aspects of computerized tasks and manual components of work in contemporary industries are also considered.

In activity theory, technology is treated as means and tools of work. Technology does not exist solely to enhance productivity but is rather utilized to perform various tasks more efficiently and easily. The technical components of a system affect a structure of activity. This raises the question of how to optimize the relationship between the human activity structure and the technical components of the system. Different structures of a technical system imply different methods of task performance. Thus, according to SSAT, if we change the configuration of an equipment, it changes the structure of activity in a probabilistic manner. Therefore, comparison of equipment configuration and activity structure is the basic principle of design in SSAT. If when a human interacts with existing equipment, the structure of his or her activity is very complex, this means that a design solution was not adequate. When workers interact with equipment, they use different strategies to perform the same task. This means that it is necessary to determine which strategies are preferable. It involves studying the mechanisms of activity self-regulation, and based on this, the most preferable strategies of task performance should be determined (Bedny and Meister, 1997). Thus, a design process includes describing the material components of a system and comparing them with a human activity structure. Therefore, task analysis is the main stage of ergonomic design.

Any task analysis and the principles of ergonomic design derived from it start with qualitative analysis of task performance that might utilize various methods such as objectively logical analysis, sociocultural analysis, individually psychological analysis, and functional analysis (Bedny and Karwowski, 2007). The first method includes traditional short verbal description of job and task performance. In the design process, this stage usually includes analysis of prototypes or any other data that can be useful in further stages of the design process. The sociocultural method takes its roots in the works of Vygotsky (1978). It focuses on the analysis of such components as a subject, an object, internal or mental tools, and their relation with external tools of activity and considers the sociocultural context under which a task is performed. Culture is regarded as a mediator between a human and technology and as an aggregation of attitudes, social norms, beliefs, standards, etc. Such sociocultural components are important, for instance, in safety analysis. Another qualitative method of study is an individual-psychological method of study. An individual style of performance is the main concept of such qualitative analysis (Bedny and Seglin, 1999a). It includes analysis of preferable strategies of task performance that depends on individual features of personality.

The most powerful qualitative method is the functional analysis of activity when activity is described as a self-regulative system (see Bedny et al., 2014). It is a systemic qualitative task analysis method that is a new qualitative approach to task analysis. Thanks to this approach, we can describe preferable strategies of flexible human activity during task performance. For contemporary task analysis when a task is performed in various ways depending on situational conditions, this method of task analysis is critical. From a functional prospective, attention is mainly paid to motives, goal of activity, mental model of a situation, subjective standard of successful result, and process of self-regulation. This allows describing activity as a goal-directed, self-regulative system. This approach is principally different from mechanistic homeostatic self-regulative systems that are utilized outside SSAT.

The process of self-regulation of activity should be described through various stages of processing information that involve different psychological mechanisms. Each stage is called a function block, because it performs a particular function in activity regulation and in formation of various strategies of its performance. A self-regulative system includes not only cognitive but also emotionally motivational mechanisms because emotionally motivational mechanisms affect the specificity of human information processing. A model of self-regulation can be interpreted as an interdependent system of windows (function blocks) from which a specialist can observe the same human activity during task performance. For example, a researcher can open a window called *goal* and, at this stage, pay attention to such aspects of activity during task performance as goal interpretation, goal formation, goal acceptance, etc. At the next stage, he or she can open another box called *subjectively relevant task conditions*. Here a researcher would study such aspects of activity as *situation awareness* (*SA*) and *operative image*. According to systemic-structural activity theory, SA is a conscious dynamic reflection of the situation during task performance. In contrast, an operative image is responsible largely for unconscious reflection of a dynamic situation. These two mechanisms interact with each other and are responsible for the creation of a dynamic mental model that is critically important for activity regulation. SSAT has developed two models of activity self-regulation. The first one is the model of self-regulation of orienting activity. It describes a stage of task performance that precedes executive components of activity. This model helps to describe strategies of dynamic reflection of a situation much more precisely than the concept of SA in cognitive psychology. The general model of activity self-regulation includes 20 function blocks or mechanisms of self-regulation. It can be used not only for analyzing strategies of dynamic reflection of a situation but also for analysis and description of executive and evaluative stages of activity regulation. Depending on the specifics of a task, some function blocks or mechanisms might be skipped all together during task analysis. All function blocks in considered models have feedforward and feedback connections and influence each other. Thus, functional analysis of activity when activity is considered as a self-regulative system is a systemic qualitative analysis of task performance. This approach helps to predict and describe flexible activity strategies and eliminate contradictions between constraint-based and instruction-based approach in design. Self-regulation is discussed in great detail in Bedny et al. (2014).

After performance of qualitative analysis, we can perform formalized methods of task analysis. It includes morphological analysis of activity during task performance. This method describes the structure of activity. SSAT offers an original psychological approach to a morphological analysis of activity. This approach describes activity during task performance as a complex structure that has a systemic organization. In other words, SSAT has developed a systemic principle of task analysis where the main units of morphological analysis of activity are cognitive and behavioral actions that consist of smaller units such as psychological operations. From a morphological viewpoint, main attention is paid to cognitive and behavioral actions as components of activity and their logical organization. Morphological analysis includes algorithmic description of activity and analysis of time structure of activity during task performance.

Algorithmic analysis of activity involves description of logical organization of cognitive and behavioral actions, which are elements and units of activity analysis. Such analysis also takes into account probabilistic features of the activity structure that unfolds in time, and therefore, the third stage includes an activity time structure description. The time structure of activity describes it as a systemic organization without reducing analysis to considering its isolate temporal characteristics. The time structure of activity is a multivariant system that unfolds in time in various ways due to its flexibility. The time structure of activity analysis can be performed after algorithmic description of task performance. Any changes in equipment configuration can change the strategy of the task performance and therefore change the algorithmic and time structure description of task performance. Thus, we can compare the structure of activity with configuration of equipment. Morphological analysis is a powerful tool in ergonomic design. All stages of task analysis utilize standardized units of analysis. Utilizing standardized principles of describing a designed object or activity is the main requirement of a design process.

The final stage is quantitative task analysis. This stage is useful for optimizing strategies of task performance. Quantitative analysis includes evaluation of task complexity and evaluation of performance reliability among other methods developed in SSAT. Such methods help evaluate the efficiency of task performance and design solutions in a precise manner. All stages of analysis are interdependent and each following stage may be a result in revision of the previous stages of analysis.

In SSAT, task analysis also outlines the parametric and systemic approaches to task analysis. A parametric approach entails the study of distinct characteristics of task performance. For example, we can measure the time of task performance, or describe characteristics of some cognitive processes that are critical for performance time. A systemic approach includes a functional and morphological analysis of activity and quantitative evaluation of task performance.

As a result of systemic analysis, various models of the same activity during task performance can be developed. These models are important tools in analyzing the efficiency of task performance and/or for the design of equipment. In our further discussions, we concentrate our effort on describing the morphological analysis of activity and its relationship with the functional analysis of activity when activity is considered as a goal-directed, self-regulative system. In the following chapters of the book, we will consider the quantitative evaluation of task performance developed in SSAT.

The SSAT approach to ergonomic design matches its meaning in engineering. The purpose of engineering design is the creation of a new product, software, manufacturing goods, etc. A design process cannot be reduced to experimentation as is done in cognitive psychology. The main objective of design is to create appropriate documentation (including creation of design models and quantitative analysis) that describes a designed object. Usually at the beginning, design involves not very well formalized procedures starting with developing a technical proposal. This is an ideation process, which is very subjective. At the further steps, design is transferred into an analytical formalized stage of design (Suh, 1990). An analytical stage of design in manufacturing involves the creation of various drawings of a designed object that is accompanied by various quantitative calculations. The next stage includes the creation of a prototype and experimental evaluation of a designed object. The obtained data are used for correcting theoretical models (drawings and quantitative calculation) and the cycle can be repeated. A design process can be presented in the following stages: not formalized (qualitative) stage → formalized stage (drawings and calculations) → experimental stage (experimental evaluation of prototype). Design models are always preserved when a designed object is modified because any new modification of a designed object often involves analysis of the history of design solutions. Project documentation includes various design models such as sketches, drawings, mathematical calculations, and text materials (soft models).

In contrast, ergonomists often eliminate the analytical stage. They are usually involved in the evaluation of physical models of designed equipment, present stimulus, observe their effect, and make suggestions on improvement of the designed equipment. Ergonomists that use the human information approach do not have a tool for description of the activity structure.

They replace an analytical stage of design with experimentation. Utilizing cognitive psychology, models are not design models because they describe mental functioning of a human brain but not designed objects. However, design models should describe the human activity structure during interaction with equipment in a standardized manner. For example, drawings always utilize standardized principles because design involves various specialists such as designers, technologists, managers, and all those who are involved in the production process. All of them utilize the same standardized language of a designed object description. Human behavior or activity is not modeled in ergonomics because there are no units of analysis or language of description for this purpose.

Thus at present, instead of design, ergonomists use experimentation. Therefore, in this book, design is referred to as "creation and description of ideal models of artificial objects in accordance with previously set properties and characteristics with the ultimate goal of materializing these objects" (Neumin, 1984; Suh, 1990). In the absence of basic analytical principles of design, this process is reduced to purely intuitive procedures. Analytical procedures involve the creation of various models of a designed object or process. The core of design is development of interdependent models of an object or process that does not yet exist. The design process can be viewed as stages of sequential refinement of design models. At the initial stage, a designer has only an ideal image or a mental model of an object being designed. During subsequent stages, this mental model is externally described using symbols and signs, which makes it available to other specialists involved in the design process.

Because activity is very flexible and unfolded in time as a process, it is extremely difficult to develop models of activity. Some scientists even reject the idea of designing human activity. For example, in cognitive psychology, Vicente (1999) suggests a constraint-based principle of design according to which a worker responds to unanticipated contingencies according to his or her preferences and existing constraints. This principle specifies what should not be done and is opposite to the principle based on the instructions that specify what should be done (Vicente, 1999, p. 68). From this follows that instruction-based task analysis cannot be efficiently applied in the design process. Further, he wrote, "Only guidance about the goal state and the constraints on actions are provided, not how the task should be accomplished" (Vicente, 1999, p. 69). Here we want to stress the fact that any type of design has to take into consideration existing constraints. However, even being aware about the constraints, a worker still needs to know how the task should be performed. Constraints are usually given in terms of external factors with respect to the structure of activity. Hence, a worker should utilize adequate strategies of performance for a situation with specific constraints.

These strategies should be described for workers as models of activity during task performance. Human activity is multidimensional and flexible, which means that strategies of activity can be described by methods that cover a coherent structure of the same activity using various languages for its description.

SSAT suggest procedures and language of description for the creation of a system of interdependent models that describe the same human activity during task performance. Obtained data are presented in a standardized and formalized manner and then these data are used as a source for the subsequent analysis and corrections of the designed equipment or method of performance. Design models are always task specific. They describe the structure of activity during task performance. The same activity during task performance should be described by several interdependent models that supplement each other, which give a better picture of the activity structure during task performance. Thus, we can compare the structure of activity with the structural configuration of equipment or software and evaluate its usability.

The design of human work activity and its associated time study allow designing equipment and software and efficient performance methods, improve safety and productivity, and develop effective training methods because there is probabilistic relationship between the structure and configuration of equipment, tools or software, and activity structure. For some psychologists (Visser, 2006), design is an effort or cognitive activity of those who are involved in the design. However, the real meaning of this term is quite different. The purpose of design is creation of documentation, according to which it is possible to produce new products, software, manufacturing goods, method of performance, etc. The main purpose of design is to create symbolic models of objects that do not yet exist in a materialized form. On the basis of the developed models, production personnel creates new materialized objects, programs, methods of performance, etc. For example, when designers design a new machine, they develop drawings of individual parts of the machine, assembly drawings, technological description, etc. Based on this documentation, as well as documentation that describes a technological process, a production process is carried out.

The model describes a designed object with some approximation especially at the early stages of design model development. During the design process, the models are refined and adjusted. Refining of design models is performed until the optimal solution is found. The final adjustment of models is done by making prototypes of designed objects and the experimental verification of their functioning. The design process can be represented as successive stages of refinement of models involving both analytical and experimental methods. Design can be presented as a continuous process of development and evaluation of obtained result.

A real object has numerous features and properties and it is not possible to include all of them in any one model. This calls for the creation of multiple models of the same object. Various models of the same object complement each other in describing the same object. All this leads to the conclusion that design is primarily working with documentation or information presented in a certain way. Experiments are conducted to test samples of a designed object and to adjust prebuilt models.

In ergonomics, it is often necessary to redesign an existing method of performance in order to create a more efficient and safe work method. Redesign includes observation and data collection, and then based on obtained data, a new project documentation is created to describe a new work method and its associated redesigned equipment. Therefore, before developing a project documentation, data are collected through observing a specific task's performance, existing documentation is analyzed, and so on. New project documentation includes drawings, calculations, and models. Thus, from the actual observation of equipment, methods of work and analysis of existing documentation a designer moves to developing a new project documentation including drawings, models, calculations, etc. An existing equipment or software is used as a prototype.

In contrast, there is no equipment to be used as a prototype for new design and observation is impossible or very limited. New design starts from the development of a technical proposal and the creation of models of an inexistent object. When a designer is involved in a totally new design process, he or she has to use analytical procedures as a main method of design. Hence, redesign is simpler than design. Artificially created prototypes play an important role in new design. For example, in engineering design, they can be mock-ups such as scaled-down models of bridges, buildings, and models of mechanized production line that are presented as 3D models of machines and structures.

6.2 Concept of Self-Regulation in Task Analysis

In this chapter, we consider an example of application of the concept of self-regulation to analyze an aviation-related task. The concept of self-regulation is discussed in greater detail in Bedny et al. (2014). Studies demonstrate that human activity is extremely flexible. Even when there is only one best way of task performance, a performer can still vary the methods of its performance within a certain range. For instance, a gymnast can perform the same movement multiple times while each movement still has some unique features that make it different from other similar ones. Bernshtein (1996) was one of the first who introduced the concept of self-regulation in physiology and psychology and utilized the phrase *repetition without repetition* wanting to emphasize the uniqueness of activity and its dependence on the situation at hand. Self-regulation influences a system in order to correct its behavior or activity (Bedny et al., 2014). The concept of self-regulation becomes useful only when the self-regulative model is developed. Such model includes various functional mechanisms or blocks and feedforward and feedback connections between them. Any psychological self-regulative system is not homeostatic but goal-directed. It provides the achievement of the conscious goal of activity by using various strategies of task performance. Strategy is a plan or program of performance that is responsive to external contingencies, as well as to the internal state of the system. In SSAT, the self-regulation process involves such functions as creation of a goal, formation of a mental model of a situation, evaluation of difficulty and significance of activity and its elements, formation of a performance program, and its realization, control, and correction (Bedny and Karwowski, 2007).

There is another reason for study activity self-regulation. In a complex activity, it is often impossible to distinguish between the mental processes of sensation, perception, memory, and reasoning. These processes are all interrelated in complex ways. For example, perception is related to reasoning. The process of perception involves memory. The process of identification includes image recognition. That is why a functional analysis of the activity suggests that the activity should be analyzed not only in terms of cognitive processes but also in terms of functional blocks, which are the main mechanisms of regulation of activity. Functional and cognitive analyses are interrelated. Each functional block integrates cognitive processes in a different way depending on the functional specificity of the block and the nature of the task. Functional analysis is qualitative systematic analysis of the activity, in which the activity is considered as a self-regulating system. Models of activity self-regulation are important tools for the analysis of strategies of human activity. These models allow analyzing and describing a strategy of task performance more accurately. In SSAT, two main models of self-regulation were developed (Bedny et al., 2014). The first model describes the *self-regulation of orienting activity*, which precedes execution, and the second one describes all stages of activity regulation. The model of self-regulation of orienting activity describes the process of dynamic reflection of the situation. Dynamic reflection of the situation, development of a dynamic mental model, and interpretation of a situation are the main purpose of orienting activity. In the self-regulation model of orienting activity, executive components of activity are significantly reduced. The *general model of self-regulation* describes all stages of self-regulation including an executive stage of activity that involves the transformation of a situation or an object of activity according to the goal of a task. Let us consider in an abbreviated manner the application of the concept of self-regulation to the task analysis of a pilot's activity utilizing the self-regulation model of orienting activity that is described in details in Bedny et al. (2014).

In emergency situations, signals with a low attractive effect may remain unnoticed by a pilot if the pilot is focused on aircraft control and is not expecting a failure. The value of the attraction effect determines how quickly the process of information reception and interpretation starts. If the attraction effect is sufficient, then it causes the voluntary switching of attention to unexpected signals. Therefore, the initial stage of this process is the orienting reaction (something happens). The functioning of this mechanism is conveyed by responses such as involuntarily turning of eyes or head toward a stimulus, altering sensitivity of different sense organs, and changing blood pressure. This mechanism plays an important role in the functioning of involuntary attention. The orienting reflex provides automated turning to external influences and influences on the general activation and motivation. The *mechanism of orienting reflex* is the first functional mechanism in the model of self-regulation that should be considered in the study of this type of emergency situation.

However, an attractive effect of the signal does not determine the next stage of human information processing that is involved in the interpretation of the meaning of information. The function block *meaning of input information* and *goal* is responsible for this stage of information processing. These two blocks are involved at the next stage of information processing. At the first stage, the pilot's activity can be described as a process of receiving information, interpreting its meaning, and creating a goal of tasks. Hence, at the second stage of analysis, such functional mechanisms or function blocks as *meaning of input information* and *goal* are particularly important. For example, the same information can be interpreted by a pilot in a totally different way. He or she can formulate different goals of task in the same objectively presented situation. Not only cognitive but also emotionally motivational mechanisms can influence the interpretation of the situation. For example, analysis of pilot eye movement demonstrates that the strategies of the pilot's gaze depend on the subjective importance of information. Therefore, such a functional mechanism or block as *assessment of sense of task* or its elements (their significance for pilot) can change the strategies of selection of information and creation of a dynamic mental model of a situation. The function block *sense* is responsible for evaluating the significance of a task or situation. Thus, interpretation of information by the pilot depends not only on cognitive mechanisms but also on emotional and motivational mechanisms. This factor is practically ignored in cognitive psychology. In the pilot's activity, the goal of various tasks very often is formulated independently. A pilot formulates a new goal and performs executive actions for its achievement based on an orienting stage of activity when a goal of task and a mental model of a situation are developed.

Functional analysis distinguishes between the objective complexity of the task and the subjective evaluation of task difficulty. A subject can evaluate the same task as being more or less difficult depending on the complexity of the task, the past experience, and even the temporal state. The more complex the task, the more probable it is that the task will be evaluated as difficult. It is important to find out how a subject evaluates the task difficulty. For example, a subject can overestimate the task difficulty and select a cautious strategy of task performance and cannot finish the work in a required time limit. Therefore, in study activity self-regulation, specialists very often have to pay attention to the function block *assessment of task difficulty*.

At the executive stage of activity self-regulation, the following function blocks are important: *formation of a program of task performance, making a decision*, and *performance of a program*. After a pilot makes a decision and executes the corresponding cognitive and behavioral actions, a result of these actions is evaluated. This is an evaluative stage of activity regulation where such blocks as *subjective standards of successful result* and *subjective standards of admissible deviation* are critical ones.

The correct interpretation of emergency situations in time limit conditions is a critical factor for pilots to choose the correct actions. At the first stage, we should distinguish the physical characteristics of signals in an emergency situation, which can be with high, average, and low attraction effect. For example, a high attraction effect is attributed to physically strong noninstrumental signals (angular rotation of an aircraft with acceleration exceeding $10°/s^2$, vibration of an aircraft, and shrill sound). Instrumental signals such as siren, ring, buzzer, and voice also can have a high attraction effect. The initial stage of receiving such information is the orienting reaction. In our model of self-regulation, such reaction is attributed to the functional block *mechanism of orienting reflex*.

This functional mechanism or block interacts with other mechanisms of activity regulation and specifically with the function block *formation of level of motivation*. The last block is tightly connected with such blocks as *assessment of task difficulty* and *assessment of task sense* (subjective significance). Under the influence of emotionally motivational factors, the function block *subjectively relevant task conditions* creates a dynamic mental model of a flight. This block includes two subblocks.

One of them is called *operative image* and the other is called *situation awareness* (conscious reflection of a situation). It not only provides a reflection of the current situation but also anticipates the near future and infers from the past. Not only the logical or conceptual components of activity but also the imaginative components provide a dynamic reflection of reality. Imaginative reflection of the situation can be largely unconscious or easily forgotten due to difficulty in its verbalization. In contrast, SA is involved in the conscious dynamic reflection of the situation, which can be verbalized. Imaginative and conceptual subblocks partly overlap. The pilot is conscious of the information being processed by the overlapping part of the imaginative subsystem. It has been discovered that a dynamic model of a flight is affected by an operator's goal, set, and those aspects of a flight situation that are subjectively significant for pilots.

All these mechanisms are critical in an emergency situation. Thus, in an emergency situation, interaction of instrumental and noninstrumental information is an important factor that determines the strategies of activity during task performance. The other important factors that influence pilot's strategies of task performance are interaction of conscious and unconscious levels of activity regulation and interaction between cognitive and emotionally motivational mechanisms of activity regulation. Whereas signal detection depends first of all on the attraction effect of the signals and therefore the unconscious level of activity regulation, the second cognitive step of information processing depends on the content of the signal, its certainty, and its ability to be interpreted correctly.

Let us consider some basic characteristics of a pilot working in an emergency situation with uncertain information (Ponomarenko and Bedny, 2011). Very often in an emergency situation, there is a combination of uncertainty information with its high attraction effect. An example of an emergency situation with a high attraction effect and ambiguity of information is autopilot failure. Each autopilot failure results in wheel deviation and aircraft rotation around the x- or z-axes. Failure is not displayed, but the pilot feels the physical influences of angular acceleration, which is perceived as jerking, pulling of the handles out from the hand, bumpy flight, etc. In experimental study the failure was associated with autopilot functioning. In real experimental flight conditions, an instructor induced malfunctioning and create an emergency situation with ambiguous (uncertain) noninstrumental information that has a high attraction effect and a pilot detects signals almost instantly. Flight instructors did not intervene in pilots' action and did not comment on flight mode changes, not until pilots reported about faults (instructor intervention was allowed only if there was a danger for flight). Correctness and timeliness of a pilot's actions were recorded. Objectively gathered data were compared with data obtained by observations and interviews of pilots. Pilots with different backgrounds were involved in the experiment. Their strategies of task performance were compared.

After a pilot reported the fault, the instructor interviewed a pilot in the air with a questionnaire designed by experimenters. The purpose of the interview was to find out what signs the pilot used to identify the fault, what difficulties the pilot met in the aircraft control and in fault recognition, and what other faults could this fault be confused with. The experimenter also interviewed the pilot right after the flight to find out if the pilot had any past experience in aircraft control with refusal of position stabilizer, what signals (physical influence that appeared at faults) the pilot considered most typical, what suggestions about the cause of the situation arose during decision making, how the pilot checked the accuracy of suppositions, and what other malfunctions resembled the malfunction that arose in this fault. In order to get additional information about the pilot's tension, biochemical analysis of urine and blood was conducted. Based on such study, five identification and interpretation strategies of fault situation were described.

The first strategy is instantaneous identification when the pilot's dynamic mental model of a situation coincides with the content of the current situation. Signals are compared with the mental model already developed in the past and the *situation is identified instantly*. The pilot experienced a feeling of familiarity and perceives noninstrumental signals as definite information. Analysis of the situation flows unconsciously. Hence in such situation, an operative image of the flight is particularly important.

The second strategy is identification and interpretation of the situation after the mental search between alternatives is evaluated based on operative thinking. Operative thinking provides development of a mental model that unfolds over time in an internal mental plane without addressing to an external stimulus. The pilot has an adequate dynamic mental model and operative image of the flight in proper time. However, inside of the model, similar signals are not differentiated sufficiently.

The third strategy is when for evaluation of the content of incoming information and development of a mental model of a flight, the pilot needs additional information. The pilot can identify information only after addressing the equipment, that is, identification and interpretation rely on additional signals and perceptual and thinking actions. A dynamic model and image of the situation derived from past experience are not completely enough, and therefore, identification and interpretation of information cannot occur in an internal mental plane.

The fourth strategy is identification and interpretation on the basis not only of perceptual and thinking actions but also of motor actions. A pilot cannot develop an adequate dynamic model and operative image of a flight based on the past experience and mental analysis of the situation. In addition, she or he needs to use the trial-and-error method, which includes motor actions and analysis of their consequences. Motor actions perform explorative functions.

The fifth strategy can be considered as identification and interpretation but very conditionally. It adjoins with the fourth mode and the difference is that it does not lead to correct interpretation of the situation.

Analysis of such function blocks as *mechanism of orienting reflex, formation of level of motivation*, and *assessment of sense of task* (significance), *goal*, and *subjectively relevant task conditions* was very useful in discovering such strategies and their precise description.

The experimental data allowed us to distinguish *certain, contradictory, and uncertain* information in emergency situations. Certain information let us recognize a situation accurately, and this is done simultaneously with signal detection. Certainness of the information is technically provided with signal representation on a display, or as a voice message in earphones. Contradictory information are usually presented by signals associated with the event indirectly. An example of such signal is mismatches in displays, indicating a result of a failure in one of them. This contradictory signal did not show the cause of mismatches directly and it hinders signal interpretation and decision making. Most of the noninstrumental signals are characterized by contradictory information; to detect them, a pilot has to search for the definite information actively and to use his or her experience and knowledge about similar signals. Uncertain information cannot be unequivocally interpreted, which is characteristic of noninstrumental information.

A pilot's information processing depends not only on signal characteristics but also on the pilot's mental preparedness and, first of all, on the content of his mental dynamical model. The function block *subjectively relevant task conditions* is responsible for the creation of such model. Functional analysis of activity allows us to divide emergency situations into five classes:

1. *Conflict situation*. In this situation, a pilot chooses to form opposite but subjectively equal significant decisions. The choice is made without clear prediction about consequences of each decision.
2. *Situation with unexpected results*. In this situation, a pilot makes purposeful action but meets unexpected results. In most cases, such

      situation is a result of an uncertain though physically intense stimulus (noninstrumental signal). The situation may be aggravated by a pilot's insufficient training for emergencies.

3. *Situation with time and information deficit.* In this situation, a pilot has to make a decision correctly and promptly in spite of the lack of information. This situation is the most complex. The pilot's ability to act safely depends only on his or her heuristic, creative decisions.

4. *Ambiguous situation.* In this situation, a pilot misinterprets controversial signals and guides his actions based on the misinterpretation. This situation is more prolonged and errors are revealed more gradually than in clause 2.

5. *Definite (certain) situation.* The pilot knows what to do and results match his predictions.

The objective complexity of these five classes is different. However, the objective evaluation of complexity in the described five classes and the subjective evaluation of their difficulty are not the same. The latter is influenced by the pilot's performance. For example, a situation with uncertain signals may be easiest for a trained pilot (falls into the fifth class). These circumstances often make interrelation of information model quality and the pilot's reliability and safety performance latent. The described analysis of emergency situation demonstrates that the pilot's interpretation of information depends not only on cognitive but also on emotionally motivational components of activity.

Even a short description of presented experimental data demonstrates that functional analysis is an efficient tool in task analysis. Usually, functional analysis utilizes experimental procedures in combination with observation and subjective judgment of experimental events by different subjects involved in this experiment. It is useful when such subjects have different experiences and individual features. Discrepancies between obtained data are an important source of information for the analysis and discussion of obtained data. The major units of analysis in this situation are functional mechanisms or blocks. Each function block can integrate different cognitive processes in a variety of ways. Hence, we have multiple representations of the same activity during task performance, and these representations can be compared with each other. During any particular study, the most important function blocks can be selected for task analysis. Therefore, when studying activity during task performance, cognitive analysis should be combined with functional analysis. Functional analysis becomes possible when it is based on the developed SSAT models of activity self-regulation where each function block can be considered as a stage of information processing with a particular purpose in activity regulation.

These stages have a nonlinear loop-structured organization with multiple feedforward and feedback interconnections. Usually, functional analysis utilizes experimental procedures in combination with observation and subjective judgment of experimental events expressed by subjects involved in the experiment.

6.3 Description and Classification of Cognitive and Behavioral Actions

In morphological analysis of activity during task performance, the major units of analysis are cognitive and behavioral actions. Therefore in this chapter, we describe in an abbreviate manner principles of action classification and description. From the activity theory standpoint, the task is a logical organized system of cognitive and motor actions. So their standardized description is of fundamental importance for design. The duration of actions is also important because activity is a process. The term *action* in activity theory is understood as an element of activity and its main building block. An action can be defined as a discrete element of activity that is directed to achieve a conscious goal of an action. Actions can be further divided into unconscious operations. Achievement of a goal of an action and assessment of its result are the end points of an action that separate one action from a following action.

From the activity standpoint, cognition is not just a system of cognitive processes. It also is a system of cognitive actions and operations. A standardized description of cognitive and behavioral actions is necessary for the description of cognitive components of an activity structure. Motor and cognitive actions are tightly interconnected and motor actions include cognitive components.

Cognitive actions have a certain analogy with motor actions according to a number of features. They are goal directed, have a beginning and an end, function according to the principle of self-regulation, and so on. Motor actions presuppose existence of material objects with which a subject interacts. Cognitive actions transform not material objects but information. More precisely, cognitive actions manipulate not with material objects but with operative units of information (OUI) or operative units of activity. Operative units of activity (image, concept, statements, etc.) are the symbolic entities that are used by cognitive actions during task performance. These units of information perform functions that are similar to the ones material objects have for motor actions. Such internalized operational unit of cognitive actions should be regarded as an internal mental tool of activity that is required for the reflection of a situation, constructing idealized objects and/or mental models of a situation, etc. According to Rubinshtein and Leont'ev, meaning is a result of mental actions and operations. Development of meaning calls for integration of various psychological processes with the leading role of thinking. Meaning and signs should be treated as psychological tools of mental actions (Bedny and Karwowski, 2004). Meaning is a product of actions that, in turn, become tools of cognitive actions. Actions may be formulated in terms of an object of an action, tools, a goal of an action, and a subject of an action.

Cognitive actions sometimes have a very short duration and it is often not easy to extract mental operations out of the content of cognitive actions. Therefore, in our further discussion, we are offering a standardized psychological description of holistic cognitive actions. The conscious level of activity self-regulation involves deliberate development of goals and planning actions. Highly automated actions entail goals that are involuntarily triggered by stimuli, which in turn guide subsequent cognitive and behavioral operations. A stage of deliberate development of a goal or its involuntarily triggering is a starting point of cognitive or behavioral actions. Evaluation of the action's result based on its goal is the end point of the action. This allows to distinguish between direct connection actions and transformational cognitive actions. Direct connection mental actions unfold without distinctly differentiated steps and require less attention than transformational cognitive actions. These actions are less consciously directed and experienced subjectively as instantaneous. For example, recognition of a familiar object is an example of such actions. Classification of cognitive actions should always be complemented by analyzing their duration. The duration of cognitive actions can be obtained by using some experimental procedures. We developed such procedures that are described in our works (see, e.g., one such method, Bedny et al., 2011, 2014). Mental actions can be classified based on the dominant cognitive process and on their ultimate purpose as follows (Bedny and Karwowski, 2007; Bedny and Meister, 1997; Zarakovsky, 2004). We describe only one group of such actions as an example.

> *Direct connection actions*—unfold without distinctly differentiated steps and require a low level of attention. They can be further distinguished as follows:
>
> *Sensory actions*—detection of noise or decision about a signal at a threshold level; obtaining information about distinct features of objects such as color, shape, and sound.
>
> *Simultaneous perceptual actions*—identification of clearly distinguished stimuli well known to an operator that only requires immediate recognition and perception of qualities of objects or events (recognition of a familiar picture).
>
> *Mnemonic (memory) actions*—memorization of units of information, recollection of names and events, etc. Direct connection mnemonic actions include involuntary memorization without significant mental efforts.
>
> *Imaginative actions*—manipulation of images based on perceptual processes and simple memory operations (mentally rotating a visual image of an object from one position to another according to a specific goal).
>
> *Decision-making actions at a sensory-perceptual level*—operating with sensory-perceptual data like decision making that requires selecting from at least two alternatives (detecting of a signal and deciding to which category it belongs out of several possible categories).

There are other two groups of cognitive actions. They are (1) mental transformational actions (deliberate examination and analysis of a stimulus [perception of an unfamiliar object in a dimly lit environment], exploration of a situation based on thinking mechanisms, etc.) and (2) higher-order transformational actions (which include a complex combination of thinking and mnemonic actions or creative actions). We do not discuss these actions in this book.

The detail description of such actions is presented in Bedny and Karwowski (2007), Bedny et al. (2014). It is necessary to distinguish between two ways of describing cognitive and behavioral actions. In the previous example of action descriptions, we utilize a standardized psychological terminology. If we have time of performance of such actions, therefore, we have all necessary information about performed subject cognitive actions. Such action description is performed by using psychological units of analysis.

Another way of describing cognitive actions involves utilizing technological terms or terms that describe some task elements associated with a considered action (typical elements of a task). This is a method of describing cognitive actions by utilizing technological units of analysis. Taking a reading from a pointer or a digital display is an example of perceptual actions that is described based on technological principles. Depending on the distance of observation, illumination, and constructive features of a display, the content of mental operations and the time of action performance can vary. Based on such description as *taking a reading from a pointer on a display*, we do not know exactly what action is performed by a subject because conditions of reading can vary. A combination of psychological and technological methods of action description and knowledge of their time performance is a powerful method of task structure description.

We developed a method of extracting cognitive actions based on eye movement data. Existing data demonstrate dependence of eye movement strategies on the features of the interface. Observations of natural behavior have demonstrated the highly task-specific nature of eye fixation patterns (Henderson, 1993). Activity theory researchers demonstrated that eye movement is an indicator not only of perceptual but also of higher cognitive functions. Mental activity involves transformation of images, searching information in memory, logic operations, and so on. Based on the analysis of eye movement, it is then necessary to determine the content of perceptual, cognitive, mnemonic, thinking, and other actions and operations (Zinchenko et al., 1973).

The thinking process is involved in problem solving and is associated with two types of eye movements. The first type is external, eye movements with a relatively high amplitude. These motor eye movements and their associated sensory components are integrated into perceptual visual actions. Formation of perceptual images of the situation is facilitated by these visual perceptual actions. In the second stage, at the time of fixation, vicarious actions are accompanied by a small amplitude of eye micro-movements.

This kind of eye movement is involved not in the perception of information but in the mental transformation of the situation needed to solve a task at hand. This is a system of vicarious actions that are involved in transforming an image of a situation. Such vicarious actions are components of the thought process. The previous data explain that during *blind fixations*, a subject performs mental operations that can be highly automated and unconscious. A subject looks at the stimulus and does not see it. Understanding the nature of these micromovements of the eyes or vicarious actions is important for correct interpretation of the eye movement record (Kamishov, 1968).

Traditional methods of eye movement interpretation use a cumulative scan path length (in pixels), cumulative dwell time or average fixation time, general number of fixations, or number of saccades. Such method is not adequate for the analysis of cognitive actions. SSAT suggested a division of tasks into relatively independent fragments. Usually each fragment traces eye movements between two clicks.

In our proposed approach, it is necessary to determine the duration of perceptual actions. If the duration of fixation exceeds the duration of perception, additional time for fixation is attributed to the more complex cognitive processes. Therefore, it is important to properly determine the duration of visual perceptual action. The duration of mental actions that begin after completion of a perceptual action is determined by the following formula:

$$T_{ment} = T_{fix} - T_{per}$$

where

T_{ment} is the duration of higher mental actions (mnemonic, thinking, decision-making actions)

T_{fix} is the duration of fixation

T_{per} is the duration of perceptual action

The content of T_{ment} is determined based on qualitative analysis of activity during this time period. We identify what information was known to a subject at the time, if he or she was aware of the course of events that preceded the fixation period, what type of cognitive and behavioral actions were performed before eye fixation took place, what actions should be performed after receiving information, and what type of cognitive and behavioral actions were really performed by a subject after fixation was completed. The ability of a subject to forecast future events in a considered time period is also important for determining the content of eye fixation time. It is important for such analysis to understand the logic of a task as a whole and the logic of its performance during a particular step in task performance. Based on such data, the action classification table can be developed (Bedny et al., 2008).

Let us consider briefly the motor action description. Motor actions can be described as a combination of standardized motions that are integrated by a single action's goal. This makes it possible to use the MTM-1 system for standardized descriptions of motor actions. According to SSAT, the MTM-1 system utilizes the psychological unit of analysis of motions. Thanks to such method, behavioral actions that are described by utilizing technological units of analysis (typical elements of a task) can be described using psychological units of analysis (typical elements of activity). Therefore at the first stage, we apply a traditional method of motor action descriptions by using technological units of analysis and then transfer them into a psychological unit of analysis. Psychological units of analysis describe elements of activity in a standardized manner that allows for unified and unambiguous interpretation of what a performer does. We want to stress that the MTM-1 system does not use such terms as motor actions and psychological or technological unit of analysis. Basic units of analysis in MTM-1 system are motions that are described in a standardized manner.

Let us consider a simple example. Suppose an operator performs the motor action *release switch and move hand to red button and press it*. This is a description of an action in technological terms. We cannot precisely understand what the worker did. Only the specialist who observed the worker's performance can understand this description. This is explained by the fact that for action description, we utilize a common language and describe a considered element of task in technological terms. However, we can also describe the same motor action by using the MTM-1 system. This action can be described in the following way: *RL1 + R26B + G5 + AP2*. Everybody who knows the MTM-1 system can understand this action. According to SSAT, the MTM-1 system utilizes psychological units of analysis for motion description, and the previously considered motor action is also described using psychological units of analysis. Moreover, based on this description of motor action, we can determine the duration of such motor action. In our example, *RL1* means normal release of switch by opening fingers; *R26B* means reach to single object (button) in location, which may vary slightly from cycle to cycle when the distance of the movement is 26 sm; *G5* is motion *grasp*, which is a simple contact with the object (button) and such performance of motion does not require time for performance (this motion is used for standardized description only); *AP2* means apply pressure, which does not require a significant effort and usually is performed after *G5*. A description of a considered motor action in technological and then in psychological terms gives us a clear picture about the performed motor action.

Thus, utilization of technological and psychological units of analysis is a powerful method of action description in a standardized manner. We provided the method of action description in a very abbreviate manner. This method will be demonstrated further using various practical examples. The next stage involves the description of the logic of action organization. This can be performed using algorithmic task description. We describe this method in the next section.

7

Morphological Analysis of Work Activity

7.1 Algorithmic Description of Activity and Task Analysis

The purpose of algorithmic analysis of activity is the subdivision of activity into qualitatively distinct psychological units and the determination of their logical organization. Algorithmic description of task performance gives an opportunity to describe very flexible human behavior. It is an important stage of morphological analysis of work activity. Logically, organized elements of activity are called members of algorithm. Usually, members of algorithm include one to four interdependent homogeneous actions (only motor, only perceptual, or only decision-making actions), which are integrated by a higher-order goal into a holistic system. Subjectively, a member of such algorithm is perceived by a subject as a component of his or her activity (mode), which has a logical completeness. Usually, the amount of actions in one member of an algorithm is restricted by the capacity of short-term memory. According to systemic-structural activity theory (SSAT) rules, motor actions can be performed simultaneously and cognitive actions should be performed sequentially. Cognitive actions can be combined with motor actions. This depends on the level of attention concentration during the performance of different actions and their elements (Bedny and Karwowski, 2007). Members of algorithms include operators and logical conditions. Operators represent actions that transform objects, energy, and information. For example, we can describe operators that are implicated in receiving information, analysis of a situation and its comprehension, shifting of gears, levers, etc. Logical conditions are members of algorithm that include a decision-making process and determine the logic of selecting the next operator. Actions as units of analysis constitute one of the most important distinctive features of a human algorithm, from different kinds of flow charts widely used to represent human performance.

Each member of the algorithm is designated by a special symbol. For example, operators can be designated by the symbol O and logical conditions by the symbol l. All operators that are involved in the reception of information are categorized as afferent operators and are designated with superscript α, as in O^α. If an operator is involved in extracting information from long-term memory, the symbol μ is used, as in O^μ. The symbol $O^{\mu w}$ is associated with keeping information in working memory, and the symbol O^ε is associated with the executive components of activity, such as the movement of a gear. Operators with the symbol O^ε are depicting efferent operators. From the previous description, one can see that, for example, O^ε cannot include any cognitive actions. Similarly, O^α can include only perceptual actions. If an operator is involved in extracting information from long-term memory (only mnemonic actions), the symbol μ is used, as in O^μ. Sometimes after receiving the information (performance of O^α), it is impossible to use it immediately. A worker keeps this information in memory, and therefore, the symbol $O^{\mu w}$ is used. This symbol describes elements of activity involved in keeping the information in working memory.

Mental actions can be complex and require a combination of several cognitive processes. For example, decision-making can be combined with memory functions. In such situation, we use a symbolic description of logical conditions such as l^μ, where μ designates the memory functions that complicate decision-making.

Often, thinking actions can be exercised based on externally provided information (e.g., mental manipulation of externally presented data) or made with reliance on the information held by or retrieved from memory (manipulation of data in memory), or thinking actions require keeping intermittent data in memory. In this case, we describe thinking operators as $O^{\alpha th}$ or $O^{\mu th}$ (α means that thinking operator is performed based on external, e.g., visual, information, and μ means that such operator requires complicate manipulation in memory). Such symbolic description described previously is a critical factor for performance of the considered members of algorithm.

The symbols l for a logical condition has to include an associated arrow with a number on top that corresponds to the number associated with its logical condition. For example, logical condition l_1 is associated with a number on top of arrow $\overset{1}{\uparrow}$. An arrow with the same number but reversed has to be presented in front of a corresponding member of the algorithm to which the arrow refers, $\overset{1}{\downarrow}$. Thus, the syntax of the system is based on a semantic denotation of a system of arrows and superscripted numbers. An upward pointing arrow of the logical state of the simple logical condition l, where $l = 1$, requires skipping all succeeding members of algorithm until the next appearance of the superscripted number with a downward arrow ($\overset{1}{\downarrow}$). So, the operation with the downward arrow with the same superscripted number in front of it is the next to be executed.

Complex logical condition has multiple outputs. For example, $L_1 \overset{1(1-6)}{\uparrow}$ indicates that this is the first complicated logical condition that has six possible outputs: $\overset{1(1)}{\uparrow}, \overset{1(2)}{\uparrow}, \overset{1(3)}{\uparrow} \ldots \overset{1(6)}{\uparrow}$. Arrows after logical conditions ($\overset{(1)}{\uparrow}$) demonstrate transition from one member of an algorithm to another ($\overset{1}{\uparrow} \overset{1}{\downarrow}$). This means that the logical condition according to the output addressed from the upward to the downward arrow is associated with the particular member of the algorithm. Therefore, human algorithm can be deterministic as well as probabilistic (Bedny, 1987). Deterministic algorithm has logical conditions with only two outputs with values 0 and 1. Probabilistic algorithm has more than two outputs with various probabilities or two outputs that can have any value from 0 to 1. In some cases, logical conditions can be a combination of simple ones. These simple logical conditions are connected through *and, or, if-then,* etc., rules. Analysis of the principles of the algorithmic description of activity reveals two groups of actions: cognitive and motor actions. Cognitive actions are described based on the analysis of cognitive processes involved in their regulation. The second group of actions is classified and described based on the analysis of motor action regulation and data that are obtained in the MTM-1 system. The methods-time measurement system has been presented in an abbreviate manner in Barnes (1980). Later we will analyze this system from SSAT perspectives. Here we only would like to underline the fact that this system is used in the modern industry. For example, Brecher et al. (2013) demonstrated the possibility of its use for the analysis of the interaction of a human operator with the robotic system. According to these authors, MTM-1 is a recognized system worldwide that is used for the description of human motions. It is interesting to note that these authors apply this system to describe a robot's movements. Moreover in their studies, they utilize the principles of motion classification without considering the performance time of a robot's motions. It is necessary to point out that when a robot's movements should be coordinated with motor actions of a human operator, MTM-1 time standards can be useful for the analysis of temporal parameters of a robotic system. In SSAT, the MTM-1 system is used in a totally different way where any motor action includes several motions that are integrated by a conscious goal of action.

Let us consider a hypothetical example. A worker needs to check a digital indicator. If the even number is lit, then the worker should turn the two-position switch up. If the uneven number is lit, he should turn the two-position switch down. If an indicator shows zero, then a worker presses the red button. Suppose that the appearance of even and uneven numbers has the same probability, $P = 0.4$, and probability of zero is $P = 0.2$. Therefore, this is a probabilistic algorithm. In deterministic algorithm, logical conditions have only two outputs with equal probabilities. Table 7.1 describes this algorithm.

The algorithm should be read from top to bottom. A symbolic description of a member of the algorithm in a standardized form in the first column on the left is an example of psychological units of analysis because they have clearly defined psychological characteristics. A verbal description of a member of the algorithm in the right column is an example of technological units of analysis. These units of analysis describe elements of work that do not possess a clearly defined psychological description. Each member of the algorithm can be described in terms of actions or operations that are smaller

TABLE 7.1

Description of the Algorithm *check a digital indicator*

Members of Algorithm	Description of Members of Algorithm
O_1^α	Take reading from a digital indicator.
$l_1 \overset{1(1-3)}{\uparrow}$	If the even number is lit, then perform O_1^ε. If the odd number is lit, then perform O_2^ε. If the number is zero, go to O_3^ε.
$\overset{1(1)}{\downarrow} O_1^\varepsilon$	Turn two-position switch up.
$\overset{1(2)}{\downarrow} O_2^\varepsilon$	Turn two-position switch down.
$\overset{1(3)}{\downarrow} O_3^\varepsilon$	Press red button.

units of analysis. At this stage of analysis, we can transfer technological units into psychological units for a more detailed description of a task. It is also an important step in describing the time structure of activity. For example, the suggested system of description and classification of the cognitive action *take reading from digital indicator*, when the duration of such reading is 0.3–0.4 s, can be considered as a technological unit of analysis and can be described as *a simultaneous perceptual action*. This description has clearly defined psychological characteristics. Such combination of technological and psychological units of analysis gives a clear understanding of what is involved in task performance. The technological description of the action *turn two-position switch down*, if the distance to and specificity of the switch are known, can be decomposed into three standardized motions: Reach (*R30A*), Grasp (*G1A*), and Move (*M2.5A*). In this example, *R30A* means reach object in a fixed location at a distance of 30 cm, *G1A* means easily grasp, and *M2.5A* depicts moving an object 2.5 cm against a stop with little effort. All these motions require a low level of attention concentration. This is also an example of psychological units of analysis because these are standardized and clearly defined activity elements. Only after transforming technological units of analysis into psychological units and determining their duration, it is possible to describe the time structure of activity and evaluate the complexity of activity during task performance. Experts in task analysis do not use such concepts as technological and psychological units of analysis that lead to an ambiguous description of activity. As the first example, let us consider the descriptions of some micromotions presented in the simo chart (Barnes, 1980, p. 152). Left-hand movements: (1) move to clamping lever and grasp knob; (2) move lever to extreme left; (3) hold lever in this position, etc. The right hand at the same time performs the following actions: (1) insert piece in fixture and release it; (2) reach a forming lever and grasp knob; (3) form first end of link; (4) return forming lever and release it, etc. The second example describes lathe work in manufacturing. Usually, production operations in this field begin with the following movements (Gal'sev, 1973): (1) move right arm to the bin with parts; (2) grasp part; (3) move part close to chuck; (4) orient the part into a correct position, etc. Such a description of movement is absolutely correct when the expert is involved in the direct observation of performed work but this description is not sufficient in cases when design issues are resolved and the specialist is working with documentation and has limited ability to carry out observation of human activity and necessary measurements. Moreover, the specialist who works with documentation might not be the one who was involved in the observation of real performance and the process of development of documentation. The specialist has received documentation that has been developed by others. When reading this documentation, the specialist must clearly understand the cognitive and behavioral actions performed by the operator on the basis of the documentation received. Usage of technological units of analysis in combination with psychological units of analysis is necessary in such cases.

We will consider such scenario using tasks with predominantly motor activity. Vast experience has been accumulated in analyzing manufacturing operations. Not only traditional methods but also the MTM-1 system can be used to describe motor activity. According to SSAT, this system utilizes psychological units of analysis. We have to remember that the MTM-1 system does not use such concepts as technological and psychological units of analysis. Due to lack of a clear distinction between the considered units of analysis of activity, their effective combined application has not been achieved. When such methods as experiment, observation, and timing dominate, these shortcomings to some extent can be compensated by empirical data, but for design where analytical methods usually are the main methods of study, such disadvantages are significant. When the role of cognitive processes increases the combined use of technological and psychological units of analysis (standardized elements of activity), it is especially important. It should also be taken into account that in SSAT, units of analysis have a hierarchical organization. The following units of analysis are used for the algorithmic description of activity: member of the algorithm (it consists of one or more of the same type of cognitive or motor actions), action, and mental or motor operation (movement).

As an example, let us examine the algorithmic description of the lathe production operation. Here, members of algorithm are presented as sequential units that have a hierarchical organization. For such production operations, cognitive activity is not separated from motor activity (see Table 7.2). So, in our further example, we describe only motor activity, but specificity of cognitive regulation of motor actions and movements is not ignored. Each movement is described by special microelements of the MTM-1 system. They can be classified based on their corresponding level of concentration of attention, which means that the complexity of cognitive regulation of motor actions and movements or motions is taken into account.

In Table 7.2, the first column on the left depicts all members of algorithm as psychological units of analysis. The next two columns on the right depict

TABLE 7.2

Turning Part in Chuck[a]

Members of Algorithm	Description of Members of Algorithm	Description of Actions	Description of Motor Actions and Operations (Motions)
Psychological Units of Analysis (High-Level Description)	Technological Units of Analysis High-Level Description (Column 2); Low-Level Description (Column 3)		Psychological Units of Analysis (Low-Level Description)
O_1^ε	Put part into an air operating chuck	1. Take part	Motor action 1 (a) Move right arm to the bin with parts ($R35A$ Reach, Case A) (b) Grasp part by fingers ($G1A$ easily grasp)
		2. Install part in an air operating chuck	Motor action 2 (a) Move the part closely to chuck ($M30B$) (b) Orient the part in correct position ($P1SE$) (c) Move the part to exact position ($M10C$ Move, Case C)
O_2^ε	Fix part in an air operating chuck	1. Take part	Motor action 1 (a) Move left arm to the lever's handle of an air operating chuck ($R30A$ Reach, Case A) (b) Grasp handle ($G1A$ easily grasp) (c) Turn handle ($T90M$) (d) Release handle ($RL1$)
O_3^ε	Start lathe	1. Grasp lever's handle	Motor action 1 (a) Move left arm to the start lever ($R25A$ Reach, Case A) (b) Grasp handle ($G1A$ easily grasp)
		2. Move lever into start position	Motor action 2 (a) Move lever ($M10A$ Move, Case A) (b) Release handle ($RL1$ Normal release)

(Continued)

TABLE 7.2 (*Continued*)

Turning Part in Chuck[a]

Members of Algorithm	Description of Members of Algorithm	Description of Actions	Description of Motor Actions and Operations (Motions)
Psychological Units of Analysis (High-Level Description)	Technological Units of Analysis High-Level Description (Column 2); Low-Level Description (Column 3)		Psychological Units of Analysis (Low-Level Description)
O_4^ε	Move cutting tool to start position	1. Grasp handle of cross slide and simultaneously perform	Motor action 1 (a) Move left arm to the handle of cross slide (*R15A*) (b) Grasp handle of cross slide (*G1A*)
		2. Grasp handle of longitudinal slide and perform by two arms simultaneously	Motor action 2 (a) Move right arm to the handle of longitudinal slide (*R15A*) (b) Grasp handle of longitudinal slide (*G1A*)
		3. Move cross slide by rotating handle in approximate position	Motor action 3 (a) Perform several rotations (*C20* turn handle, minimum effort)
		4. Move longitudinal slide by rotating handle in approximate position	(b) Perform several rotations (*C20* turn handle, minimum effort)
O_5^ε	Set instrument to the required starting position	1. Move cross slide by moving handle (install cutting tool in precisely required position)	Motor action 1 (a) Move handle ($\approx 1/3C20$) (b) Release handle (*RL1*)
O_6^ε		2. Move longitudinal slide by moving handle (install tool in precisely required position)	Motor action 2 (a) Move handle ($\approx 1/3C30$) (b) Release handle (RL1)
O_7^ε	Turn on longitudinal feed.	1. Grasp handle by left hand	Motor action 1 (a) Move left arm to the handle (*R30A* Reach, Case A) (b) Grasp handle (*G1A*)
		2. Move handle	Motor action 2 (a) Move handle (*M10A*) (b) Release handle (*RL1*)

[a] In the right column, numbers designate distance in centimeters (distances were selected as an illustration).

all members of algorithm using technological units of analysis. The second column gives a general description of the member of the algorithm and the third column describes individual actions involved in performing this member. Members of algorithm are described with various levels of decomposition. The far right column describes motor actions utilizing psychological units of analysis where motor actions decomposed into motions and is presented in a symbolic form using the MTM-1 system. Only the comparison of such description of the units of analysis facilitates a clear understanding of what is involved in the operator's activity.

The MTM-1 system does not use hierarchically organized units of analysis, the concept of action and members of algorithm, and some other important concepts in its description of human work. In SSAT, each member of the algorithm is described as hierarchically organized elements such as member of algorithm–actions–operations. For a description of these elements, we use technological and psychological units of analysis. We describe algorithmically the performance of the whole production operation but consider in a detailed manner only installation of a part into an air operating chuck (only two members of algorithm) (see Table 7.2).

Table 7.2 demonstrates that each member of the algorithm (motor actions) was performed sequentially. This means that production operation is performed utilizing only one best method of performance. Such production operations exist in mass production. In this example, all members of algorithm have linear organization and each member of the algorithm includes hierarchically organized elements. There are no logically organized elements of activity. For example, there is no if-then rule that would determine the sequence of activity elements.

This example of production operation is useful for discussing some aspects of algorithmic description of work activity. Table 7.2 includes hierarchically organized units of analysis. The left column utilizes standardized psychological units of analysis (in symbolic form). This column describes activity with high-level description. The second column describes activity by using technological units of analysis by using high-level description. It includes verbal description of the whole member of the algorithm and then decomposes it into separate actions (technological units of analysis with low-level description). For describing the member of the algorithm, professionals used common language. This description is ambiguous and can be interpreted in different ways by different specialists especially when they do not have the ability to directly observe the work of the performer (they only read documentation). The fourth column on the right describes human actions in terms of standardized motions—psychological units of analysis (low-level description). Utilization of MTM-1 allows a clear understanding of the substance performed by worker actions at the lower level of hierarchical description.

From such description, one can see that each member of the algorithm consists of actions and actions in turn consist of movements or operations. Each member of the algorithm has one goal that integrates several actions into the hierarchical subsystem of activity. The goal of a member of the algorithm is higher in hierarchy in relation to the goal of individual actions. Members of algorithm have a certain logical organization. In our example, they simply follow one another in a certain order. Usually, the amount of actions that can be included in one member of an algorithm is restricted by the capacity of the working memory. Subjectively, a member of an algorithm is perceived as a completed stage of activity. A member of an algorithm also has relatively constant material components such as objects and tolls of activity.

For instance, O_1^ε (psychological unit of analysis in symbolic form) depicts a motor component of activity that consists of two actions: *put part into an air operating chuck* is the description of the member of the algorithm, and *take part* and *install part in an air operating chuck* are the descriptions of actions (technological units of analysis).

These members of the algorithm are described further as psychological units of the algorithm. Motor action 1: move right arm to the bin with parts and grasp it (*R35A* + *G1A*), where *R35A* depicts Reach, Case A and G1A depicts grasp part by fingers (easy grasp); motor action 2: move the part close to chuck, orient the part in correct position, and move the part to exact position (*M30B* + *P1SE* + *M10C*), where *M30B* describes moving an object 30 cm to an approximate location; *P1SE* depicts orienting a symmetric object into a certain position, no pressure required, easy to handle; *M10C* describes moving an object into an exact position. Such combination of units of analysis gives a precise understanding of what this member of algorithm involves.

After completion of the first member of the algorithm O_1^ε performed by the right hand, a worker switches his or her attention to the execution of the second member of the algorithm O_2^ε performed by the left hand. When a worker completes performing the second member of the algorithm, subjective perception of completion of a certain part of the task coincides with technological completion of this part of the task, but after completion of the first member of the algorithm, technological completeness of this portion of the task is not observed. If a worker interrupts his or her actions after executing O_1^ε and releases the part, it will fall on the lathe bed, which would not be observed after the second member of the algorithm O_2^ε is performed, when the part is fixed in the chuck. Thus, technological completeness and psychological completeness of the task do not always coincide. Separation of a technologically completed portion of a production operation into two members of algorithm is also explained by the preferable strategy of attention reorientation. A worker concentrates on actions that involve the right hand and then on using the left hand. Such a strategy reduces the workload on working memory. At the first step, information that is related to execution of the first member of the algorithm is actualized in memory and then this information is eliminated from the working memory and information related to execution of the second member of the algorithm is actualized in memory. Thus, when dividing the task into members of an algorithm, one should take into account interaction of technological and psychological factors and use technological and psychological units of analysis. Actions that are included in a member of an algorithm as its components also have their goal. However, it is hierarchically a lower level goal. When a worker achieves the goal of a separate action, in most cases, it does not provide completeness of a part of a task. For example, execution of the first action, *take part*, provides attainment of a goal of an action but a performed action is not a completed part of a production operation. Only when a worker installs a part into an air operating chuck, the worker can formulate a next goal or stage of work, *fix part in an air operating chuck*. At this stage of analysis, a specialist has to take into account completeness of a step in task performance and compare it with subjective completeness of activity. The other factor is capacity of working memory (approximately 1–4 actions can be included in one member of an algorithm). Complexity of actions is also an important factor. The more complex the actions are, the less number of actions can be included in one member of an algorithm. The notion of *goal* is important in allocating such units of activity analysis as a member of an algorithm and an action.

A hierarchical description of each member of an algorithm allows comparing actions and their motions with utilized tools and equipment. Based on analysis of performed actions and operations, we can improve the design of tools, equipment, or software. If actions and their motions require a high level of concentration of attention, design features of equipment or software should be changed. For example, during performance of O_1^ε, the second action *install part in an air operating chuck* includes a motion (c) move a part to an exact position (*MC*). We can introduce a special limiter, and movement will be performed against a stop, changing this motion to *move part against stop (MA)*, which simplifies this movement and the concentration of attention shifts from the third category of complexity to the first one. Hence, hierarchical description of activity helps us to evaluate more precisely a structure of production operation and task performance in general. Moreover, even when describing motor activity, we still pay attention to the cognitive aspects of motor action regulation.

It should be noted that in our manufacturing operation, all members of the algorithm are of similar type and belong to motor activity, which makes it difficult to break the same type of activity into separate hierarchically organized units. This aspect of algorithmic description needs a detailed discussion because extraction of separate members of an algorithm that belong to the same category of classification is the most complex part of algorithmic description. For example, several members of an algorithm that describe only motor or only perceptual types of activities should be extracted, which makes it difficult to find a clear border between members of the algorithm. In the previous example, all members of the algorithm describe motor activity, which made it difficult to find a border between members of the algorithm O_1^ε and O_2^ε. When separating this part of the manufacturing operation into two members of the algorithm, we took into account such factors as awareness of goals, at least for a short period of time, and principle of their organization, strategies of shifting attention, capacity of working memory, feeling of completeness of the same portion of the task, and objective or technological completeness of a portion of production operation. Finishing work with the right hand ends a certain part of the operation and a worker shifts his attention to another part of the operation associated with working with another hand (new goal).

In our example, a worker utilizes a preferable strategy, which clearly demonstrates that in normal conditions of task performance, we can extract members of algorithm O_1^ε and O_2^ε and divide them into units depicted in Table 7.2. Any description of real activity is always performed with some approximation. If a worker changes a strategy of performance, a structure of activity also changes. Sometimes several strategies of activity have to be described such as task performance in a dangerous situation or under stress. In some cases, identification of the most preferable strategies of performance is possible only after functional analysis of activity during task performance when activity is considered as a self-regulated system and the main unit of analysis is a function block.

After algorithmic description, the next stage of analysis is description of the time structure. We intentionally focused our efforts on description of production operations that included only motor actions that allowed us to consider a rather complicate example of dividing a production operation or task into separate members of algorithm of the same type and to demonstrate the method of description of motor actions by using elements of task or technological units and standardized elements of activity or psychological units. The considered example also demonstrates the principle of hierarchical description of members of an algorithm. Detailed analysis of manufacturing operations is particularly relevant when actions of the same type are repeated during a long period of time and automation of a technological process is not cost-effective. Further, we will also consider in a detailed manner the tasks that are performed in automated and semiautomated systems and include a combination of motor and cognitive actions.

7.2 Algorithmic Description of Tasks in Automated and Semiautomated Systems

In this chapter, we will consider very flexible tasks performed by an operator in semiautomatic and automatic systems, where cognitive and motor components of activity prevail. Manual control is important even in automatic systems. In a malfunctioning situation, an operator has to transfer from automatic to manual control and intervene in controlling automatically functioning system. An operator is facing emergency situations very rarely, but such situations are critical. In abnormal situations, an operator usually depends on presented information to perform a logically organized sequence of actions. In this section, we demonstrate an algorithmic description of such an activity by utilizing laboratory experiment.

For conducting our laboratory study, we developed a special control board. On one side of this control board, there was a panel for the participant, and on the other side a panel for the experimenter. The experimenter panel allowed the experimenter to set the program, which would present the participant with different versions of the task. The duration of the performance of various versions of the task was registered automatically through timers (exact to the 0.01 s). In addition to the main experimental panel, the participant was also instructed to use another panel, which he or she believed to be part of the experiment, but which simply served to increase task complexity and did not register any measurements. In one series of experiments, the participant worked only on the main panel, in the other series the participant worked simultaneously on two of the panels. Each experimental trial lasted 1 h and 30 min. Subjects performed a number of tasks in this period of time. They did not know when exactly each task can start. During breaks between tasks performed on the panels, the participants perform supplementary activities, such as coping texts, performing simple arithmetic tasks, or performing attention tests. A sound signal indicates that the participants should cease the supplementary activities and switch to major task with the panels. The work with the panels was presented to the operator as *emergency conditions*. The participants were told that they needed to perform quickly and reliably. As a result, the participants could not completely concentrate on the work with the main panel. This manipulation of introducing a second panel and supplementary activities resulted in a more naturalistic simulation of work conditions required for evaluation of work pace. The participants were shown various versions of task performance. After this, the participants were trained to work on the panels. The operator panel has the following instruments and controls (see Figure 7.1). Figure 7.1 shows the numbered 1–5 instruments and numbered 6–10 controls.

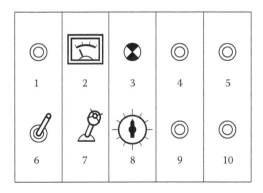

FIGURE 7.1
The operator panel's instruments and controls.

The panel of the participant had (1) a signaling bulb that lit up either number 1 or 2; (2) a pointer indicator; (3) a digital indicator, which showed 1 of 10 possible numbers; (4) a signaling bulb that lit up as green; and (5) a signaling bulb that lit up as red. All the indicators were located on a vertical panel, which was slanted according to ergonomic standards. The controls were located under these indicators on another horizontal panel. Figure 7.1 schematically depicts the utilized devise. All the indicators and controls were situated in order from left to right and had linear organization. The controls were as follows: (6) a four-position switch; (7) a hinged lever, which could be moved to four perpendicular positions (up, down, left, and right). This lever has a button on the top of the lever's handle, which could be pressed with the thumb. Only following the depression of the button could the hinged lever (7) be moved. The next control was (8) a 10-position switch. Furthermore, the panel had (9) a red button and (10) a green button. In sum, there were five indicators and five controls. Each organ of control was located under the corresponding instrument. Consequently, movement of the eyes from instrument to instrument and the movement of the right hand, which was used to manipulate the controls, had linear organization. In order to make the task more complicated, green button 10 had been installed under red bulb 5. Red button 9 has been installed under green bulb (4). Therefore, color was used as an interfering factor. In real work conditions, when an operator performs a variety of tasks, this kind of interfering becomes critical. The work on the panel was an imitation of a logically organized system of mental and motor actions, the completion of which was done under conditions of constrained time. The incorrect sequence of actions or exceeding required time of performance of a task was followed by an unpleasant sound. After the completion of a task on the panel, the participant returned to the interrupted task supplementary work. The system of signals presented to the participant using 5 indicators allowed to present 110 different versions of an algorithm.

The work on the main panels had the following logic. The operator receives information from instrument 1 at the main panel that can demonstrate (1) or (2). If this instrument presents (1), the subject is to turn the switch down, if (2) is presented, the switch is to be turned up. Then he or she uses a lever that can be moved into one of four directions, depending on the information that was presented on pointer display. However, before the subject can do it, he or she depresses the button at the top of the handle with his or her thumb. After the hinged four-position lever (7) was moved in the required position, the digital indicator (3) presents a corresponding number. Depending on the presented number, the operator can turn the 10-position switch in the required position. Green or red bulbs may be illuminated. Depending on which indicator is lit, the subject presses the green or red button. After pressing the corresponding button, the task on the main panel was completed. We can describe a more general version of task performance. In this version of the task, all instruments and controls are used during task performance. In other versions of the task, some instruments are not activated and they, together with associated controls, were not involved in the performance of this particular version of the task.

Seven male subjects were involved in an experimental study. A warning signal that informs about emergency conditions and necessity to work on main control panels had been presented to subjects in a random fashion. In this situation, subjects have to drop the supplementary activities and switch to major task with the panels. During experimental trials, performance time was registered for only three preselected versions of the task on the main panel. These versions of the task had been combined with other versions of the task. Time performance of the other versions of the task was not measured. All versions of the task had been presented in random fashion. Subjects did not know that only the time performance of three selected versions of the task were recorded. All versions of the task (selected as basic and supplementary) were equally important subjectively for the subjects.

Thus, the task performed on main control board has 110 versions of realization. All versions can be with the same approximation divided into three groups of complexity. In the first, more complicated version of the task, all controls were used. In average group complexity, one of the controls was not used (hinged lever 7). In the simple version of the task, two controls were not used (additionally, the digital indicator 3 is not used).

Flexibility of task performance can be explained by the variability in the signals and thus in many ways performed responses to their presentation. In our studies, we adhere to the principle known in mass production. Uniqueness in shape and size of individual parts during mass production is taken into consideration by utilizing the concept of range of tolerance. Parts that vary in size and shape in a range of tolerance are considered to be identical. Similarly, we can describe very flexible human activity in automated or semiautomated man–machine systems. Based on excepted range of tolerance, there is no need to measure execution time of each of the 110 versions of the algorithm. It is enough to measure several most representative versions of the algorithm. With this approach, all the other versions of the algorithm are considered as variations of selected versions. Their variations are considered as acceptable because they are in a range of tolerance in comparison to an adequate standard version of the algorithm. Three of the most representative versions of algorithm have been chosen for analysis. All other possible variations of an algorithm are considered to be close to one of the selected standard options according to the performance time criterion. If there are more rigorous requirements for analysis, we can select more versions of an algorithm as a standard.

In the first version of the task performance the subjects utilized all the indicators and controls. This version of the task included the following steps. After signal bulbs (1) went on, the subject was to turn on the four-position switch (6) to the required position. After checking the instruments and making sure that none of the ones were turned on, the subject moved his or her hand to the hinged four-position lever (7), grasped the handle, and pressed the button on the handle using his or her thumb. As a result, pointer of indicator (2) assumed one of the four possible positions. The subject then moved the four-position hinged lever (7) to one of four possible positions. Then, the digital indicator (3) displayed the number 5 (possible numbers that can be presented by this indicator varied from 0 to 9). A subject turned the multi-positioning switch to position 5. After that, the red bulb (5) is turned on and the subject moved his or her hand toward the green button (10) and pressed it.

In the second version of the task pointer indicator 2 has not been used and this version consisted of the following steps. After signal bulb (1) was turned on, the subject was to turn on the four-position switch (6) to the required position. After making sure that none of the instruments were turned on, the subject moved his or her hand to the four-position hinged lever (7) and in the same way pressed the button on the handle by the thumb finger. As a result, pointer of indicator (2) assumed one of the four possible positions. The subject then moved the four-position hinged lever (7) to one of four possible positions. Then, the digital indicator (3) displayed the number 0 and simultaneously the red bulb was turned on. After that, the subject pressed the green button (10).

In the third version of the task a subject did not use indicator 2 and related to it four-position lever 7, and digital indicator 3 and multi-positioning switch 8. This version of the task has the following steps. The subject switched position switch (1) after the bulb went on. The digital indicator showed the number 0 and simultaneously the green bulb turned on. In response to this signal, the subject was to press the red button. We measured time performance for only these three versions of the task, which were presented randomly between other versions.

It should be noted that we also studied one more version of the task. However this version is excluded from this consideration to simplify our discussion. One of our goals in conducting this experiment was to compare the experimental data in time performance of considered versions of the task with analytical estimation of task performance. Analytical method includes algorithmic description of task versions, calculation of probabilistic characteristics of selected version of tasks, and calculation of their time performance by using temporal characteristics for cognitive activity existing in activity theory. Duration of motor activity was evaluated by using the MTM-1 system in combination with analysis of activity strategies during task performance. Comparison of experimental and analytical methods has shown that they gave approximately the same results (Bedny and Karwowski, 2007).

This experiment as well as other studies has shown effectiveness of the suggested approach for determining time parameters of activity during task performance (Bedny, 1979). At the first step of task analysis, the task considered earlier is described algorithmically. All motor actions in this task are performed by right hand. Algorithmic description of the task begins when the right hand is in the start position. An algorithmic description of the task is presented in Table 7.3. This is a general algorithm that can depict all possible versions of task performance. Algorithmic description of the task performed on this panel adheres to the previously covered rules and principles.

TABLE 7.3

General Algorithmic Description of Task Performed on an Experimental Control Board

Members of Algorithm	Description of Member of Algorithm
O_1^α	Look at first the digital indicator.
$\overset{1}{l_1}\uparrow$	If the number 1 is lit, perform $_1O_2^\varepsilon$; if the number 2 is lit, perform $_2O_2^\varepsilon$.
$_1O_2^\varepsilon$	Move the two-position switch 6 to the right.
$\overset{1}{\downarrow}_2O_2^\varepsilon$	Move the two-position switch 6 to the left.
O_3^α	Determine whether the digital indicator 3 or the signal bulb 4 or 5 is turned on.
$L_2\overset{2(1-3)}{\uparrow}$	If neither the digital indicator 3 nor the signal bulb 4 or 5 is turned on ($L_1 = 0$), perform O_4^ε; if the digital indicator 3 presents numbers 1–9 ($L = 1$), perform L_4; if the digital indicator 3 presents the number 0 and bulb 4 or 5 is turned on, perform l_5.
$\overset{2(1)}{\downarrow}O_4^\varepsilon$	Move right arm to the four-position hinged lever 7, grasp the handle, and press the button with the thumb.
$O_5^{\alpha w}$	Wait for 3 s.
O_6^α	Determine the pointer's position on the pointer indicator 2.
$L_3\overset{3(1-4)}{\uparrow}$	If the pointer position is 1, perform $_1O_7^\varepsilon$; if 2, perform $_2O_7^\varepsilon$; if 3 perform $_3O_6^\varepsilon$; if 4, perform $_4O_7^\varepsilon$.
$\overset{3(1)}{\underset{1}{\downarrow}}O_7^\varepsilon$	Move the hinged lever 7 to the position that corresponds to the number 1.
………	……… ……… ………
$\overset{3(4)}{\downarrow}_4O_7^\varepsilon$	Move the hinged lever 7 to the position that corresponds to the number 4.
O_8^α	Determine whether the second digital indicator 3 or the signal bulb 4 or 5 is turned on.
$\overset{2(2)}{\downarrow}L_4\overset{4(1-10)}{\uparrow}$	If the digital indicator 3 displays the number 1, perform $_1O_9^\varepsilon$; if number 2, perform $_2O_9^\varepsilon$; if ……… number 9, perform $_9O_9^\varepsilon$; if the digital indicator 3 presents number 0 and bulb 4 or 5 is turned on, perform O_{10}^α.
$\overset{4(1)}{\underset{1}{\downarrow}}O_9^\varepsilon$	Turn multiposition switch 8 to position 1.
………	……… ……… ………
$\overset{4(9)}{\underset{9}{\downarrow}}O_9^\varepsilon$	Turn multiposition switch 8 to position 9.
O_{10}^α	Determine which one of the two bulbs 4 or 5 (red or green) is turned on.
$\overset{2(3)4(10)}{\downarrow}\overset{5}{\downarrow}l_5\uparrow$	If the red bulb 4 is turned on ($l_5 = 0$), perform O_{11}^ε; if the green bulb 5 is turned on, perform O_{12}^ε.
O_{11}^ε	Move the arm to the red button 9 and press it.
$\overset{5}{\downarrow}O_{12}^\varepsilon$	Move the arm to the green button 10 and press it.

In the left column, we present a symbolic description of the task. In this column, we utilize psychological units of analysis (typical elements of activity). These symbols provide a more general description of the task by using standardized units of analysis. In the right column, we present technological units of analysis. As we already know, this description of the task in common language or using technological terminology is not precise. If we compare the data in the left and right columns, we can understand better what specific type of human activity each member of the algorithm belongs. For a more detailed description of human algorithm, we need to describe human standardized action, which is used for the performance of each member of the algorithm. However, here we restrict ourselves only to the description that is presented in Table 7.3. This example clearly demonstrates the possibility of utilizing algorithmic method for analysis of flexible tasks in automated and semiautomated systems.

All the described algorithms belong to the most widely used group of algorithms known as the algorithms of transformation. These algorithms describe processes geared toward transformation of material objects or information. At the end of the chapter, we briefly dwell on algorithms of identification. This is a small group of algorithms that is nevertheless of particular interest. Algorithms of identification provide an effective tool for describing diagnostic tasks that are very important in studying certain types of tasks. Such tasks are associated with classifying objects into a particular group based on existing criteria, solving diagnostic problems, and so on. As an example, such problems are encountered in the military when it is necessary to recognize enemy armored vehicles from a long distance (Keebler et al., 2008). The process of recognition should be distinguished from identification. Recognition is a perceptual process when a subject relates a presented object to the ones he or she has been previously exposed. A subject matches the current stimuli with the images that are stored in memory. In recognition, perceptual processes are integrated with a function of memory. Identification always suggests dividing all presented stimuli into two classes: those that are identical by all features to templates stored in memory (positive identification) and those that are not identical to templates by at least one feature (negative identification). Sometimes template can be presented externally. Recognition is different from identification because it additionally includes categorization (Zinchenko, 1981). From activity perspectives, perceptual process and classification of objects into various categories can be considered as a system of perceptual, mnemonic, and thinking actions that has a logical organization. Therefore, all perceptual tasks that are involved in recognition of ambiguous objects and their classification can be described algorithmically (Bedny and Meister, 1997). These are related to diagnostic tasks. In simple situations, diagnostic tasks can be expressed as follows: if an object X has attributes A and B or D, then it belongs to class M; if it has attributes A and B and does not have D, then it belongs to class P. If a subject can clearly define and relate a perceived object to a certain category of objects or phenomenon based on a logically organized system of cognitive and behavioral actions, such process can be described by an identification algorithm.

In more complex cases, diagnostic tasks can be described using probabilistic algorithms. When solving diagnostic task problems, it is important to correctly identify positive, negative, and irrelevant features of an object or phenomenon. It is also important to clearly define what cognitive and behavioral actions a subject should perform with these features.

Algorithmic description of the task is used for the analysis of flexible human activity. It also used for development of the time structure of activity. The time structure is a new concept in the area of time study and ergonomics. Time structure description is a systemic principle of time study which was developed in SSAT. Rather than describing separate temporal characteristics (parametrical characteristics) of activity such as time of task performance, reaction time, and time of performance of separate actions, as elements of activity, it described temporal characteristics of all elements of activity as a system, which unfolds in time.

7.3 Time Structure Analysis of Activity during Task Performance

Time structure of activity is a new concept in task analysis. This is not a description of separate temporal data of activity but rather a description of temporal characteristic of activity as a system. Rather than considering separate parametrical characteristics of activity, such as time of task performance, reaction time, reserve time, and pace of performance, we attempt to develop the holistic time structure of an activity. We define time structure of activity as a logical sequence of activity elements, their duration, and possibility of their performance simultaneously or sequentially. Algorithmic analysis of activity and development of the time structure of activity are two basic methods of morphological analysis of activity during task performance. Without the development of the time structure of activity, we cannot perform quantitative assessment of task complexity. This is explained by the fact that activity is a process and we have to evaluate the complexity of this process. At this stage of analysis, all activity elements are translated into temporal data that demonstrate the duration of standardized elements of activity. When we design a time structure activity, technological units of analysis should be transformed into psychological units of analysis.

The description of activity time structure is important in the study of human–computer interaction (HCI) and in the design of tools and equipment for the operator. The main idea is that changes in equipment configurations probabilistically change the time structure of activity. The specialist can evaluate and change the equipment characteristics based on time structure analysis. The time structure of activity helps the specialist evaluate the efficiency of the performance of production operations; thus, it can be used in the evaluation of safety and training. The time structure of activity cannot be developed until we determine strategies of task performance and possibility to perform activity elements simultaneously or sequentially. It is important also preliminarily to determine logical organization of elements of activity. Therefore, before developing a time structure of activity, it is necessary to perform an analysis of activity self-regulation to determine strategies of task performance and then to describe task algorithmically. The following are the stages of time structure development:

1. Determine the content of activity with the required level of decomposition for defining their elements (psychological units of analysis).
2. Determine the duration of elements while considering their mutual influence on each other.
3. Define the distribution of activity elements over time, taking into account their sequential and simultaneous performance.
4. Specify the preferable strategy of activity performance and its influence on the duration of separate elements and the total activity.
5. Determine the logic and probability of transition from one temporal substructure to another.
6. Calculate the duration and variability of activity during task performance.
7. Define how strategies of activity change during skill acquisition, and estimate what is intermediate and final about the time structure.

A critically important step in the development of a time structure of activity is determining what elements of activity can be performed simultaneously and what elements can be performed only sequentially. This stage of analysis starts with a qualitative analysis, including analysis of the mechanisms of activity self-regulation for determining a strategy of task performance. For example, in a dangerous situation, when actions have a high level of significance, an operator performs them sequentially, even if they are simple. However, in a normal situation, where the consequences of error are not severe, the same simple actions will be performed simultaneously. The strategies of activity also depend on the logical components of the work process and the complexity of separate elements of activity.

During the design of time structure, one should distinguish cognitive elements that are independent components of activity (cognitive actions) and those cognitive elements that are components of motor actions (microblock of programming and correction of motor motions of activity). The last cognitive components are not independent elements and will be related to motor activity. We will consider further the following situations: (1) possibility of combination of motor components of activity; (2) possibility of combination of cognitive components; (3) possibility of combination of motor and cognitive components of activity.

In SSAT, the possibility to perform cognitive and motor actions or their elements (operations) simultaneously or sequentially depends on the level of concentration of attention during the performance of these elements. According to SSAT the higher is the level of attention concentration during performance considered element of activity the more complex is considered element of activity. We already consider this question before in our work (Bedny and Karwowski, 2007). Therefore, we simply present here basic rules that determine the possibility to perform elements of activity simultaneously or sequentially.

A. Motor components of activity

1. Two motor actions that require high levels of concentration of attention and visual control can be performed simultaneously only after development of high-level automatic skills and if two motor actions are performed in normal visual field.

2. Two motor actions that require high level of concentration of attention and visual control along all trajectories of actions and are performed outside of normal visual field should be performed only in sequence.

3. Two motor actions that require low and average level of concentration of attention can be performed simultaneously.

4. Two motor actions when one of them requires high level of concentration of attention and the other requires low or average level of concentration of attention can be performed simultaneously.

B. Cognitive and motor components of activity

1. The simultaneous recognition of different stimulus is possible if they are well structured and the number of stimuli is not greater than 3–4 (based on working memory capacity), and the stimuli are familiar. In other cases, input information should be received sequentially.

2. If an operator recognizes well-known stimuli in a familiar situation, mental actions can be simultaneously combined with motor actions whatever the level of attention required.

3. If an operator recognizes unfamiliar stimuli in unfamiliar situations and the system of expectation does not coincide with ongoing information, motor actions that require only a lower and average level of concentration of attention can be performed simultaneously with perceptual actions.

4. The decision-making actions and motor actions that require a high level of attention should be performed sequentially.

5. Simple decision-making (e.g., choice between alternatives) can be simultaneously performed with motor actions that require a lower or average level of attention.

6. Cognitive components should be performed sequentially.

7. Simultaneous performance of activity elements that might result in working memory overload (e.g., require simultaneous keeping in memory different data items) should be performed sequentially.

8. In stressful situations, or when personnel are not highly skilled, all activity elements requiring high levels of attention should be performed sequentially.

Let us consider several examples. Two motions R30C (move two hands 30 cm to an exact position) should be performed sequentially because each of them requires a high level of concentration. If a subject performs two motions and one of them is R30C and the second one is R30B (average concentration), then they can be performed simultaneously. Two simple decisions should be made in sequence. At the final stage of the design process, analytical models are tested experimentally and some correction is possible. These are common steps not only for ergonomic but also for more engineering design. Between design activity models and real performance are probabilistic relationships. The more subjects perform the same tasks, the more closely the subjects' activity approaches developed models.

While analyzing the task performance, simultaneously performed actions and combined actions should be distinguished. Each of the simultaneously performed actions has its own goal. For example, two motor actions can be performed simultaneously (the subject moves two arms simultaneously and grasps the objects). An example of combined action can be *move arm with an object and turn it in a vertical position at the same time*. This is one action because it has one action's goal. If a subject moves an arm with an object in the given position and then turns it, this would be one action with two sequentially performed motions that has one goal. Cognitive actions can also be combined (in thinking actions, a subject uses external visual information or information from memory). Without combining perception with thinking or memory with thinking, such actions cannot be performed. Cognitive actions that have their own goal very seldom can be performed simultaneously (Bedny and Meister, 1997).

In the succeeding text, we start to consider some examples that are derived from laboratory and field studies. In Table 7.3, we presented the method of the algorithmic description of the task performed on the control board. In this section, at the first stage we demonstrate the principle of activity time structure description of the same task. There are tabular and graphical forms of time structure description. At the first stage, we present a tabular form describing the temporal structure of the task. For this purpose, it is necessary to introduce additional information in algorithmic description of the task (see Table 7.4 and compare with Table 7.3).

TABLE 7.4

Time Structure of Task Performance on Experimental Control Board
(First Version of Algorithm)

Members of Algorithm (Psychological Units of Analysis)	Description of Elements of Tasks (Technological Units of Analysis)	Description of Elements of Activity (Psychological Units of Analysis)	Time
O_1^α	Look at first the digital indicator.	Simultaneous perceptual operation	0.15 s
l_1	If the number 1 is lit, turn the switch left (perform $_1O_2^\varepsilon$); if the number 2 is lit, turn the switch right (perform $_2O_2^\varepsilon$).	Simultaneous perceptual operation	0.15 s
$_1O_2^\varepsilon$ or $_2O_2^\varepsilon$	Move the two-position switch 6 to the right or move the switch 6 to the left.	M2,5A	0.14
O_3^α	Determine whether the digital indicator 3 or the signal bulb 4 or 5 is turned on.	Simultaneous perceptual operation	0.15 s
L_2	Decide to move an arm to the hinged lever 7 and press button 8 (perform O_4^ε);	Decision-making operation at a sensory-perceptual level	0.15 s
O_4^ε	Move right arm to the four-position hinged lever 7, grasp the handle, and press button 8 with the thumb.	RL1 + R13A + AP2 + G1A	1.15
$O_5^{\alpha w}$	Wait for 3 s	Waiting time	3.00 s
O_6^α	Determine the pointer's position on the pointer indicator 2.	Simultaneous perceptual operation	0.15 s
L_3	Decide how to move hinged lever 7 (if the pointer position is 1, perform $_1O_7^\varepsilon$; if 2, perform $_2O_7^\varepsilon$; if 4, perform $_4O_7^\varepsilon$).	Decision-making operation at a sensory-perceptual level	0.15 s
$_1O_7^\varepsilon$ ⋮ $_4O_7^\varepsilon$	Move the four-position hinged lever 7 to the position that corresponds to the number of pointer indicator.	M5B	0.27
O_8^α	Determine whether the digital indicator 3 displays the number 5.	Simultaneous perceptual operation	0.15 s
L_4	Decide to move multiposition switch to position 5 (if the digital indicator 3 displays the number 5, perform $_1O_9^\varepsilon$).	Decision-making operation at a sensory-perceptual level	0.15 s
$_5O_9^\varepsilon$	Turn multiposition switch 8 to required position 5.	RL1 + R13A + G1A + T150S	1.12

(Continued)

TABLE 7.4 (*Continued*)

Time Structure of Task Performance on Experimental Control Board
(First Version of Algorithm)

Members of Algorithm (Psychological Units of Analysis)	Description of Elements of Tasks (Technological Units of Analysis)	Description of Elements of Activity (Psychological Units of Analysis)	Time
O_{10}^{α}	Determine whether bulb 5 (red) is turned on.	Simultaneous perceptual operation	Overlapped by motor activity (0.15 s)
l_5	Decide to press green button 10 (if the green bulb 5 is turned on ($l_5 = 1$), perform O_{11}^{ε}).	Decision-making operation at a sensory-perceptual level	Overlapped by motor activity (0.15 s).
O_{11}^{ε}	Move an arm to the green button 11 and press it.	RL1 + R26B + G5 + AP2	1.46
Total	Working time is 3.73 s; waiting time is 3 s		6.73 s

In the right side of the table, we need to enter two additional columns. In the third column from the right, we describe the cognitive and motor actions by using psychological unit of analysis. For this purpose, the method of cognitive and motor action descriptions is presented in Section 6.3. Cognitive actions usually have standardized verbal description according to their classification principles presented in SSAT. If there are data about the content of cognitive actions (description of their mental operations), these information are also introduced in the description of cognitive actions. Motor or physical actions are described with the aid of motions described in MTM-1. According to SSAT, motor action includes, in its content, motions that are integrated by a goal of the motor action. Typically, each member of the algorithm includes 1–4 similar types of actions. In the last column to the right, the time performance of each member of the algorithm is given. If it is possible, the duration of separate cognitive actions should also be presented in this column. The duration of motor actions can be determined by using the MTM-1 system in combination with some procedures developed in SSAT. Here, we maintained that system MTM-1 does not use the concept of motor actions.

As has been shown in Figure 7.1, there are 110 versions of task realization on the control board. However, the number of studied versions should be much less. This can be explained by the fact that a significant number of versions of task performance are similar in their content and temporal parameters. For example, the difference in realization time of two different versions of the task (versions of realization algorithm) that differ only in switcher position (9) on one to two positions can be ignored. Therefore, we need to select the most representative versions of the task algorithm realization.

As an example, we describe time structure of the first version of task performance in the succeeding text. The first version corresponds to the situation when one uses all instruments and controls (see Figure 7.1) including a hinged lever (7) and a multiposition switch (8). A hinged lever can be turned into one of four positions (up-down and right-left). In our experiment we only consider the case when it can be turned only up-down. However, before subjects can do it, they must press the button with the thumb (button was installed directly into the lever's handle). Movements in one to four positions were similar and required the same time for performance. Multiposition switch, according to this version of algorithm, should be turned into position 5 according to presented information from digital display (3). After that, the red bulb is turned on and the subject moves his or her right arm to the right direction and presses the green button (10). Temporal parameters of the considered version of algorithm were measured. Here, we want to mention that in the experimental analysis, this version of algorithm realization is presented between other versions of algorithm in a chance manner. When we design a time structure of activity without using experiments, it is necessary to know the performance time of separate cognitive or motor actions or operations. In the considered task, the subjects received signals, evaluated them, made decisions, and performed actions. Mental actions consisted of simple recognition and decision-making, retrieving from long-term memory well-known information. This kind of activity is common for operator's activity in semiautomatic systems when they perform familiar tasks. Therefore, in this study, we will limit ourselves to using the MTM-1 system or data from engineering psychology handbooks.

In contrast to traditional methods of using MTM-1, when specialists immediately divide a task into its constituent motions in SSAT, the following steps of MTM-1 application are followed: According to SSAT, specialists have to (1) conduct qualitative task analysis and discover preferable strategies of its performance; (2) describe them algorithmically; (3) describe actions in each member of algorithm; and (4) describe a list of motions in each motor action by using the MTM-1 system. Each the following steps requires reconsideration of the previous one. The duration of perceptual and mental actions or operations can be determined based on a technique presented in a handbook of engineering psychology (see, e.g., Myasnikov and Petrov, 1976) or based on chronometrical methods developed in cognitive psychology (Sternberg, 1975).

SSAT presents a number of additional requirements during performance of the chronometric measurement. For example, the duration of cognitive actions may be measured at their isolated execution, when they integrated in the structure of holistic activity, in the presence of various conditions and level expectations. The MTM-1 system has microelement *EF* (Eye Focus), which can be considered as a simple cognitive action. The MTM-1 system also takes into consideration cognitive processes during performance of different motions. For our example, we selected the following data from a handbook of engineering psychology (Myasnikov and Petrov, 1976): reading pointer display 0.4 s; recognition of simple signal 0.4 s; decision-making at sensory perceptual level 0.29; retrieval of information such as simple, well-known signal 0.25–0.35 s; *EF* recognition and making decision (*yes–no or if-then*) 0.27 s (≈0.30 s).

For determination of duration of motor actions, we used the MTM-1 system. The decision-making involved in this task is the simplest one. It merely requires the operator to recognize the initiating stimuli and to remember the (usually) binary rules associated with it. Because of this, stimulus recognition is conflated with decision-making. According to our classification of cognitive actions, it is decision-making at the sensory-perceptual level. The MTM-1 system assigns 0.27 s for this type of action. In our task, there are also more complicated decision-making actions at the sensory perceptual level. They require simultaneous receiving of information from several instruments and making decision based on it (2–3 stimuli can be perceived as one operative unit of information that requires no more than one fixation of eye). For this kind of decision-making at the sensory-perceptual level, we can assign 0.3 s. According to the rules that are described in Bedny and Karwowski (2007), time for the simplest recognition and decision-making operations at the sensory-perceptual level can be determined by dividing element *EF* into two. One half of *EF* is related to sensory-perceptual operation and the other to decision-making operation. This rule will be applied in our example. However, one should understand that this is simply a conventional rule that helps us pay attention to perceptual and decision-making mental operations that are components of one unitary cognitive action. Eye travel time in our study had not been taken into consideration due to the fact that the control board was relatively small. The first version of the algorithm of realization was almost the same as the general algorithm of task performance. The only difference was that the arm was always turned to multiposition switch position 5 and at the final stage the subject always presses the green button (10).

Table 7.4 presents the time structure of task performance for the first version of the algorithm. However, suggested method allows identifying performance time for any version of task, including the ones performed with average, maximum and minimum task performance time. In the presented tables, time will be given in most cases in units that correspond to 0.01 min. In those cases when time will be given in seconds, it will be shown in the tables.

Attention should be drawn to the fact that the main units of analysis are those that belong to the category of psychological units of analysis. Technological units of analysis are an additional means of describing the time structure of activity. Comparison of technological and psychological units of analysis facilitates an understanding structure of activity during completing a task. Translation of technology units in the psychological unit of analysis must be mentioned as a required step in the design of the time structure of activity.

The task execution time that was determined by the chronometrical study had a similar value to the experimental data. We present earlier the time structure of activity in table form. However, in situations when there are complex combinations of activity elements (activity elements are performed simultaneously), the most informative is a graphical form of time structure description. This method of presentation usually can be done after the table form is developed. Figure 7.2 presents the graphical model of the time structure of activity during performance of the earlier-described task (first version).

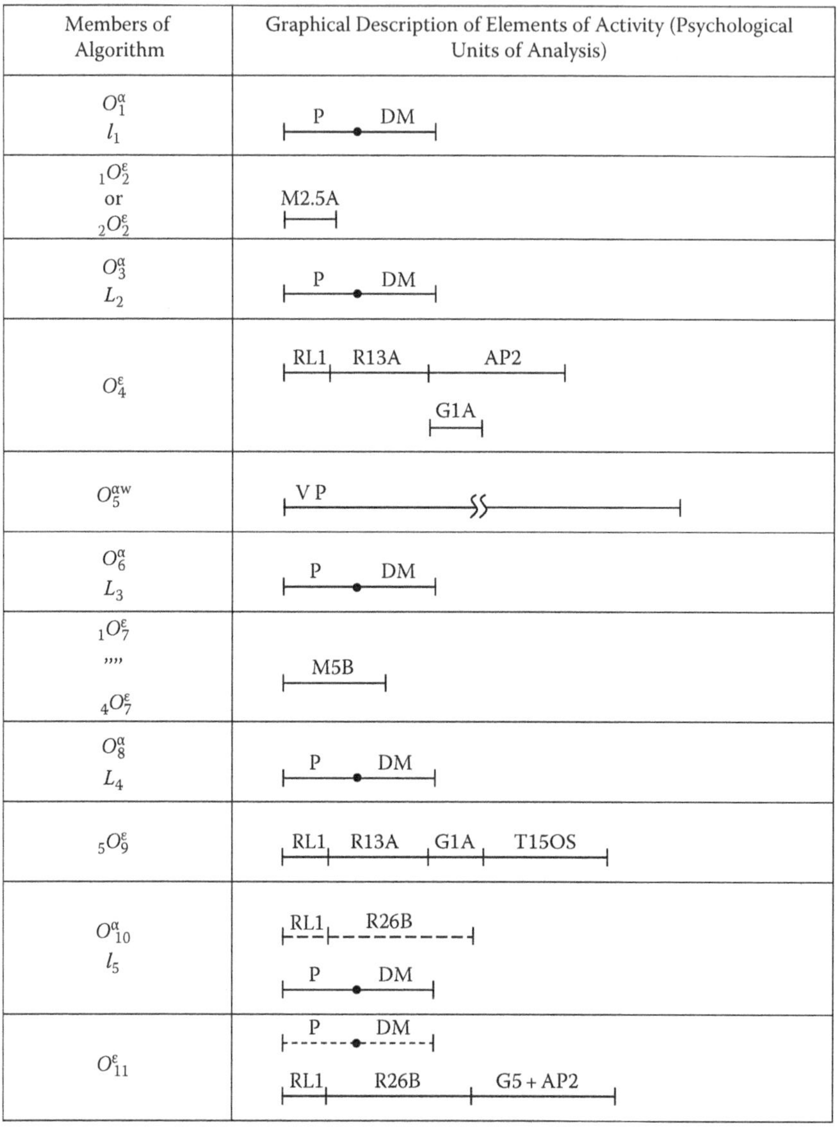

Members of Algorithm	Graphical Description of Elements of Activity (Psychological Units of Analysis)
O_1^α l_1	P DM
$_1O_2^\varepsilon$ or $_2O_2^\varepsilon$	M2.5A
O_3^α L_2	P DM
O_4^ε	RL1 R13A AP2 G1A
$O_5^{\alpha w}$	V P ʃʃ
O_6^α L_3	P DM
$_1O_7^\varepsilon$,,,, $_4O_7^\varepsilon$	M5B
O_8^α L_4	P DM
$_5O_9^\varepsilon$	RL1 R13A G1A T15OS
O_{10}^α l_5	RL1 R26B P DM
O_{11}^ε	P DM RL1 R26B G5 + AP2

FIGURE 7.2

Graphical presentation of time structure of task performance on the experimental control board (the first version of algorithm).

In the presented graphical model of the time structure of activity, individual elements of activity are presented in a horizontal line. The elements are specified by symbols above the segments. Microelement *EF* describes the perceiving of signals as simple decision-making at sensory perceptual level that includes *yes–no* or *if-then* decisions. The segment under *EF* designates the duration of this kind of mental action. In the same way, duration of any other segment designate the duration of other elements of activity. According to the introduced rules (Bedny and Karwowski, 2007), for purposes of determining durations of simple perceptual operation and decision-making operation, which are components of unitary decision-making action at sensory-perceptual level, we divide *EF* (0.3 s) into perceptual and decision-making mental operations (½ *EF*).

In more complicated situations, the duration of decision-making actions can be evaluated experimentally or required data can be taken from other sources. For example, O_6^α and L_3 are also involved in decision-making process at the sensory-perceptual level. We can define duration of this element based on data from the *Handbook of Engineering Psychology* (Myasnikov and Petrov, 1976) and divide this action in the same way as *EF* into two mental operations. Some members of the algorithm overlap. The considered in the left column member of the algorithm is depicted by a solid line while the overlapping member of the algorithm is depicted by a dashed line (see Figure 7.2). For example, O_{10}^α and l_5 are overlapped by O_{11}^α. Because of that, for O_{10}^α and l_5, we did not assign time for performance when we calculated the duration of the whole task. One can make conclusions about possibilities to perform actions simultaneously or sequentially, not only based on analyses of separate actions or operations but also based on analysis of possible strategies of task performance. For example, if a performer is very skilled and consequences of wrong action are not important, then actions can be performed simultaneously. If actions are not automated and errors undesirable, they should be performed sequentially.

Some symbols used in Figure 7.2 require additional explanation. *M2.5A* means *move object against stop when distance is 2.5 cm*. Letters P and D over segment mean *perception* and *decision-making*. *RL1* means *normal release performed by opening fingers*. *R13A* means *reach to the object in fixed location when distance is 13 cm*. *AP2* designates *apply pressure with effort less than 15 kg*. *G1A* means *easily grasped*. This element is overlapped by AP2. The letters W and P mean *waiting period*. *M5B* designates *move an object 5 cm to approximate location (requires an average level of concentration of attention)*. *T180 S* designates *turn 180° with small effort (from 0 to 1 kg)*. In a similar way, other elements of activity are designated.

Let us consider units of analysis that are used during algorithmic description of the task and temporal analysis of activity. The first two members of algorithm (O_1^α and l_1) are the result of artificial dividing of element EF into two separate mental operations, which are related to different members of algorithm. This was performed for the purposes of distinguishing, in future analysis, members of algorithm associated with decision-making at a sensory-perceptual level from members of algorithm that are comprised of simultaneous perceptual actions (operation). This is why in this chapter we introduced an artificial rule according to which, in some situations, we can divide decision-making actions at a sensory-perceptual level into operations and relate them to different members of algorithms, when one member of the algorithm is associated with perceptual stage and others with decision-making stage. In all other situations, we divide tasks into separate members of algorithms according to the recommendations described in Section 7.1. For example, a member of algorithm O_2^ε contains one motor action: *move arm to the lever, grasp it, and simultaneously press the button with the thumb*. This action, in turn, is comprised of the following motor operations (motions): *move arm, grasp the handle*, and *press the button with the thumb*. All these operations are integrated by the goal of motor action. In a similar way, other members of algorithm are described.

Very often, the more complex time structure of activity can be encountered in situations when relatively simple cognitive actions are combined with various complexities of motor actions. This is explained by the fact that cognitive actions cannot be performed simultaneously. Cognitive actions should be performed in sequence (see previously presented material).

In the following, we present an example of a time structure of activity when the subject in experimental conditions performs two elements of task (member of algorithm), *grasping the pins* and *installation of pins in pin board*. In front of the subject, there is a pin board that contains 30 holes for metal pins. Behind the pin board was a box containing pins. There were regular (without a flute) pins and fluted pins. The pins are put in the holes according to specific rules.

1. If pins are regular (without a flute), they can be installed in any position.
2. If a fluted pin is picked up by a subject's left hand, it must be placed so the flute is below the hole.
3. If a fluted pin is picked up by a subject's right hand, it must be placed so the flute is above the hole.

These are complex logical conditions that we designate by the letter L. Logical condition for when fluted pins are absent in both the left and right hands can be designated as $L_1 = (l_1 = 0$ and $l_r = 0)$. Logical condition for when fluted pins are presented in the left hand and absent in the right hand can be designated as $L_2 = (l_1 = 1$ and $l_r = 0)$. The last logical condition can be designated as $L_3 = (l_1 = 0$ and $l_r = 1)$. There are also logical conditions when both pins are fluted and therefore can be designated as $L_4 = (l_1 = 1$ and $l_r = 1)$. It was a sufficiently complex manual combinatory task because it included various logical rules and motor manipulations with pins. Below we consider an example of time structure analysis of the fragment of activity when both pins have flutes.

We can present these two elements of activity in a simplified way. The first element consists of moving two hands simultaneously and grasp two pins (see Figure 7.3). Figure 7.4 depicts installation of two pins into the holes simultaneously according to described rules. The time structure of activity is presented in Figure 7.3. Figure 7.3 describes a time structure of activity when subjects move two hands simultaneously and grasp two pins. Figure 7.4 describes a time structure of activity when subjects move the pins and install them into the holes of the pin board. LH and RH means left and right hands. According to MTM-1 system in Figure 7.3, R32B means reach to single object in location, which may vary slightly from cycle to cycle (average or the second level of attention concentration), and G1C1 means grasp nearly cylindrical object with interference for grasp when diameter is more than 12 mm.

The time structure of activity when subjects install two pins with flutes is presented in Figure 7.4.

When hands move the pins, cognitive components of activity are overlapped by motor components. In Figure 7.4, P is simultaneous perceptual operation and DM is decision-making operation at sensory-perceptual level. PDM has the fourth category of complexity because decision is made based on information extracted from memory when a subject decides *how to turn a pin*. Motor actions are performed by left and right hands

FIGURE 7.3
Graphical model of activity time structure *move hands and grasp pins.*

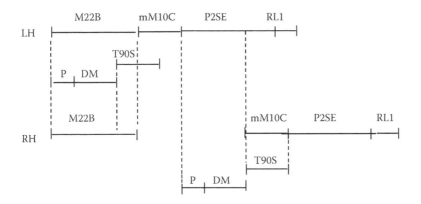

FIGURE 7.4
Graphical model of activity time structure *move hands with two pins and install them into the holes.*

(symbols LH and RH). They consist of the following motions: *M22B*, move hand with part when distance is 22 cm (average or the second level of attention concentration), and *mM10C*, continue hand movement with part to an exact position (high or the third level of attention concentration); the second decision-making action, PDM, has the fourth and motor motion mM10C has the third category of complexity. Therefore, they cannot be performed simultaneously. From this, it follows that the second PDM can be performed only after completion of the first PDM and *mM10C: P2SE*, installation of part into the hole, with low pressure and easy to handle (average or second level of attention concentration); *RL1*, release a part by opening fingers (low or first level of concentration of attention). Two motor actions start simultaneously and decision-making is performed sequentially. Thus, the performance of the motor action by right-hand movement is interrupted until the second decision is completed. From this example, we can see that cognitive actions are performed in sequence and they can be combined with motor components of activity. The combination of motor and cognitive elements of activity in time is determined by rules described earlier in this chapter. The combination of activity elements is also determined by equipment configuration. Changes in the equipment or interface design lead to changes in the time structure of activity. If the time structure of activity is very complex during interaction with equipment, this means that the method of performance should be changed or equipment is not designed efficiently. Hence, efficiency of ergonomic design and efficiency of task performance can be evaluated based on analyzing the activity time structure. In the succeeding chapter, we will demonstrate how to evaluate the complexity of task performance. The developed method can be used as a purely analytical one or in combination with simplified experimental procedures. Design should not be reduced to purely experimental methods as it is done in cognitive psychology. Analysis of time structure of activity is the very useful method when studying the skill acquisition process (Bedny, Meister, 1997, p. 315) because time structure of activity changes during skill acquisition. It has been discovered that some elements of activity during the skills formation process are temporal and are not included in final structure of activity. The more complex the task is the more intermediate strategies are utilized by trainees. Each strategy has its own specific time structure of activity.

The time structure of activity should be distinguished from a timeline chart that consists of lines or bars, whose length is proportional to the amount of time necessary for performing a particular task or task element (Kirwan and Ainsworth, 1992, p. 136). This method does not take into account the structure of activity, because it does not distinguish between psychological and technological units of analysis, the possibility of combining elements of activity in time, and the probabilistic characteristics of activity. Let us consider a hypothetical example. Suppose a subject has to take two pins by left

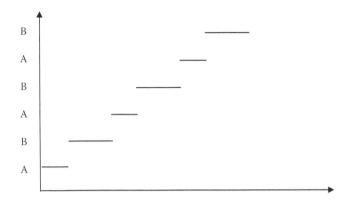

FIGURE 7.5
Timeline chart of the task *take two pins and install them into the holes three times* (vertical axis). A—move two hands and grasp pins; B—install the pins into the holes; performance time in seconds (horizontal axis).

and right hands and put them into the holes three times. Figure 7.3 shows a time structure of activity when the subject picked up two pins with both hands and installed them into the holes only once. The timeline chart when the subject performs these components of the task three times is presented in Figure 7.5.

The length of the line for element *A* depicts the time scale for the element *move hands and take two pins* and the length of the line for element B reflects the time scale for the element *install the pins into the holes*. Each type of lines is repeated three times, because the installation of the pins is performed three times. If the subject performs installation of pins only once, only two lines (line A and line B) would be present, in Figure 7.5 Such a comparison of the temporal structure of activity (see Figures 7.3 and 7.4) and the timeline chart (see Figure 7.5) clearly demonstrates the difference between them. As can be seen, the timeline chart describes the duration of some task elements without describing the subject's activity structure during these periods of time. In this method of analysis, the specialist utilizes technological units of analysis. When the specialist develops a time structure of activity, he or she utilizes psychological units of analysis. SSAT introduced the concept of the technological unit of analysis (typical elements of task) and the psychological unit of analysis (typical elements of activity.) The relationship between them is of fundamental importance in the design of activity. Using the psychological unit of analysis is also of fundamental importance in the design of the equipment based on the analysis of the structure of activity. We used the examples based on laboratory studies. This was done in order to explain the proposed methods of analysis easier. In the following sections, examples that are based on real studies will be mainly used.

The previous discussion allows us to draw some conclusions. Models of activity time structure obtained by an analytical method reflect an idealized description of activity. Actual models can be developed by repetitive experimental measurements of the task performance time and the following averaging of obtained data using experimental methods of analysis. Real data obtained during experimental study approaches an idealized model as a result of repeated executions. The suggested method of time structure development is totally different from the traditional methods of time study. Rather than considering separate parametrical characteristics of activity such as task performance time, reaction time, reserve time, and pace of performance, we have developed a holistic method of analysis of the time structure of activity. At this stage, all activity elements (psychological units of analysis) are translated into temporal data that demonstrates duration of standardized activity elements. The duration of performed task elements that are described in technological terms may be useful if they can be compared with corresponding typical elements of activity or psychological units of analysis. In cases where task elements are performed sequentially, the time structure of activity can be presented only in tabular form. If there are a lot of task elements that are performed simultaneously, then the time structure of activity should also be presented in graphical form. This allows us to describe more clearly the complexity of the time structure of activity during task performance. It is especially important in quantitative assessment of task complexity.

The proposed approach of analysis of the temporal characteristics of the work activity is important not only in ergonomics and work psychology but also in economics. This approach allows us to address the issues of time study of production operations with a high level of variability. It is well known that existing methods of time study and efficiency of performance are basically adapted to analyze only those tasks (production operations) that have strictly defined the sequence of executed task elements (one of the best methods of performance). An algorithmic description of the work activity and the construction of its time structure is the basis of morphological analysis of human performance. This approach allows us to eliminate the contradiction between the normative approach to task analysis and the variability of work activity.

8

Design and Time Study

8.1 Time Study as an Important Aspect of Ergonomic Design

What is the relationship between design of activity and time study? At first glance, these are two independent problems. However, in reality, they are closely interrelated because activity or behavior is a process. In essence, when we set ourselves on designing activity, it is necessary to understand that we have to design a process that unfolds over time. More precisely, activity is a complex structure that consists of various elements that are unfolding over time. In Section 7.3, we demonstrated that the time structure analysis is a critically important stage of ergonomic design. However, there are a number of other issues related to the time study in ergonomics. These issues will be discussed in this section.

Psychological aspects of time study are important aspects of task analysis. Time study can be used not only for analysis of efficiency of work but also for evaluation of cognitive processes and external behavior (Bedny, 1979, 1981).

Traditionally, time study is used to determine the time required for the performance of a particular task when a well-trained operator works at a normal pace (Barnes, 1980; Gal'sev, 1973; Karger and Bayha, 1977). This determined time is called a standard performance time for a task or production operation. Time study is used for cost estimating, planning, and scheduling work; developing a wage incentive plan; evaluation of labor cost; measuring productivity; etc. There are a variety of methods of time measurements such as using a stopwatch, videotaping, and recording using software. All these methods provide chronometrical analysis of work. A detailed analysis and usage of chronometrical methods can be found in Barnes (1980) and methods of time study in ergonomics were described by Drury (1995). In systemic-structural activity theory (SSAT), time study includes new nontraditional methods of analysis of the temporal characteristics of activity. This approach allows us to describe the time structure of the holistic activity and determine the basic time parameters of variable activity. Such data are critically important in solving design problems. The proposed approach allows us to introduce new methods of time study in the analysis the operator performance in automated control systems, including computerized systems.

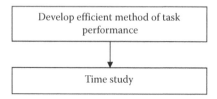

FIGURE 8.1
Relationship between stages of time study (traditional approach).

These questions are important not only in ergonomics and psychology but also in economics that study the efficiency of human work.

Traditionally, time study is divided into two stages. The first stage is associated with determining an efficient method of performance and the second stage involves time study or determining standard performance time. These stages are presented in Figure 8.1.

As we will show later, this scheme is not entirely accurate. The method of task execution can be determined at the first stage only preliminarily. At the next stage, it is necessary to determine the temporal structure of activity and hence the performance time of the whole task. After obtaining this new information, it becomes possible to develop a new adjusted method of task performance. Thus, these two stages, according to the SSAT approach, have a loop structure as shown in Figure 8.2. In complex situations, the cycle can be repeated several times.

Chronometric measurements (timing) of an entire operation are applied very rarely, when precision of the time study is low. This method can be applied when production operation is used only for a very short period of time. In most cases during time study, production operation or task is divided into separate elements and their duration is measured separately. The beginning and end points (chronometrical points) for each element should be specifically indicated (Barnes, 1980; Drury, 1995; Gal'sev, 1973; etc.). The standard time is determined for each element of task. After that, the total standard time for each task is calculated. During chronometrical study, handling time should be separated from machine time. Machine time is calculated by technologists. Time standards for the manual component of work are based on the principles developed in the area of time and motion study. In this area, we have some rules that help us to determine which motions can be performed in sequence and which can be performed simultaneously (see, e.g., method–time measurement (MTM)-1 system).

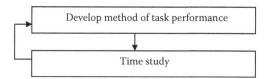

FIGURE 8.2
Relationship between stages of time study (activity approach).

We developed a more advanced rule to help determine which cognitive and motor actions can be performed in sequence and which can be performed in parallel (Bedny and Meister, 1997). It is also necessary to determine the possibility of combining the machine and manual elements of a task. Only then can the total time of task performance be determined. Thus, the idea that analysis of labor and determining of working time are reduced to a simple summation of time for separate elements of task or motions is incorrect. All these ideas are important for the study of any kind of human work. As can be seen, the principle of dividing human work into separate elements or, according to the SSAT terminology, into separate units of analysis is an important aspect of time study. Psychologists Schultz and Schultz (1986) point out that one of the shortcomings of the time and motion analysis is job simplification that can lead to monotony, boredom, etc. However, the purpose of contemporary time study is not simplification but rather optimization of work and determining the required time standard for task performance. One drawback of the traditional time study is that it is not fully adapted to the measurement of mental components of activity. Methods of analyzing the performance time of cognitive components of activity will be discussed further in this section.

Some psychologists wrote that one drawback of traditional time study and design of human performance was associated with ignoring workers' individual differences. The criticism of some psychologists in this area (Schultz and Schultz, 1986, p. 410) is partly true. In activity theory, this aspect is known as individual style of activity (Bedny and Seglin, 1999a). Individual style of activity may be understood as a strategy of activity that derives from personal features of a performer. In considering this problem, one should take into account the following limitations associated with the individual style of activity in the field of time study and ergonomic design. Any production process has certain normative requirements for task performance. It has some standardized rules of performance and prescribed technological procedures. Hence, any individualization of work methods is restricted by the range of tolerance in variation of job performance. If the individual style of performance contradicts the normative requirements, it might lead to the violation of safety, increased equipment wear, etc. This means that individualization of work methods can be implemented with some stipulations.

Individual style of activity in the production environment can be accepted if it can provide a required level of productivity and safety. Individual strategies of performance, their description utilizing probabilistic algorithms, and determining their adequacy to standardized requirements and safety of task performance should be identified (Bedny and Seglin, 1999a). If individual style of activity performance is in conflict with productivity and safety requirements, there is a problem with individual selection for the job.

Time study is a broader concept than time and motion analysis. Moreover, the term time and motion analysis is not an adequate term from contemporary time study perspectives. This is explained by the fact that activity has logical and hierarchical organization. For example, motions are components of motor actions, and activity during task performance has cognitive components. At the same time, motor actions and their constituent element motions are important components of any contemporary tasks. The most powerful method of time study and efficiency of performance for manual tasks is the MTM system. However, this system should be adapted for contemporary task analysis. This system has been developed by Maynard et al. (1948). Later this system was significantly improved and is known now as the MTM-1 system. Currently, there are MTM associations in multiple countries. MTM-1 can be applied for manual operations with a standardized method of performance. Its basic principle is to divide a method of performance into basic motions and to assign to each motion a predetermined time standard. Each motion and its performance time depend on conditions under which it is performed.

The system has rules of combining individual movements over time, depending on such factor as attention concentration. Therefore, the opinion of some psychologists that the total time of operation is determined by a simple summation of mean values of time for separate element when the MTM-1 system is used is incorrect. Any analytical design method allows for certain errors. They are adjusted by means of experimental verification. MTM-1 also suggests a possibility of errors. Such errors, as in any design, are corrected in the following process of theoretical and experimental verification.

The system proved itself in the application of designing repetitive manual manufacturing operations and in various types of industry where tasks are usually performed in a standardized way. We will further consider a more detailed analysis of this system and its application to the design of contemporary work. We will argue that the MTM-1 system can be applied in modern conditions for study motor components of activity. The MTM-1 system is a specific method of time study. It can be applied at the design stage when there are no real manufacturing operations and chronometrical analysis and observation in general cannot be conducted. One of the important advantages of the MTM-1 system lies in the fact that this system provides a standardized language for motor activity description. It is very useful for solving design problems. Standardized motions can be considered as one important type of *units of analysis* when studying motor behavior. Such concept as *units of analysis* is important in activity theory. Time study in general and MTM-1 in particular are important areas of studying human work not only for motor tasks but also for cognitive tasks. For this purpose, we have to know not only the duration of motor motions but also the duration of cognitive actions. Cognitive actions have a very short duration and extraction of separate cognitive operations is not always possible.

The MTM-1 system is defined as a procedure that analyzes any manual operation or method into basic motions required to perform it and assigns to each motion a predetermined time standard that is determined by the nature of the motion and the conditions under which it is made (MTM 1 Analyst Manual, 2001). The task performance time for motions that are performed in sequence is determined by the summation of the individual time standards for separate motions. The MTM-1 system takes into consideration an ability to perform some motions simultaneously. From SSAT perspectives, the classification and description of motions according to their purpose is a powerful method of standardizing the description of motor action because each motor action includes several motions. This permits utilizing MTM-1 in contemporary task analysis for the description of manual components of work in a totally different way. Another advantage of this system is that it takes into consideration some aspects of cognitive regulation of motions. The application of MTM-1 is usually restricted to the analysis of routine tasks such as assembly-line jobs where motor activity dominates. However, further, we will demonstrate how this system can be adopted to contemporary task analysis and equipment design.

Most traditional methods of time study are based on the idea that it is necessary to find *the best way* of performing a production operation or task to increase efficiency and productivity (Barnes, 1980; Gal'sev, 1973). Such idea has some shortcomings in its application to modern industry. For example, in contemporary industry, the proportion of mental work and the variability of strategies of task performance are significantly increased. SSAT offers an algorithmic method of task analysis. A combination of this method with MTM-1 allows describing very flexible motor components of work that will be demonstrated later in this book. Although the MTM-1 system was derived from ideas of Gilbreth (1911), it is a new, much more powerful system. Critical comments of Gilbreth's system are not directly related to the MTM-1 system. We presented a detailed critical analysis of MTM-1 in the first volume of the book. We showed that, in combination with newly developed methods in SSAT, the MTM-1 system can be an effective tool for the analysis of a motor component of tasks.

Time not only reflects the distinguishing features of external behavior but also the specifics of the internal psychic process. Hence, chronometric studies play an important role in cognitive psychology (Sperling, 1960, Sterenberg, 1969). One of the first and the most widely known methods is measurement of reaction time, which includes perceiving a stimulus and transferring this perception into a well-learned response. There is a simple reaction time, choice reaction time, and reaction time to a moving object when, for example, a lathe operator cutting a piece of metal has to stop the machine's cutting tool in the exact position. The reaction time depends on the reaction type. It was also discovered that reaction time depends on stimulus modality, stimulus intensity, temporal uncertainty, stimulus–reaction compatibility, etc. However, in most cases, reaction time does not reflect the specifics of a task performance time. Human activity cannot be represented as a set of independent reactions performed with maximum speed. Cognitive and motor actions or operations have a certain logical organization; they influence each other and can be performed sequentially or simultaneously. The pace of performance of such actions is significantly different from the speed of reaction. All these issues are mostly ignored in ergonomic time studies. Therefore, time study of separate reactions and time study of cognitive actions in the structure of activity have certain specificity. At the same time, methods of chronometrical analysis developed in cognitive psychology can be adapted for analysis of cognitive components of activity. It should be noted that the attempt was made to create a predetermined time standard for cognitive components of activity (Van Santen and Philips, 1970). However, this system of time standards has certain disadvantages. Its time standards are excessively detailed and it is difficult to use them in practice.

Temporal characteristics of mental actions can be determined utilizing experimental procedures developed in cognitive psychology and systemic-structural activity theory. The most important factor is what should be measured and how chronometrical data should be obtained and described in a standardized manner. It is important to give a clear description of a beginning and an end of an activity element under chronometric study. A verbal description of an activity element should be accompanied by its graphic description that would assist in using the time standards in further applied studies with understanding of what specific cognitive action or several cognitive actions were performed by a subject during a measured period of time. Performance time of cognitive actions should always be combined with a standardized description of cognitive actions and various conditions in which they are performed. Only after that can obtained time standards be used as reference data by other specialists. For example, when time standards for perceptual action are developed, it is important to present the types of indicators that were used, verbally describe a beginning and an end of perceptual actions, and describe the perception time of isolated data or in the context of complicate activity. In some cases, it is important to indicate which strategies are used by a subject when he or she perceives information. When developing time standards, researchers can use various instructions. For instance, an instruction can be given to perform an activity element with maximum speed or with optimal pace. A standard description of data about the performance time of cognitive activity elements (usually separate cognitive action) is critical for further understanding of what was done in the considered time period (Bedny, 1987; Bedny and Karwowski, 2007). The pace of performance in such studies is a critical factor.

A number of temporal characteristics of cognitive processes in activity theory and cognitive psychology are useless because of the lack of a clear and standardized description of chronometrical data. In most cases, only professionals who directly perform chronometric measurement can understand their description of the measured elements, and using these results as reference data by other specialists is very difficult. Moreover, there are also certain inaccuracies that are critical in constructing the time structure of activity and subsequent assessment of task complexity.

In some cases, it is necessary to perform simplified experimental studies for obtaining a specific time standard for cognitive actions. For example, Zarakovsky and Pavlov (1987) have conducted research to analyze the performance time of recoding actions. An example of such recoding actions is translating from one language to another. This type of actions is encountered in operators' work. In determining translation time from one language to another, it is necessary to consider the level of knowledge of the foreign language. Similarly, we need to take into account a skill level during the performance of any recoding actions by the operator. Zarakovsky studied recoding actions when an operator had to transform numeric data into symbolic form, and vice versa. Information presented to subjects imitated situations specific for pilots' tasks. The time of such recoding actions was measured. It was discovered that this time depended on the skill acquisition process. The range of recoding actions' performance time varied from 1 to 2 s with average time being 1.7 s. It was an interesting fact that during the development of skills, the content of recoding actions changed. Using the SSAT terminology, we can say that subjects changed their strategy of recoding actions' performance. The content of the internal operations of the recoding actions has changed. Therefore, we need to know the most preferable strategies of action performance. All this suggests that, in determining the duration of cognitive actions, it is important to not only describe an action's name and point out time of its execution but also indicate the most preferable strategies of its performance and describe the beginning and end points of actions, the common situations when they were utilized, and their performance time. Conditions of the same cognitive action can vary and therefore various versions of the same cognitive action should be clearly described. The performance time of each version of action should be presented. The language of activity description should be standardized. SSAT suggests principles of standardized description of cognitive and behavioral actions.

When analyzing temporal characteristics of a task, one has to distinguish between time standards for performing elements of a task (technological units of analysis) and time standards for elements of activity (psychological units of analysis). Their differences and ways to use will be the subject of further discussion.

It is necessary to distinguish between the time study in the analysis of individual tasks and the time study during job analysis. The latter is carried out in a more general manner in comparison with the time study of individual tasks or production operations. The time study of a job is usually carried out during the entire work shift or a specified work period when work is divided into stages that have a clear qualitative difference. It is important to indicate a beginning and an end point of each stage. The duration of each stage is usually measured in minutes. In contrast, in the chronometrical study of production operations or tasks, time measurement is performed in seconds or even fraction of seconds. In some cases, the results of time study during job analysis may be compared with the physiological or psychological measures. For example, we can define work and rest periods during the shift and their specific location in time. These data then can be compared with physiological indicators of fatigue. A combination of physiological indicators of fatigue with chronometrical analysis is a useful method of determining the cost-effectiveness of an ergonomic intervention (Bedny et al., 2001). This method is useful in economics when efficiency of any interventions that are directed to reduce physical fatigue should be evaluated.

In ergonomics, a method that sometimes is called event/time record provides information on the sequence of events, duration of events, and their frequency (Drury, 1995). In our further discussion, we will concentrate on time study in task analysis.

Time study is critically important not only in studying traditional work in mass production but also in the design of any type of human work because activity is a structure that unfolds in time. We cannot design human activity if we do not know its time structure. The term time structure is a new and important concept in the area of time study and work design. This concept was considered in Section 7.3.

Another aspect of time study is analysis of the system reserved time. Time during which a man–machine system is transferred from the initial to the required state is called *time of the regulation cycle*. The task performance time of the operator very often constitutes the substantial part of the cycle of time regulation, which is an important system characteristic that influences the system reserved time (Kotik, 1974; Siegal and Wolf, 1969). Reserved time is defined as a surplus of time over the minimum that is required to detect and correct any deviations of system parameters from allowable limits and to bring the system back into tolerance. Thus,

$$T_{res} = T - T_0,$$

where
   T is the time that cannot be exceeded without peril to the system
   T_0 is the cycle regulation time

From the activity self-regulation point of view, it is necessary to differentiate between objectively existing reserved time and operator's subjective evaluation of this time, which are often not the same. This may lead to an inadequate evaluation of the situation and, more importantly, to the inadequate behavior of the operator in a critical situation.

A decrease in reserved time can often produce various kinds of tension. In activity theory, one distinguishes two kinds of tension (Nayenko, 1976). One is called operational and the other is emotional tension. Operational tension is determined by a combination of task complexity and lack of available task performance time. In SSAT, emotional tension is determined by personal significance of a task for an operator. The concept of significance serves an important functional purpose, which we already discussed in the context of activity self-regulation. It should be noted that both kinds of tensions are tightly interrelated and under certain conditions can be transferred into each other.

Subjective perception of reserved time influences the cognitive components of activity and the emotionally motivational state of an operator. Psychic tension can emerge even when objectively there is plenty of time for the task performance. Subjective perception of reserved time is an important component of the dynamic mental model of the situation. In general, the relationship between objective and subjective reserved time is an example of application of the concept of self-regulation in studying the temporal parameters of task performance. The emotionally motivational state of an operator in time-restricted conditions is an important aspect of functional analysis of activity.

8.2 Pace of Performance and Time Study

Activity is a process embedded in time. Therefore, task analysis cannot be performed without taking into consideration the concept of work pace. Unfortunately, work pace analysis in ergonomics is reduced to study separate reactions. Let us consider some examples. According to the Hick–Hyman law (Hick, 1952; Hyman, 1953), the reaction time increases by a constant amount each time the amount of information in the stimulus is increased by one bit. This law can be expressed by the equation $RT = a + bH$, where the constant b reflects the amount of added processing time that depends on each bit of stimulus of information to be processed and the constant a depicts the processing latencies that are unrelated to the reduction of uncertainty (see also Wickens and Hollands, 1999). This formula has a very restricted application. It has been revealed that the speed of information processing does not change when selecting from alternatives if a number of stimuli is more than 8–10. Moreover, it is almost impossible to determine the amount of information in a real situation. The speed of information processing is also not constant and depends on the holistic structure of activity. For example, it has been revealed that the time of the second choice reaction that is performed immediately after the first choice reaction depends on the complexity of both of these reactions. This means that the reactions are not independent and affect each other (see Bedny et al., 2014). Hence, the performance time of separate reactions or actions is changing in the holistic activity. All this is not taken into account in the Hick–Hyman law. Fitts' (1954) law is used when the focus is on motor responses in performance of manual tasks. This scientist investigates the relationship between time, distance, and accuracy of motor movements. The formula suggested by Fitts' law is obtained when a subject hits two targets with maximum speed or hits a single target from the start position. Such conditions are seldom in the production environment. There are other aspects of task performance time such as logically organized sequence of simultaneously and sequentially performed cognitive and motor actions. Subjects develop complex strategies of task performance based on the mechanisms of activity self-regulation. Fitts' law, the same as the Hick–Hyman law, ignores the concept of pace in task performance. A subject never performs a task with maximum speed or multiple times reputes the same actions with maximum speed as it is done in Fitts' experiment. When a subject used four targets instead of two, the pace of such movements changed significantly (see Bedny et al., 2014, Section 5.1). The pace of performance also depends on the duration of work, the significance and precision of task, etc.

Shannon's measures, which are derived from statistical analysis of information, are not sufficiently adequate for psychological studies. In activity theory, it was shown that quantitative measures of information defined by a purely statistical Shannon's approach do not take into account qualitatively structural characteristics of information that are derived from the set-theoretical approach (Vekker, 1976; Vekker et al., 1993). Hence, Shannon's measures cannot be used for the time study of human work.

Operators perform the tracking function where they have to anticipate future errors and develop complex strategies that cannot be predicted by transfer functions. As we already discussed, Zabrodin and Chernishov (1981) discovered additional harmonics that were not anticipated by models that describe the tracking dynamic functions.

Analysis of data obtained in studies of simple reaction time, choice reaction time, and performance time of positioning actions is not helpful for the solution of real-world tasks. All these methods consider human activity as a summation of independent responses that are performed with the maximum speed. A subject is considered as a reactive system that responds to stimuli. In reality, a subject formulates goals, regulates actions, changes his or her strategies, etc.

Suggested methods ignore important data that were obtained in the traditional area of time study. A comparison of traditional work in time study and the study of temporal parameters in operators' performance demonstrates that they have both common and distinctive features. The first approach concentrates on studying production operations and efficiency. The second approach is used for studying the operator's performance in time-restricted conditions, evaluation of safety, etc. However, in both cases, the temporal parameters of human work activity are considered. These two aspects of time study are interdependent and should not be studied separately.

The maximum speed of single or repeated responses cannot be the basis for determining the task performance time. According to such approach, pace depends on the interstimulus interval (force-paced) or response–stimulus interval (self-paced) stimulus rate. In the last case, the frequency with which stimuli appear depends on the latency of a worker's response (Wickens and Hollands, 2000). A worker is considered as a reactive system and pace is described as a number of unrelated responses initiated by various stimuli. However, the pace of performance depends on the holistic structure of activity. Even in force-paced conditions, a worker has an opportunity to regulate his or her pace of performance in a certain range. Pace is not a result or consequence of reactions but is a result of self-regulation of activity. External and internal contour of self-regulation can be identified in the process of pace regulation. The external counter of self-regulation provides a comparison of pace with some external signals, which are actively selected by a subject, that are used as a *subjective standard of a successful result*. The internal counter of pace self-regulation is based on the comparison of pace with the prevailing internal time standards. Pace is formed and maintained at a given level due to the mechanisms of self-regulation, rather than due to the principle of human's reactivity.

Without understanding the concept of work pace, one cannot determine the time of task performance. In studying the pace of work, it is necessary to distinguish two main situations. In one case, we consider the time study of blue-collar workers (traditional time study), and in another situation, it is a question of determining the task performance time, when an operator interacts with the complex technical systems. When it comes to traditional time study, one should take into account that a worker performs the same task multiple times when it is necessary to maintain the same pace during the work shift or significant periods of time. In the second situation, an operator functions as a monitor of the complex system. The role of mental components of tasks and complexity of task increases. An operator does not perform the same task multiple times. Rather, an operator performs different kinds of tasks. This significantly increases uncertainty about some aspects of task performance. Variability of activity during task performance significantly increases emotional tension. As with the traditional time study and time study of the operator's work, when serving complex technical systems, the maximum pace cannot be sustained during the workday.

There are a lot of difficulties in studying work pace. There is no precise definition of work pace. Barnes (1980) defines work pace as the speed of operator's motions. However, this definition is unsatisfactory because it ignores the cognitive components of activity and the logical organization of cognitive and behavior actions. Pace can be considered as the speed of performing various components of activity that are structurally organized in time. Hence, the pace of performance can be defined as an operator's ability to sustain a specific speed (below maximum) of holistic activity structure that unfolds during task performance. This pace should be sustained during the work shift and subjectively evaluated by an operator as an optimal pace. It has been discovered that the slowest workers' pace of performance can be two times slower than the fastest blue-collar workers' pace (Barnes, 1980). Hence, in a large group of workers who perform the same task by using the same method, the fastest operator would produce approximately twice as much as the slowest operator. Our study showed that performance time in vocational school can vary in average from one to four (Bedny and Zelenin, 1989). This has been uncovered when students worked without time standard requirements.

There is a lot of difficulty in pace evaluation. One widely used method of pace evaluation in industry is based on subjective judgment. This method is called rating. Rating is a process during which a specialist compares the pace of a blue-collar worker's performance with the observer's own concept of normal or standard pace. The latter can be understood as an average worker's pace that can be maintained during a shift without excessive mental and physical effort, assuming that the quality of work would be within the assigned standard.

An average person walking on a level grade at 3 miles/h (4.8 km) along a straight road is used to represent a normal walking pace. This criterion has been supported by physiological studies. It is a traditional type of activity that is also easy to compare with subjective feelings and psychophysiological measurements. Physiological studies demonstrate that energy expenditure per unit of covered distance is minimal if the speed of walking is between 4 and 5 km/h (Frolov, 1976). In evaluating the pace of performance, experts use methods that were developed in psychophysics. These methods are based on a subjective evaluation of such phenomena as subjective scaling for evaluation of noise and brightness. Similarly, this method may be carried out for the subjective evaluation of pace.

There are several different rating scales for the evaluation of work pace. For example, there is a scale where the standard or normal pace is 100. If the actual pace of performance is less than normal, it thus receives a number less than 100, and if actual pace is higher than standard, it receives a number above 100. These kinds of scales are based on psychophysical methods. Pace is designated by numbers. The last number that is assigned to the real pace of performance should be "0" or "5" (70, 75, 80, etc.). Pace evaluation can be done for individual elements of task whose duration is no more than 30 s (Barnes, 1980). After evaluating the pace of performing the elements and measuring performance time, the standardized performance time for each element of task is determined using the following formula:

$$S = T \times P,$$

where
  S is the standardized time for an element of task
  T is the time obtained during chronometrical measurement
  P is the coefficient of pace performance (it defines the relationship between evaluated by expert pace of performance and standardized pace of performance)

For example, a real task element performance time is 0.30 min; the pace of performance is 90. Therefore, $S = 0.30 \times 90/100 = 0.27$ min.

The other method based on physiological evaluation of performance pace is an experimental one. In cases when a practitioner evaluates medium and heavy physical tasks, physiological evaluation of performance pace is possible. Oxygen consumption in calories per minute and heart rate in beats per minute can be utilized. It is more difficult to evaluate the pace of performance when cognitive components of activity dominate in the task.

Expenditure of energy at 4.17 kcal/min is equivalent to a pulse rate of 100 beats/min. Analysis of the publications (Lehmann, 1962; Rozenblat, 1975) demonstrate that a pulse rate of 100 beats/min or 4.17 kcal/min should be used as the benchmark for the boundary between acceptable and unacceptable strenuousness of work. It corresponds to the boundary between low and heavy physical work intensity according to Rozenblat's classification. In work conditions when the pulse rate increases beyond this standard, an additional break time is recommended.

In vocational schools when physical components of work dominate energy, expenditure should not exceed 3.7 kcal/min for boys and 3.2 kcal/min for girls (Kosilov, 1979). This means that the standard or normal pace assigned for teenage students based on subjective judgment should be equivalent to 70 units instead of 100 units assigned for adult workers. Such psychological methods as analysis of error rate, subjective evaluation of pace, and observation of external symptoms of fatigue can be used for the evaluation of students' pace.

Subjective judgment of a performer about his or her pace is also valuable. If a worker evaluates his pace as not optimal, the quality of work can deteriorate. Chebisheva (1969) had conducted the following laboratory experiment. Subjects had to sort wooden sticks with different colors matching the pace of the metronome strokes. At the beginning of the experiment, metronome strokes were set on slow pace. Gradually the pace of metronome strokes increased. Hence, the students should sort the sticks with differed pace. The following levels of performance pace have been discovered during this study:

1. Very low pace that was evaluated as uncomfortable
2. Optimal pace
3. Effortful or intensive pace
4. Difficult to achieve pace
5. Unachievable pace

It has been discovered that transition from very slow pace to optimal one reduces the amount of errors. This pace is conveyed by the most positive emotional state of subjects during task performance. However, further increase of pace causes increase in error rate. The effortful pace that exceeds optimal level is evaluated as emotionally tensioned and more difficult. The difficult to achieve pace is considered as excessive and can be sustained only during a very short period of time. The error rate is an important criterion for pace evaluation. It was discovered that the optimal pace activates subjects and motivates them to seek the most efficient task performance strategies. Gradual increase in pace is possible. After acquisition of optimal pace, it is possible to perform with higher pace. Task performance with the pace that insignificantly exceeds the optimal pace stimulates better performance. Therefore, the concept of optimal pace during training can be changed accordingly. Training with gradual increasing speed of task performance is known as *above real-time training*. This method has been applied in the air force pilot training (Miller et al., 1997).

The pace of experienced workers is relatively stable. In vocational training, the pace of performance is changing during skill acquisition. It has been discovered that transition to a higher level of performance cannot be reduced to increase the speed of performance. The ability to perform a task with a higher pace is accompanied by changes in the structure of activity (Bedny, 1979, 1981), which to a significant degree is a new kind of skills. It takes special training to prepare students to work with a required pace.

The concept of professional pace in ergonomics is important not only for training of blue-collar workers but also for the study of operator work in semiautomated and automated systems. However, the concept of pace of performance has not been studied in this field. Trying to transfer the result of reaction time studies to the work environment, one can assume that each operator's action is performed at the maximum pace and that each action does not influence the previous or the subsequent action. However, it is important to know not only the speed of isolated reactions but also how much time is needed for performance of the total task and particularly when it is performed in emergency conditions. The task is not a sum of independent reactions but rather a system of logically organized actions integrated according to a set goal. An operator never performs the task with the speed that is equivalent to the speed of the isolated reactions. For instance, it has been discovered that when a subject has to hit four targets instead of two, the pace of performance slows down (see Bedny et al., 2014).

The speed of cognitive actions mostly depends on their content because the pace of cognitive processes is less regulated voluntarily. A person can widely voluntarily regulate a speed of a motor action but the speed of cognitive actions depends primarily on the composition of mental operations within the cognitive actions. For example, a simultaneous perceptual action can be performed during 0.3 s. A successive perceptual action might require 0.8 s. The difference in execution time is not due to the speed of performing considered actions but primarily because successive action includes a number of additional mental operations. For complex perceptual action, the speed of unfolding of individual operations over time may be approximately the same as for the simultaneous perceptual action. The degree of automaticity with which the actions are performed depends on past experience and complexity of a task. The more complex the task is, the less is the probability that this task can be performed with the high level of automaticity and the pace of performance will be lower as well.

In the MTM-1 system, the pace of performance is equivalent to the walking speed of 5.7 km/h (Smidtke and Stier, 1961). However, according to physiological data, the standard pace for a physical job should be 4.8 km/h. This pace guarantees that energy expenditure does not exceed 4.17 kcal/min or a workload that is equivalent to a heart rate of 100 beats/min. These are physiological criteria that are considered as a border between acceptable and unacceptable workloads during performance of physical work. The MTM-1 system was developed for mass production, assembly work as in electronic industry, etc., where one cannot observe substantial physical efforts.

For such work, the pace of MTM-1 is considered to be optimal. However, according to experimental data, the pace offered by the MTM-1 system is too high even for mass production (Smidtke and Stier, 1961). Gal'sev (1973) recommends to use coefficient 1.1–1.2 to reduce the pace of performance. Only after this correction, physiological costs of performed work can approach the standard physiological levels. The level of automaticity of task performance in mass production is higher than during performance of tasks in automated or semiautomated system. Our study demonstrated (Bedny, 1979) that one has to consider three levels of work activity pace: *very high*, *high*, and *average*. A *very high pace* is slightly slower than the operator's reaction time to various stimuli. This pace is possible only in those cases when an operator reacts to isolated signals, using discrete actions in highly predictable situations. For example, an operator can have a high level of readiness to push a button or throw a switch when a particular signal appears. A *high pace* is that in which an operator performs a sequence of logically organized mental and physical actions in response to the appearance of various signals. It is essentially the same pace as the one offered by the MTM-1 system for motor activity. Such pace should be used when an operator works in emergency conditions and performs not isolated reactions but various tasks. The pace of performance for mental actions should be determined based on analysis of strategies of their performance in a particular situation. This refers us to the functional analysis of activity that will be discussed later. Conditions when an operator performs actions in a logically organized sequence lower the degree of his or her readiness to perform particular actions. An *average pace* is that in which an operator performs tasks at his or her own subjective time scale (when there are no time constraints).

8.3 Pace Formation Process and Mechanisms of Activity Self-Regulation

According to cognitive psychology, pace can be described in terms of stimulus and reactions. "The pacing factor defines the circumstances under which the operator proceeds from one stimulus to the next" (Wickens and Hollands, 2000, p. 371). In SSAT, the pace factor is considered from an activity self-regulation perspective. Therefore, the study of the speed of performance of the isolated reactions is not adequate for analyzing the pace formation process. It is important to understand how a subject can change the pace of his or her holistic activity when performing a logically organized sequence of actions. The purpose of this study is to investigate the temporal structure of activity rather than the temporal characteristics of isolated reactions or actions because the pace is determined by characteristics of the activity time structure and not by speed of performance of individual actions or movements. When it comes to the time structure of the overall activity, it is important to determine the duration of individual elements of activity and of activity in general, specifics of elements' distribution in time, logic of transition from one element to another, and so on. Hence, the goal in front of us is to study the pace formation process when subjects have to perform a sequence of logically organized actions for attainment of the goal of task.

In the experiment described later, we used the same device as the one in Section 6.4 of Bedny et al. (2014). However, the purpose of this study was different. In Figure 8.3, we present only the allocation of the start position for the index finger and the buttons that have been utilized by the subjects during the experiment. The device had horizontal and vertical panels for the subject on one side of the device. On the other side of the apparatus, there was the experimenter's panel. At the subjects' horizontal panel, there was a start position for the index finger (button 1), a middle button (button 2), and two right-edge buttons (buttons 3 and 4). One right-edge button had red and the other green color as shown in Figure 8.3.

There were two stopwatches on the experimenter's panel that were used to register the performance time of the first and second actions. After the signal is presented on the subject's vertical panel, he or she can move an index finger to an intermittent button (button 2) and then to the peripheral buttons (buttons 3 or 4) and press one of them depending on the color of the signaling bulb. This device allowed us to measure not only the whole task performance time but also the performance time of the separate actions. Here we also applied functional analysis of activity considering activity as a self-regulative system.

At the vertical subject's panel, there were one digital and four colored bulbs. In this experiment, we have utilized only green and red bulbs (the

FIGURE 8.3
Description of subject panel. 1, start position; 2, intermittent button; 3, red button; 4, green button.

same color as two device's edge buttons). Two stopwatches on the experimenter's panel allowed to measure the performance time of the first and second action in combination with their corresponding cognitive components. The summation of performance time of these actions provides information about the whole task's performance time.

Three groups of subjects were selected from the university student population. In each group, there were five male subjects. All of them had approximately the same physical characteristics and similar age as the subjects in the previous study. Each group performed a different task. Each subject performed the same task 50 times in 1 day. The subjects knew how many trials they were to perform every day. This information can influence the pace of activity. For example, Konopkin (1980) showed that when the subjects knew about a significant increase in the number of trials, this information led to increased reaction time. Subjects performed the same task for 3 days. In the previous chapter, when subjects performed a similar task, the task performance time was not specified. In these studies, the required task performance time is always indicated. Subjects received information about their task performance time (performance time of two actions) in all trials.

The first group performed the following task. After a red bulb is turned on, a subject removes his or her index finger from the start position and presses an intermittent button and then moves his or her finger to the red edge button and presses it. If the green bulb is turned on, a subject presses the green button. Subjects should perform this task in 0.9 s.

The second group performed the same task with the following differences. If a red bulb is turned on, they should press an intermittent button and then move their finger to the green edge button and press it. If a green bulb is turned on, they should press an intermittent button and then move their finger to the red edge button and press it. This task was more difficult because the displayed color of the bulb does not match the color of the button that the subject had to press. If a subject from the second group made a mistake, he or she received an electric shock. In this experiment, the time standard for task performance was also 0.9 s.

The third group performed the same task as the second group. However, their performance time was 0.8 s. In all these experiments, we did not measure a reaction time to an isolated stimulus when a subject reacts with a single action as it is done in traditional reaction time measurement procedures. Subjects had to perform a logically organized sequence of actions. The second difference included a requirement not to react with maximum speed but to perform the task according to the time standard requirements. The results of measurements carried out with an accuracy of up to 0.01 s. We present later only the average data for all three groups (see Tables 8.1 through 8.3). Let us consider the result of the experiment with the first group where the time standard was 0.9 s and electrical shock was not used. Only information about the performance time of two actions is presented.

TABLE 8.1

Average Time of Task Performance in the First Group of Subjects (Seconds)

Subjects	First Day			Second Day			Third Day		
	First Action	Second Action	Two Actions	First Action	Second Action	Two Actions	First Action	Second Action	Two Actions
1	0.53	0.34	0.87	0.58	0.31	0.89	0.55	0.32	0.87
2	0.52	0.43	0.95	0.46	0.37	0.83	0.48	0.34	0.82
3	0.54	0.35	0.89	0.57	0.33	0.9	0.53	0.29	0.82
4	0.45	0.33	0.78	0.35	0.29	0.64	0.35	0.27	0.62
5	0.48	0.37	0.85	0.44	0.36	0.8	0.43	0.31	0.74
Σ	0.50	0.36	0.86	0.48	0.33	0.84	0.47	0.31	0.77

TABLE 8.2

Average Time of Task Performance in the Second Group of Subjects (Seconds)

Subjects	First Day			Second Day			Third Day		
	First Action	Second Action	Two Actions	First Action	Second Action	Two Actions	First Action	Second Action	Two Actions
1	0.54	0.32	0.86	0.49	0.30	0.79	0.50	0.31	0.81
2	0.61	0.42	1.03	0.42	0.32	0.74	0.40	0.30	0.70
3	0.67	0.47	1.14	0.48	0.40	0.88	0.46	0.35	0.81
4	0.62	0.45	1.07	0.58	0.50	1.08	0.55	0.43	0.98
5	0.63	0.41	1.04	0.50	0.43	0.93	0.42	0.40	0.82
Σ	0.61	0.41	1.03	0.49	0.39	0.88	0.46	0.36	0.82

TABLE 8.3

Average Time of Task Performance in the Third Group of Subjects (Seconds)

Subjects	First Day			Second Day			Third Day		
	First Action	Second Action	Two Actions	First Action	Second Action	Two Actions	First Action	Second Action	Two Actions
1	0.40	0.58	0.98	0.32	0.43	0.75	0.37	0.40	0.77
2	0.46	0.29	0.75	0.39	0.33	0.72	0.43	0.29	0.72
3	0.49	0.38	0.87	0.29	0.38	0.67	0.29	0.39	0.68
4	0.84	0.45	1.09	0.47	0.35	0.82	0.43	0.31	0.74
5	0.60	0.36	0.96	0.44	0.33	0.77	0.45	0.34	0.79
Σ	0.52	0.41	0.93	0.38	0.36	0.74	0.39	0.35	0.74

Experimental data for the first group that performed the simplest task demonstrate that this group achieved the time standard requirements on the first day (0.86 s). On the following 2 days, the task performance time reduced negligibly, and on the third day, the performance time of the task was 0.77 s decreasing by 0.09 s. To assess the statistical significance for within the group comparison, we used one-way within-subject ANOVA ($F(2,8) = 6.66; p < 0.05$). Post hoc Tuckey t-test for day 1 is significantly slower than for day 3 ($p < 0.05$). The difference between performance times on the first and second days was statistically insignificant.

The obtained result demonstrates that subjects already achieved their time standard requirements on the first day of experiment. Hence, the pace of performance has been developed in the first day for this group of subjects. Observation, discussion with subjects, and analysis of performance time of separate actions and of the whole task help to understand strategies that were utilized by the subjects during the task performance. Subjects selected the following strategy. If their performance time was more than time standard requirements, they increased the speed of performance significantly. After a number of trials, they selected one subjectively preferable result. This result was considered as a subjective standard of success. Subjects compared their results not with objectively given time for task performance but with a subjectively selected standard of successful result. The selected standard was slightly below the established time for task performance. The standard has been selected in such a way that, despite the variation in the results, it guaranteed that the required time of task performance would be achieved and subjects did not spend more time than has been required. Subjective standard includes not only criteria that a subject is well aware of but also some ambiguous subjective feelings. Thanks to repeated trials, such criteria become sufficiently accurate. The pace of performance that corresponded to the subjective standard of success was considered by subjects as optimal and the task performance time that is slightly less than the selected standard is considered as a successful result. Sometimes subjects slightly corrected their subjective standard of a successful result. Therefore, the purpose of the strategy was not simply to increase the speed of performance but to stabilize the result in relation to the subjective standard.

An activity goal that is accepted or formulated by the subject does not always determine the exact result of activity. This can be explained by various factors. For example, a goal often does not include all necessary information about the required results of activity. Moreover, mental representation of a desired result often can be developed only during the performance process. Different subjects can formulate a different mental representation of a desired result when they have the same goal. Thus, subjective standard can be gradually developed and deviated from the established goal.

Depending on the motivational factor and significance of the task, which are other mechanisms of activity self-regulation, subjective standard of successful result may be above or below the externally given goal of task. Thus, sometimes a subjectively accepted standard of successful result can be in contradiction with the task requirements that were determined by the goal of task. In our study, subjects were motivated to perform the task successfully and they gradually formed an adequate standard that guarantees to meet a required goal.

The other interesting aspect of the strategy of task performance was the fact that the first motor action from the start position to the intermittent button (button 2) required more time than the following movement with the same distance to the green or red button. The difference between the first and the second actions was 0.14, 0.15, and 0.16 s for the first, second, and third days. This difference was statistically significant (for the first day, within-subject t-test $t(4) = 6.67$, $p < 0.01$; for the second day, $t(4) = 3.35$, $p < 0.05$; and for the third day, t-test $t(4) = 5.16$, $p < 0.01$) (see Table 8.1).

The difference between performance times for the first and second actions was similar and was consistent within all 3 days of experiment, meaning that the activity strategy did not change during all these periods of time. Cognitive functions dominated during performance of the first action. Therefore, subjects developed a program of performance for two actions.

The second action was largely performed automatically. Subjects develop their own strategy that was not completely predetermined by instructions given to the subjects. The differences in performance time of the first and second actions can be explained by the fact that decision to press the green or red button was made during execution of the first action. We want to stress the fact that the experimental conditions in this study are different from the conditions of the experiment in Section 6.4 of Bedny et al. (2014). In contrast, in the present experiment, two actions were to be carried out within a certain period of time (0.9 s) but not with maximum speed when psychologists measure the reaction time.

In the experiment with the second group, subjects (more complex task) could receive an electric shock if they made a mistake. The subjects perceived the task as not only more difficult but more significant, since the untimely and improper execution of the task leads to the receiving of electric shock. This leads in changes of task performance strategies and in increase in task performance time (see Table 8.2).

On the first day of the second experiment, the difference between performance times for the first and the second action was 0.2 s, and on the second and third days, it was 0.1 s. The difference was statistically significant (for the first day, within-subject t-test $t(4) = 21.10$, $p < 0.001$; for the second day, $t(4) = 4.72$, $p < 0.05$; and for the third day, $t(4) = 3.99$, $p < 0.01$).

For the first group of subjects (see Table 8.1), the difference between the first and the second actions was similar for all 3 days. Therefore, the performance strategy during the 3 days did not change for the first group of subjects. In the second group, subjects gradually changed their strategy of task performance under the threat of electric shock. This was clearly manifested when subjects made an error and actually received an electric shock.

The performance time for the first day of the experiment was 1.03 and exceeded the required time standard that was 0.9 s. In the experiment, subjects had two factors that were determining their success. One of them was the time factor and the second one was precision, violation of which was punished by electric shock. At the first stage, the precision factor had the higher level of significance for subjects. Therefore, subjects sacrificed the time factor and improved the accuracy of performance. The purpose of such a strategy was to avoid electrical shock.

On the second day, three subjects performed the task according to the time standard requirement. An average performance time matched the required time standard. On the third day, the pace of performance slightly increased. Therefore, the required time standard has been achieved mainly on the second day. The difference between task performance times on the first and the second days was 0.15 s and on the first and the third days was 0.21 s. The difference between task performance times in the second and the third days was only 0.06 s. The difference between task performance times during 3 days was statistically significant (one-way within-subject ANOVA, $F(2,8) = 9.11$, $p < 0.01$). Post hoc Tuckey HSD t-test for day 1 is significantly slower than days 2 and 3 ($p < 0.05$).

Comparison of performance time on the third day for the first and the second group demonstrated that the pace of task performance was approximately the same. However, the dynamics of the pace of performance was different. In the first group, the required pace has been achieved on the first day, but in the second group, where the task was more complicated, this pace was achieved only on the second day. This can be explained if we consider complexity of the task and strategies of task performance.

The task was perceived as more difficult by subjects. The second action became more significant for them. The subjects made a decision about selecting the second action during the first action performance, and this decision has been now double-checked before starting the second action. As a result, time differences between the first and the second actions were reduced. If the first group performed the second action largely under automatic control, in the second group, subjects controlled the second action also consciously.

The second group selected a cautious strategy based on the precision factor in the first day. The subjective standard of successful result according to time parameters exceeded a predetermined time standard. Subjects sacrificed such factor as time to increase precision. Only on the third day, the subjective standard of successful result became sufficiently lower than the time standard, which was 0.9 s. There were other changes in strategies of task performance. According to instruction, the second group had to use the bulb color and the button color for decision making. However, in the second group, subjects gradually abandoned these distinguishing features and started using the space position of the red and green buttons instead.

The space position is a more complicated distinguishing feature than the color, but the interference in color features during decision making made this feature more difficult to use during decision making. As a result, all subjects gradually started to ignore color and use only the space position of the buttons for their decision making. This strategy eliminated the interference of the colors in the second group of experiments. Hence, subjectively relevant task conditions or mental representation of task has changed. This transformation in the strategy of performance was achieved at the end of the second day by all subjects. The performance time was reduced even on the third day. This shows that various components of the strategy have different dynamics of their formation. Not all components of performance strategies are changed consciously. For example, when we asked subjects in these two groups which action requires more time, seven said that the second action required more time than the first one.

In the third experiment, when subjects had more rigorous time standard (see Table 8.3), there were further changes in strategies of task performance.

The second action became more consciously controlled. Differences in performance time of two actions kept decreasing. Two subjects even changed their strategies in such a way that the performance time of the second action became greater than the performance time of the first action because the increased time standard (0.8 s instead 0.9 s) and the contradiction between the bulb color and the color of buttons made the second action a more significant component in the task performance. As a result, the second action became more cognitively controlled. The difference between task performance times during the 3 days was statistically significant (one-way within-subject ANOVA, $F(2,8) = 15.11$, $p < 0.01$). The post hoc Tuckey HSD t-test for day 1 is significantly slower than days 2 and 3 ($p < 0.05$). The difference between the performance times of the first and the second actions during all 3 days was not statistically significant (for the first day, within-subject t-test $t(4) = 1.55$, $P < 0.2$; for the second day, $t(4) = 0.36$, $P < 0.73$; and for the third day, $t(4) = 1, 01$, $p < 0.37$). This study demonstrates that externally given instructions do not exactly predetermine strategies of the task performance. Subjects develop their own understanding of a goal, develop a mental model of a situation and subjective standard of success, evaluate significantly different elements of the task, etc. As a result, dynamic strategies of activity performance can be developed. All of these are a result of a complicate process of activity self-regulation. Hence, the pace of task performance is actively developed during the self-regulative process. Information presented to subjects about temporal parameters of activity is critically important. The duration of the pace acquisition depends on the efficiency of the mechanisms of self-regulation. The more complicate the task is, the more stages of strategy transformation are needed. As a result, the duration of the pace formation process increases. A mental model of a situation constantly changes depending on the stage of the learning process. Allocation of attention between different elements of activity is changing as well. Some actions become more automatic, and the other more cognitively controlled. Elements of activity that become automatic very often are performed simultaneously. As a result, the pace of activity performance can be increased without increasing the speed of separate actions. This in turn influences the pace formation process. The functional analysis of activity is the foundation for the self-regulative concept of learning. Learning is considered as a process of strategy transformation during activity performance. The more complicate the skills are, the more intermittent strategies are utilized by the students. Formation of the pace of performance can be explained only through analysis of mechanisms of activity self-regulation and its derived strategies of activity performance utilized by the learners during their skill acquisition process. In the following section, we will consider some principles of pace regulation when a subject performs tasks of various complexities. We want to draw attention to the fact that the same instructions were given to all three groups. However, the relationship between speed and accuracy was not predetermined by these instructions. Similarly, other aspects of developed strategies were not defined by instructions. The subjects choose their strategy based on analysis of the objective conditions of the task performance, their past experience, evaluation of the results of performance, significance of the task for subjects, and their individual characteristics. The obtained result demonstrates that instructions do not uniquely determine possible strategies of the subjects in any experimental conditions. Therefore, mechanisms of self-regulation and strategies of activity derived from them should be taken into consideration in all experimental studies. This aspects of human activity has not considered in the experimental study of cognitive psychology. In conclusion, we note that in the analysis of strategies of task performance, we have used such functional blocks of self-regulation as a goal, subjectively relevant task conditions (dynamic mental model), subjective standards of successful results, assessment of task difficulty, assessment of sense of task (significance), and formation program of performance (see Sections 3.2 and 3.3 of Bedny et al., 2014).

8.4 Pace Regulation and Task Complexity

Complexity is one of the most important cognitive characteristics of a task that may affect the strategies of its performance. Subjects are always trying consciously or without clear awareness to optimize their activity in accordance with the level of task complexity. Difficulty is the subjective equivalent of the objective complexity of the task. This means that, according to the perceived difficulty of the upcoming task, subjects develop their strategy and allocate their efforts over time. In this regard, the purpose of our study was to find out how the complexity of the task affects the ability of a performer to regulate his or her pace of performance.

We intended to analyze not only the performance pace of isolate actions but also the sequence of logically organized cognitive and motor actions that made the experimental task closer to real ones. The experimental tasks varied in their complexity.

The one useful way of studying the ability of pace regulation is analysis of the difference threshold (Krilov, 1970; Zalkind, 1966). Typically, the difference threshold is measured in situations when subjects perform simple rhythmic motions multiple times in the same sequence. For example, subjects had to perform the tapping test, rotate a crank multiple times, and so on. Then they were asked to slow down the rate of movement in the faint magnitude (just-noticeable difference). Changes in the performance pace of multiple actions performed in sequence and the pace of regulation of the whole task have not been studied.

However, it is important to know how subjects can regulate their pace during the performance of various tasks. One of our assumptions was that with the increasing complexity of the task, the difference threshold will be more significant for the more complex task. Next, we were interested in the factors involved in the regulation of pace when subjects do not perform the same actions multiple times but execute a complete task. For example, subjects can adjust their pace based on estimating performance time of the whole task or some of its elements. Thus in our study, we measured the difference threshold for the performance pace of sequentially executed motor actions that had some logical organization and therefore required sufficiently complex cognitive regulation. Five subjects took part in this experimental study. For this experimental study, we used the stand that has been depicted in Figure 8.3. Subjects perform actions in training sessions with maximum speed. During the first part of training, they received information about their performance time. In the final stage of training, they alternated performing tasks with and without information about their execution time. After relative stabilization of actions' performance time at maximum speed without being informed about the performance time, training was completed.

Subjects were trained to perform three tasks with different levels of complexity that were used in the main experiment. Training procedures were carried out in 2 days. On the third day, the subjects were involved in the main experiment where the subjects performed five training trials before each new task. In these trials, they received information about the performance time. Only after this, they were involved in the main experiment, when they worked without information about the performance time, but their preliminary trials allowed them to recall how fast they performed the same tasks in the training sessions.

The main experiment consisted of three series. In the first series, the subjects performed a simple task. This task required performing a single action with maximum speed. When the white bulb was turned on, subjects moved their index finger from the start position to a middle button and pressed it. Performance time was recorded. The results of measurements carried out with an accuracy of up to 0.01 s.

Ten trials were used in the main part of the experiment. The average performance time with maximum speed was calculated. After that, subjects received instructions to perform the same action 10 times reducing the speed at just-noticeable value. In both conditions, information about performance time was not given. The average performance time was calculated. The average performance time of one action with maximum and just-noticeable speed reduction is presented in Table 8.4 (see Task 1).

The second series of experiments included the task that consisted of a sequence of actions (average complexity). This task included a simple decision to choose the appropriate button. When the red bulb is turned on, a subject moves his or her index finger from the start position to the middle button and presses it. Then he or she moves his or her finger to the red edge button and presses it. If the green button is turned on, a subject performs the same sequences of actions and presses the green button. All actions should be performed with maximum speed. In five preliminary trials, subjects were informed about their actions' performance time. In the main experiment, actions' performance time has not been presented to the subjects. Subjects performed 20 trials in the main experiment. Then they were instructed to slow down actions' speed at just-noticeable value and perform another 20 trials. The average performance time with maximum and slower speed was calculated (see Table 8.4; Task 2).

TABLE 8.4

Performance Time of Actions with Maximum Speed and a Just-Noticeable Slower Speed (Only Average Data Are Presented)

Performance Time of One Action (Task 1)	
Performance time of one action with maximum speed	Performance time of one action with just-noticeable slower speed
0.402	0.497

Performance Time of Two Actions with One Logical Condition (Task 2)					
Performance time of actions with maximum speed			Performance time of actions with just-noticeable slower speed		
1	2	3	4	5	6
First action	Second action	Two actions	First action	Second action	Two actions
0.501	0.403	0.904	0.52	0.503	1.023

Performance Time of Two Actions with Two Logical Conditions (Task 3)					
Performance time of actions with maximum speed			Performance time of actions with just-noticeable slower speed		
1	2	3	4	5	6
First action	Second action	Two actions	First action	Second action	Two actions
0.508	0.67	1.18	0.55	0.7	1.25

The third series of experiments was performed similarly and the only difference was that the task was of an even higher complexity (see Table 8.4, Task 3). This series of experiment involved using an additional digital bulb. As a result, subjects had to take into account additional logical conditions. If the white bulb is turned on and simultaneously an even number appears on the digital bulb, then subjects should push an intermittent button and, after that, push an edge green button. If the white bulb is turned on and an uneven number appears on the digital bulb, then subjects should press an intermittent button and then press a red edge button.

Thus, the subjects performed three tasks of increasing complexity. In the first task, subjects performed only one action. In the second and third task, subjects performed two actions in sequence. The subjects performed a sequence of motor actions in combination with mental operations according to a particular logic. This made the task similar to real tasks, where actions are performed in sequence according to a specific logic. We have studied strategies of the subjects' performance and their ability to regulate the pace of task performance of varying complexity. All the average results are shown in Table 8.4.

Based on obtained data (see Table 8.4), we measured the following indexes:

1. $\Delta T_1 = T_{1SL} - T_{1F}$
2. $\Delta T_2 = T_{2SL} - T_{2F}$
3. $\Delta T_{SUM} = T_{SUMSL} - T_{SUMF}$
4. $S_1 = \Delta T_1 / T_{1F}$
5. $S_2 = \Delta T_2 / T_{2F}$
6. $S_{SUM} = \Delta T_{SUM} / \Delta T_{SUMF}$

where

T_{1SL} is the execution time of the first action at a slower pace
ΔT_{1F} is the execution time of the first action at a faster pace
ΔT_{1F} is the execution time of the first action at a faster pace
T_{2SL} is the execution time of the second action at a slower pace
T_{2FL} is the execution time of the second action at a faster pace
ΔT_{1F} and ΔT_2 demonstrate the difference in performance time of considered actions when they are performed in a slower and faster pace
T_{SUMSL} is the execution time of the two actions at a slower pace
T_{SUMF} is the execution time of the two actions at a faster pace
ΔT_{SUM} is the difference in performance time of two actions when they are performed in a slower and faster pace
$S_1; S_2; S_{SUM}$ is the ratio of the minimum change of the time performing the actions to the initial time of their execution at a faster pace

The results of measurements carried out with an accuracy of up to 0.01 s.

Mental actions were combined with the motor ones. Therefore, the time of cognitive actions was not specifically extracted. So when describing the experiment, we will call movement from the start position to the intermediate button the first action, and movement from the intermediate button to the red or green edge buttons will be called the second action.

Measured indexes and statistical analysis of obtained data are presented later. The statistical significance of the observed mean difference in all three tasks was checked by using the paired t-test.

In the first task, when the subjects had to voluntarily slow down, only one performing action (the easiest, Task 1) yielded the following results. In accordance with formula 1, we have determined ΔT_1:

$$\Delta T_1 = T_{1SL} - T_{1F} = 0.49 - 0.40 = 0.09 \text{ s}$$

Then utilizing formula 4, S_1 has been defined as follows:

$$S_1 = \Delta T_1 / T_{1F} = 0.09/0.4 = 0.23 \text{ or } 23\%$$

Statistical analysis demonstrates the following.

When performing only one action, the performance time was faster when participants were instructed to work with maximum speed than when participants were instructed to perform the task a little slower (mean max = 0.402 s, SD = 0.052; mean slower = 0.497 s, SD = 0.038; paired t-test, $t(9) = 4.9$, $p = 0.0008$).

The following results have been obtained in the second task with average task complexity (see Table 8.4; Task 2):

According to formula s 1 and 2, the values of ΔT_1 and ΔT_2 were determined:

$$\Delta T_1 = T_{1SL} - T_{1F} = 0.52 - 0.50 = 0.02 \text{ s}$$

$$\Delta T_2 = T_{2SL} - T_{2F} = 0.50 - 0.40 = 0.1 \text{ s}$$

When performing two actions as fast as possible, action one (column 1) was slower than action two (column 2) (paired t-test, $t(19) = 5.18$, $p < 0.0001$). By contrast, when participants were performing two actions slower (Table 8.4; Task 2), performance time for actions one (column 4) and two (column 5) were approximately the same and there were no statistical difference between them (paired t-test $t(19) = 1.21$, $p = 0.24$). There was no difference in performance time between action 1 performed at maximum speed (column 1) and action 1 performed slower (column 4) ($t(19) = 1.59$, $p = 0.13$). Action two was faster when performed at maximum speed (column 2 vs. column 5) ($t(19) = 3.50$, $p = 0.024$).

Based on formula 3, ΔT_{SUM} has been calculated:

$$\Delta T_{SUM} = T_{SUMSL} - T_{SUMF} = 1.02 - 0.90 = 0.12 \text{ s}$$

The sum of performance times for actions one and two was less when participants were instructed to perform at maximum speed (column 3 vs. column 6, $t(19) = 3.63$, $p = 0.002$).

The value of S_1 has been determined according to formula 4 ($S_1 = \Delta T_1/T_{1F}$):

$$S_1 = 0.02/0.50 = 0.04 \text{ or } 4\%$$

Since the difference in ΔT_1 was not statistically significant, S_1 should be considered as an unreliable indicator for evaluating changes in the performance pace of the first action.

S_2 was determined according to formula 5 ($S_2 = \Delta T_2/T_{2F}$):

$$S_2 = 0.1/0.4 = 0.25 \text{ or } 25\%$$

S_{SUM} was calculated as follows:

$$S_{SUM} = \Delta T_{SUM}/\Delta T_{SUMF} = 0.12/0.90 = 0.13 \text{ or } 13\%$$

Analysis of the data obtained in the second experiment indicates that general reduction in the task execution time (performance time of two actions) was due primarily to slowing down in the second action. The first action has been performed approximately with the same speed in both cases: when instructed to perform at maximum speed and when instructed to slow down.

The following results have been obtained in the third task (the most complex task; see Table 8.4):

$$\Delta T_1 = T_{1SL} - T_{1F} = 0.55 - 0.51 = 0.04;$$

$$\Delta T_2 = T_{2SL} - T_{2F} = 0.70 - 0.67 = 0.03;$$

$$\Delta T_{SUM} = T_{SUMSL} - T_{SUMF} = 1.25 - 1.18 = 0.07;$$

$$S_1 = \Delta T_1/T_1 F = 0.04/0.51 = 0.07 \text{ or } 7\%; \quad S_2 = \Delta T_2/T_2 F = 0.03/0.67 = 0.04 \text{ or } 4\%;$$

$$S_{SUM} = \Delta T_{SUM}/\Delta T_{SUMF} = 0.07/1.18 = 0.06 \text{ or } 6\%$$

Let us consider statistical data.

Similarly, there is a statistically significant difference between the first and second actions' performance time in both conditions: complex action 1 performed with max pace vs. action 2 (column 1 vs. column 2 [mean time action 1 = 0.508, mean time action 2 = 0.67, $t(19) = 8.25$, $p < 0.0001$]) and complex action 1 performed with slower pace vs. action 2 (column 4 vs. column 5 [$t(19) = 7.85$, $p < 0.0001$]).

There is a small but still statistically significant difference between action 1 of max speed and action 1 of slower speed (column 1 vs. column 4, $t(19) = 2.81$, $p < 0.01$). There is no difference in performance time between action 2 with maximum speed and action 2 with slower speed (see action 2 max speed vs. action 2 slower speed [column 2 vs. column 5, $t(19) = 1.02$, $p = 0.32$]).

In the task with average complexity, the difference in performance time of two actions was 0.12 s (see Table 8.4). In the more complex task, this difference is reduced to 0.07 s (see Table 8.4; Task 3). However, this difference was statistically significant (sum max vs. sum slower [column 3 vs. column 6, $t(19) = 2.52$, $p = 0.02$]).

The decrease in the speed for action 2 from when the instruction was to perform with maximum speed to when the instruction was to perform with slower speed was greater when the task was simpler (Table 8.4; Task 2 vs. Task 3 [within-subject ANOVA, $F(1,57) = 4.77$, $p = 0.03$]).

In the third series of experiments when the task was the most complex one, slowdown in the pace of the task performance was only 0.07 s and less than during performance of the second task with average complexity where the difference was 0.12 s. Experimental results demonstrate that increase in task complexity at the particular level decreases the ability to voluntarily regulate the work pace, where pace regulation is regarded as subjects' ability to maintain a given speed of activity over time.

Let us consider qualitative analysis in greater detail. The instructions were for subjects to perform tasks of varying complexity at the first stage with maximum pace and then just subjectively noticeably slow down the pace. In psychophysics, just-noticeable value is known as the difference threshold. From a position of self-regulation, there are some contradictions in such requirements as decreasing speed of performance at just-noticeable value. A slower pace should be close to the maximum one but at the same time should be certainly slower than the maximum pace. There were no external indicators to assess the pace and evaluation of the two paces should be performed based on subjective criteria that performers kept in their memory. To perform the task according to presented instructions, subjects choose certain strategies. We can describe them from the standpoint of the theory of self-regulation by paying attention to the most important function blocks or mechanisms of self-regulation in this situation (see Bedny et al., 2014.; Figures 3.1 and 3.2): *stable model* (block 13), or *dynamic model* (block 9); *assessment of task difficulty* (block 8); *assessment of sense of task* (block 7); *formation of a program of task performance* (block 14); *formation of the level of motivation* (block 6); *subjective standard of successful result* (block 19). These blocks are the most important for regulating the pace during the considered tasks' performance.

Block 13 or block 9 (stable or dynamic model of the task) give subjects information about the main characteristics of the task in a particular situation. This information is only partially reflected at a conscious level. Further, this information affects the assessment of the task difficulty (block 8) and significance (block 7) and formation of the level of motivation (block 6). Interaction of these blocks is important for the formation of a program of task performance (block 14). The factor of task complexity has a particularly negative impact on regulation of pace because subjects have no external criteria for assessing the pace and adequately adapting to such requirement as slowing down the maximum speed of task performance just noticeably. Analysis of the main blocks that are important in the pace regulation demonstrates that not only cognitive but also emotionally motivational factors can influence pace regulation.

In the first task when subjects performed only one action, they voluntarily have reduced the speed of a single action by ≈23%. In Krilov's (Krilov, 1970) study, it was shown that when the subjects performed repetitive actions (the same actions performed multiple times without interruption) such as rotating the handle a certain number of times and tapping actions, slowing down the pace just noticeably resulted in 20%–22% pace reduction. Approximately the same value has been received in our first experiment, when subjects performed only one action (simple task).

In the second task when subjects performed two actions in sequence (task of average complexity), subjects performed the first action with the same speed (0.50 s for maximum speed and 0.52 s when they slowed down). The speed of the first action was practically unchanged. Subjects reduced speed first of all during performance of the second action. The slowdown of the second action in the second task S_2 was 25%. This value is equivalent to a difference threshold. The slowdown of the whole task S_{SUM} was 13% and the slowdown of the first action was 4%. The values of S_{SUM} and S_1 were less than a difference threshold for repetitive motor actions. Hence, subjects can reliably detect a difference in the pace when performing the second task only based on comparing the speed of performance of the second action. Thus, subjects only consciously regulate the speed of the second action without consciously regulating the speed of the first action. Such strategy made regulation of pace simpler and resembled the strategy of pace regulation for the first task. Subjects believed that they slowed the speed of both actions in a similar way. This strategy is more reliable since it is easier to control the pace of only the second action. Subjects are striving to optimize their strategies based on the difficulty criterion. Increasing performance time of the first action in the faint magnitude requires consideration of the obtained result during the slowing down of the second action. The subjects simplified their strategy. They left the performance time of the first action practically unchanged while decreased the performance time of the second action. This ensured a reduction in the time of whole task performance on a subtle amount. Such a strategy was formed mostly unconsciously and subjectively perceived as most convenient. Conscious regulation of the pace in this case was very fragmented and was quickly forgotten by subjects. The pace of the second task as a whole was reduced only by 13% because the first action was performed approximately at the same speed.

We expect that in more complex tasks (the third task), reducing the pace should be more significant. However, in the experiment, the opposite result was obtained. Reducing the pace in complex tasks was very minor. For subjects, it is more difficult to reliably and consciously detect a difference between the paces of task performance with maximum and slower pace for the most complex task (Task 3) (see Table 8.4). Subjects claimed that they reduced the speed of task execution in all three experiments at approximately the same value. But in reality, this value is not the same.

Increasing the complexity of the task and therefore its subjective difficulty is also accompanied by increasing the motivation level and its related emotional tension, which can inadvertently result in a higher performance pace. In other words, there is a possibility to overshoot the pace of work, or to involuntary increase it. These factors are manifested less when performing simple tasks and are more pronounced for more complex tasks. The results of the third experiment, observation, and discussions with the subjects confirmed these conclusions.

Increasing task complexity leads to a situation when subjects could not focus much attention on separate actions. The pace of the first and second actions in the third task has been reduced by approximately the same negligible amount because increased task complexity did not allow subjects to concentrate on individual actions. For the more complex and, therefore subjectively, more difficult task, the ability of conscious and voluntary regulation of the pace decreases in general. In the third task, subjects utilize simpler strategies of pace regulation without concentrating on the speed of separate action performance. As a result, the speed of performance of the first and second actions in the third task varies in various conditions in approximately the same range. There is only some tendency for reducing the speed of performance of some element of the task. The role of conscious and voluntary pace regulation decreases for more complex tasks.

With increasing complexity of the task, the subject evaluates the task as more difficult.

In such conditions, important factors influencing the pace of performance are emotional and motivational mechanisms of activity regulation.

Increasing the complexity of the task leads to elevation of emotionally motivational tension. This factor manifests itself in involuntarily raising the pace. Discussion with subjects showed that they did not notice the fact that they almost did not reduce the pace of task performance when asked to do it by the instructor. The difference in performance time of more complex tasks with faster and slower pace is reduced.

When a subject performed the same motor action multiple times (tapping test and rotation of crank), his or her pace varied within approximately 5% (Krilov, 1970). When a subject was trying to make just-noticeable difference in his or her pace, this new pace also varied in a similar range. This variation was taken into account by a subject intuitively when he or she was changing his or her pace. Therefore, when a subject changes his or her pace on just-noticeable value, such difference has to exceed considered above variation of actions. A 20%–22% change of pace at just-noticeable value provides overlapping of such variation.

However, in our example, a subject performs a logically organized sequence of actions and this explanation is not sufficient. In this case, a subject has to take into account not only the variability of activity but also a variety of other factors. A subject consciously or unconsciously creates complex strategies of activity that allow achieving a required goal with the least efforts. This becomes possible due to activity self-regulation.

Studies indicate that a factor of task complexity contributes to the search for strategies that provide an easier way to achieve a goal. The search of such strategies can be performed consciously or unconsciously. Therefore, the subjects cannot always clearly understand their strategies of performance when they regulate their pace of performance. The possibility of pace regulation depends on the listed above mechanisms of activity self-regulation.

Interesting data on the regulation of the pace were obtained by Konopkin (1980). They provide additional evidence that demonstrates the interdependence of the pace formation process and mechanisms of self-regulation of activity. Let us look at them briefly. In one experiment, subjects were presented from one to eight signals in different series. In each series of experiment, there was the same number of alternatives and only one interval between stimuli was used. After appearance of signals, subjects had to respond with maximum speed. Two intervals were used in the experiment: 1 and 3 s. The average response time was calculated for each series. Subjects received special training before taking part in the main experiment. It has been revealed that despite receiving the same instruction, *act with maximum speed*, subjects' reaction time significantly varied depending on the interval between stimuli. For 1 s intervals, the reaction time was significantly shorter than for 3 s intervals.

Similar data have been obtained when five intervals were used. Intervals between stimuli stayed the same within each series. They were as follows: 3.0, 2.0, 1.5, 1.0, and 0.75 s. A 5 min break followed every session of the experiment. Before each session, the subjects received information about the pace of incoming signals. With shortening of the intervals between presenting of signal up to 1.0 or 0.75 s, reaction time became shorter.

In other groups of experiments, Konopkin (1980) used false information about the pace of presenting the signals. The signals were presented by using only 1.5 s intervals. In one experiment, subjects were informed that there will be a 1 s interval and 2 s in another one. The reaction time was shorter in conditions when subjects were informed that 1 s interval will be used. The difference was statistically significant despite the fact that in real experiments, the same 1.5 s interval was used. The study also was conducted in conditions of gradual change of pace of stimuli presentation, when subjects did not know about it. It was found that the reaction time was changed only when the subjects were aware of a change in pace. Changes in the speed of reactions were explained by different mental representations of tasks. Such information in our model of self-regulation is associated with block 9 (see Sections 3.1 and 3.2 in Bedny et al., 2014). However, interaction with other functional blocks should also be taken into account.

Konopkin and Zhujkov (1973) discovered that the reaction time also depends on the mental representation of the work duration. The longer the subject predicted to perform some tasks with maximum speed, the lower the speed of performance. This study demonstrated that a performer can independently self-program or self-regulate his or her pace and his or her activity in general and extract information in order to do that. It should be remembered that given instructions also affect the pace of task performance. Nojivin (1974) determined that the speed of performance of even a simple reaction significantly changes depending on the instructions utilized by an experimenter.

Presented in this section, material demonstrates that in cognitive psychology, scientists focused on the issue of *speed–accuracy trade-off* and did not pay sufficient attention to mechanisms of activity self-regulation and strategies of task performance.

It was assumed that a given instruction completely determined the reaction time. However, it is not always the case. Subjects evaluate not only instructions but the task at hand and the conditions of its performance. The same instruction can produce totally different strategies of task performance in different conditions.

In contrast to studies of repeated actions that were performed many times in each trial, we studied tasks in which actions were performed only once in each trial. We have also investigated the sequence of different actions performed in each trial that made our research settings similar to real-world tasks when actions are performed in sequence. Studies have shown that the pace formation is connected with the choice of appropriate strategies of performance that depend on mechanisms of activity self-regulation.

Transition from one pace to another leads to a reconstruction of the strategies of performance associated with redistribution of attention on various elements of activity, uneven changes in duration of individual actions, possibility to combine elements of activity in time, etc. Depending on task complexity, performance strategies and the speed of action execution vary. Subjects are not quite aware of how they change the pace of performance. Analysis of changes in the speed of individual reactions is not sufficient for analyzing the pace of activity as a whole. Various reactions usually are performed with maximum speed, but in practical situations, the pace of performance in most cases is below the maximum speed. Performers carry out not just isolated and independent reactions but rather an interdependent sequence of actions. The pace of activity depends not only on the speed of performance but also on the structure of activity, subjectively formed ideas about a task-problem, individual characteristics, and past experience of a performer. Depending on these factors, workers can choose a wide range of strategies for task performance. Such activity is not performed in a pace equivalent to the reaction time. The speed–accuracy trade-off can be performed in a variety of ways. In view of this, a wide range of possible strategies of task performance that is shaping the pace of activity should be considered. To solve practical problems, it is more accurate to consider not the *speed–accuracy trade-off* but the relationship between pace and precision of task execution.

We can make the following general conclusions. When a subject regulates the pace of task performance, he or she can use various strategies. Performing simple tasks, subjects have an ability to consciously control the speed of individual actions. The more difficult a task is, the more difficult it is to control the pace of its performance and subjects switch from the conscious control of individual components of a task to undifferentiated general and less conscious regulation of the overall task's pace.

The ability of subject to change the speed of task performance depends on the adequacy of a selected strategy of pace regulation. From the position of the self-regulation of activity, the specificity of pace regulation depends on such function blocks as *goal, subjectively relevant task conditions* (dynamic mental model, which includes conscious and unconscious components), *criteria of evaluation* (pace evaluation criteria), *assessment of task difficulty, motivation,* etc. Changing the task complexity leads to different strategies in pace regulation. That is why in our studies with increasing complexity of the task, the difference in the pace of task performance did not increase but decreased. This can be explained by the fact that an ability to consciously regulate pace of performance decreases and at the same time emotional tension involuntary provokes increase in the pace of performance. In some cases, a person cannot adequately adjust the pace of task performance not because subjects were slow but because they rush.

The pace of performance can also be evaluated from physiological perspectives. The physiological method of pace evaluation is specifically useful for the evaluation of medium and heavy physical tasks. The evaluation of pace in a work situation based on energy expenditure is a very difficult method. Therefore, it is recommended to use pulse rate measurment procedures. In any work conditions in which the pulse rate increases beyond 100 beats/min, an additional break time should be introduced. The task also has to be performed at a slower pace. We suggest to extend this method in order to evaluate the cost-effectiveness of mechanization of task performance or evaluate the cost-effectiveness of work environment improvement (Bedny and Seglin, 1997; Bedny et al., 2001).

9

Error Analysis

9.1 Error Analysis and Design

Not only time study but also error analysis is important in design. Analysis of errors in the system is an important source of evaluation of correctness of the design solutions.

A lot of interesting data were obtained when analyzing human errors in cognitive psychology (Norman, 1988; Reason, 1990; Senders and Moray, 1991, etc.). In this section, we present some new aspects of human error analysis that can be utilized in man–machine and computer-based systems. The systemic-structural activity theory (SSAT) approach is the basis for this analysis. Human errors can occur in different types of tasks performed by an operator. Errors are a potential source of hazard on human health and malfunctioning of the technical components of a system. We distinguish operator errors from operator failures based on the criticality of the errors. If human actions do not lead to unacceptable deviations or variations in the functioning of the system, do not lead to personal injury, and only lead to the deterioration of the system and staff functioning, they are regarded as errors. Errors very often are reversible or one that can be corrected without significant consequences. Irreversible errors are the main reasons for deterioration of product quality and decreasing productivity.

Actions that render the operator incapable of further functioning or shut the system down or cause the system not to achieve assigned goals are considered an operator failure. Reliability refers to failures of performance and how the probability of failure can change over time or in stressful situations. Human performance can be precise but not reliable and not all errors can be considered as system failures (Bedny, 2004). Some errors can be recoverable or have a relatively small effect on the functioning of personnel or technical components of the system, while other errors are associated with hazardous accidents, nonadmissible losses of time, and so on. Only the last kind of errors can be categorized as failure. We will use the term *errors* to evaluate the precision of human performance and *failures* to evaluate human reliability. When accuracy declines and falls below acceptable level, it becomes an operator's error. If as a result of operator's errors the system cannot function and achieve its goal or goal achievement is conveyed by unacceptable losses, it is considered as a failure. Hence, the major criterion for distinguishing between errors and failures is their consequences for the system as a whole. Accuracy and reliability are two important and interdependent characteristics of human performance. Similar but not identical principles are applied to their assessment. In the analysis of errors committed by humans, there are two aspects of study, qualitative and quantitative. The qualitative method of analysis is primarily connected with the analysis of the precision of human performance and the quantitative analysis with the assessment of the reliability of human performance.

There are errors or failures caused by technical components of the system and errors caused by operator's erroneous actions. Of course, this division is relative because an operator can perform wrong actions due to the system design flaws or when the worker does not possess the required skills. A reliability specialist can conclude that failure is caused not by technical components of the system but by human factors only after uncovering operators' erroneous actions and understanding their causes.

The concept of cognitive and behavioral action is the central focus for the analysis of errors and failures and therefore for the assessment of precision and reliability. Logical organization of actions, their complexity, and the possibility of their sequential or simultaneous performance are major factors that determine the strategies of task performance. The strategies of task performance also depend on the mechanisms of activity self-regulation. One of the most important factors that determine the rate of the operator's errors is task complexity, which becomes particularly important in a stressful situation (Bedny, 2006; Bedny and Sengupta, 2005; Sengupta et al., 2008).

Earlier, we have considered some aspects of error analysis from the activity self-regulation perspective, where strategies of positioning action performance have been described (Bedny et al., 2014).

In this chapter, we consider human error analysis and its associated evaluation of precision of human performance. We also emphasize the importance of error analysis in learning and training.

Trial and error is not useless in a training process. Both correct and incorrect cognitive and behavioral actions give the learner useful information about the changes in strategies of activity during the learning process. Analysis of various consequences of correct and incorrect actions enables one to develop adequate strategies of task performance. The more complex the learning process is, the more intensive is the explorative activity, thereby increasing the number of errors. In the absence of any organized information, a learner's activity relies on the trial and error strategy. A learner tries to find the causes of erroneous actions and correct the strategies of their performance. Learning can be considered as the transformation of strategies of performance that correspond to a particular stage of learning. The more complex a task is, the longer it takes a learner to find an adequate strategy of its performance. Algorithmic prescriptions utilized by a learner have a purpose of eliminating undesirable errors and leaving only those that perform informational functions in the action regulation process. During learning, a learner uses various temporal components of activity that cannot be considered erroneous. Such components and associated errors are intermittent and are not included in the final stage of performance. Therefore, written and oral instructions should be adjusted depending on the stage of the learning process.

All previously presented material demonstrates that error analysis requires both analysis of activity self-regulation (functional analysis) and morphological analysis, which includes an algorithmic description of task performance and a design of the time structure of activity. A potentially dangerous point of task performance from the safety analysis perspective can be discovered.

Self-regulation of activity during task performance can be performed on conscious or unconscious levels. They can be partly transformed from one to another. Errors more frequently occur when the level of activity regulation does not correspond to the conditions of task performance. For example, in unpredictable situations, stereotyped methods of activity may result in errors. Due to the goal-directed self-regulative process, a worker continuously adapts his or her strategies of task performance to the dynamic conditions of a situation. The concept of strategy and continuous changes of strategies of task performance in accordance with a dynamic situation are important aspects in the study of human errors from the self-regulation point of view. A situation that provokes errors or failures becomes a problem-solving situation. Such mechanisms of self-regulation as meaning and sense (subjective significance) play an important role in the selection of adequate strategies of task performance. If meaning determines how the operator understands a situation, then sense determines the situation's significance, which can be positive or negative.

For example, in a potentially dangerous situation, without a sufficiently correct interpretation of the situation, an operator can disregard some safety requirements and select more risky strategies of task performance. Such strategy can help to pursue a more significant goal for the worker. This suggests that not only cognitive but also emotional-motivational components of activity can influence the accuracy and reliability of an operator's performance. Here we consider emotionally evaluative aspects of activity regulation in unity with human-informational processing. The factor of significance as an emotionally evaluative mechanism can be considered outside of the study of the effect of stress. This was shown in the Chernobyl tragedy, which occurred during the performance test program. It was discovered that all protection systems were turned off because these systems complicated the test program. Stress has no effect when operators perform the test program. And at the same time, test performance was a very significant task for operators. The presented example demonstrates that cognitive and emotionally motivational mechanisms always coordinate their functions in a particular way. The strategies of performance can also depend on the specific logical organization of cognitive and behavioral actions. The logical organization of actions cannot be adequate for a situation, which provokes errors.

Human errors play an important role in the regulation of activity. Subjects can regulate their activity based on feedback from errors, existing goal, subjective standard of a successful result, and significance of the errors for the subject (worker). Errors perform informational function and are particularly important during the skill acquisition process. Usually such errors can be corrected, and they sometimes can be produced by the subject intentionally for better understanding the principles of the skill acquisition process. Very often such errors are simply deviations inside of the range tolerance for normatively prescribed strategies of performance. The range of tolerance for separate actions or activity during task performance for experts is narrower than for a novice. From the analysis of mechanisms of self-regulation and their derived strategies of performance, it follows that errors cannot be totally eliminated from the skill acquisition process. Variability of human activity and separate cognitive and behavioral actions can be explained by the fact that activity and its elements are formed based on the mechanism of self-regulation. Variability inside of the range of tolerance cannot be considered as an error. For example, in the study movements involved in the tracking function, it was discovered that movements contain micromotions with additional harmonics that have not been anticipated as per tracking theory. The produced micromotions give the subject additional information that is important in the regulation of movements. These additional micromotions are considered by mathematical models as tracking errors. However, from self-regulation perspectives such errors perform explorative cognitive functions. The subject works as a self-regulative system. Not only the performance time of a specific task but also the error rate is an important criterion for learning process evaluation. Based on the analysis of task performance time and errors committed by learners, we can introduce significant corrections to learning and training.

Later we consider some new aspects of error analysis. The presented material demonstrates that error analysis is important for studying the precision of man–machine systems, their reliability when rough errors can be transferred into failures, and when studying the learning process.

The differences in the understanding of errors in cognitive psychology and activity theory can be found in the interpretation of the role intention in error analysis. According to Senders and Moray (1991), the concept of intention is central to the meaning of *error*. The notion of intention suggests that in analysis of errors, it is important to find the differences between what was intended and what was done and what should be intended and was intended. In activity theory, such terms as intention, wishes, desires, etc., are considered as motivational factors. These factors are involved in the creation of the vector *motive* → *goal*. Based on the analysis of this concept, we can say, for example, what the goal of activity was and what the real result of activity was. If deviation of the result exceeds the range of tolerance, then error has happened. It is important to understand that there is an objectively given range of tolerance and subjectively accepted range of tolerance.

In cognitive psychology, the term *risk homeostasis*, which is derived from homeostatic self-regulative mechanisms, is used for explanation of the fact according to which a worker has a preferred level of risk at which he or she likes to operate (Senders and Moray, 1991). In SSAT, the self-regulative process is goal-directed. Humans are goal-directed self-regulative systems that select the level of risk depending on the goal of the task, significance of errors, difficulty of task, etc. Therefore, the level of risk at which a subject operates can be different in various situations and depends on the strategies of task performance.

In our analysis of human errors, we utilize the terminology developed in applied and systemic-structural activity theory. For example, the term *action* in activity theory has a different meaning in comparison to other fields of psychology. SSAT gives a precise definition of actions and their classification. Instead of the term *actor*, we utilize the term *subject*. In all cases when we use such terms as *performer, worker*, and *operator*, we mean different subjects that perform human work. In all such cases, the conscious goal of a subject plays a central role in error explanation. The self-regulative process is not considered as a homeostatic but as a goal-directed process. Failure to provide the precise terminology could also have an impact on the legal arguments about blame and responsibility of workers for committing various types of errors. In cognitive psychology, behavior consists of cognitive processes such as perception, attention memory, and motor actions (Senders and Moray, 1991, p. 19). According to these authors, the study of errors is to study ordinary psychological processes and motor actions. However, how these motor actions are extracted from human behavior and how they are classified according to various criteria are different in activity theory. In applied activity theory (AAT) and SSAT, error analysis involves not only the study of cognitive processes but also the analysis of cognitive actions, because activity is not only a process but also a structure that unfolds over time. When we conduct human error analysis, erroneous cognitive and motor actions should be determined. In such situations, one should determine what the action goal is, what kind of information should be used by the worker during action execution and what information is provided in reality, what interfering factors complicated the performance of the particular action, etc. This is a parametric qualitative analysis of human actions. The analysis of erroneous action from a self-regulation standpoint (functional analysis) begins with the discovery of the goal and the motive of action, type of feedback, time of action performance, and complexity of action. The cause of erroneous action can be incorrect formation and understanding of the goal, or the incorrect selection of strategies of goal attainment.

9.2 Error Analysis in Man–Machine Systems

SSAT considers error analysis from the systemic approach perspectives. This approach includes analysis of separate cognitive processes, analysis of cognitive and behavioral actions, and analysis of the holistic structure of activity. The last method is the basis for discovering more preferable strategies of performance and their algorithmic description. The analysis of activity structure and therefore the strategies of performance are not conducted separately from technology but only during the interaction between human and technology. This is explained by the fact that the structure of activity depends in a probabilistic manner on the constructive features of equipment. Systemic qualitative analysis from the self-regulation point of view can be used as a source of information for error analysis. Sometimes such analysis can be performed in an abbreviate manner.

Let us consider an example (Ponomarenko et al., 2003). In one of the flights, the pilot was instructed to return to the airport due to the worsening weather. He formulated a goal to make one more flight circle over the airport in the afterburning regime without opening the chassis in order to get rid of the remaining fuel. When the pilot approached the landing airfield, he suddenly received instruction from the commander of the flight to release tanks and immediately land. However, before landing, he unexpectedly received the command *one more circle*, because the pilot forgets to open the chassis for landing.

The cause of the error was the unexpected changing of the goal—*immediate landing*—instead of the planned early flight without opening the chassis. The earlier stated strategy was inadequately used in the new task.

Let us consider another example (Ponomarenko, 2006). The pilot returns from the flight. Before the third turn, he decides to open the chassis. Suddenly at this point, the flight commander has requested information on the rest of the fuel. The pilot replied, "all right, the chassis was released, three green bulbs lit...," and unexpectedly tried to land the aircraft without opening the chassis. The reason for this error can be explained through analysis of the structure of the activity and strategies for its implementation. Actions in the structure of the activity are interconnected. Their organization and significance determine the strategy of attention for shifting the focus from one action to another. An unexpected question from the commander and the answer to the question violate the attention strategy related to the shift focus from one action to another. What's more, action that has the same goal can be performed in various ways. In this case, upon violation of the attention strategy, a motor action has been substituted by a verbal action. In the previously considered examples, we utilize the term *errors* but not *failures*. This is explained by the fact that erroneous human actions did not result in injuries or equipment damage. Erroneous actions have been corrected by of subjects involved in committing these errors.

The more significant and complex the tasks are for the pilot, the more the tasks interfere with each other. Landing is a significant and complex task especially in cases when a task is performed in unexpected and changeable conditions. This explains the interfering effect of the considered tasks and the emergence of errors.

The accuracy with which the pilot reads an aviation instrument often depends more on the significance (subjective importance) of this instrument than on the visual features of the instrument. For example, an aircraft's attitude indicator has a rough scale, the distance between the scale elements is about 5°, and the distance between the numbers is about 15°. This instrument is very significant for the pilot, because they learn to read the horizon of the aircraft with an accuracy of about ±1.3°, which is much higher than their precision in reading other displays, which have a more detailed scale (Kotik, 1978).

Activity in general and separate cognitive and behavioral actions are very flexible systems. The same goal of task can be achieved by using not exactly the same actions. Moreover, logical organization of actions may vary. However, in the production process, this variation has some normative restrictions, defined by safety requirements, possible wear of equipment, economical factors, etc. A separate action also can vary according to some parameters. This is due to the fact that activity is a self-regulative system. In SSAT, different methods of task analysis were developed and all of these methods were adapted to the fact that activity is a variable system. For example, variability of task performance can be described by using the algorithmic method of task description. Separate actions are also described as variable self-regulative subsystems of activity. It is possible to develop a heuristic description of task performance when unpredictable external disturbances are acting on the system. Such descriptions can be developed in a similar manner as a human algorithm of task performance (Bedny and Karwowski, 2007; Bedny and Meister, 1997). A heuristic description of task performance does not guarantee the achievement of the task goal. Error analysis also should be adapted to the fact that human activity and actions are variable systems. SSAT developed the concept of *error tolerance*. For example, the performance time of a motor action can vary in the ±0.5 s range. Any variation in the performance time in the presented limit is acceptable, and such actions should be considered as correct. The smaller is the range of activity or action variations according to particular parameters, the more precise is the activity or action. Actions can produce an output that varies on some sort of scale. For example, in manufacturing, a cylindrical part of a nominal size 50 mm needs to be machined with a tolerance in diameter from 50 to 50.05 mm. Here we have a maximum limit of size 50.05 mm and minimum limit of size 50 mm. This variation in size of the part is acceptable and considered inside of the range of tolerance. Variations inside of tolerance are considered as deviations. However, if variation in size exceeds the range of tolerance, then this deviation will be considered as an error. Any actions or method of task performance that lead to variations in the size of the part outside of the acceptable range should be considered as erroneous actions that produce errors. Of course human activity is much more complex than the production process. However, some general ideas about the analysis variability of human activity can be useful in the analysis of human errors. For example, in an actual flight, a pilot very often perceived not isolate displays but logically interconnected items of display information. It is therefore not as important for a pilot that some scale pointer is not in the exact position, but how far it deviates from the specified position, in other words, how far it exceeds the tolerance limit. The pilot compares the mental image of the required scale value with the real value, which is demonstrated by the pointer. Therefore, depending on the situation and the applied strategies, the pilot continually switches from the precise quantitative reading to the less precise, but more reliable for the goal of the task, qualitative reading. An example of qualitative reading is when an operator can still accurately read presented information even if most of the numbers are not presented on a visual display because the pointer position can be mentally associated with a range of quantitative data. A qualitative reading implies very flexible strategies and different ranges of tolerance for precise instrument readings. In such readings, the mental model of a situation and its dynamics in time are very important. Not only verbalized components of activity, which are known as situation awareness (SA), but also imaginative components of such mental models are important in considered situations. Therefore, in SSAT, we distinguished deviations in performance that are inside of tolerance and are not considered as errors and deviations that exceed tolerance and are human errors. The previously described examples demonstrate that errorless performance in emergency situations depends on selecting adequate strategies of goal attainment. A strategy of gathering necessary information is the basis for the formation of an adequate dynamic mental model of the situation. It should be considered as a system of flexible and adaptive cognitive actions and operations utilized by a subject for the correct interpretation of the situation. From the SSAT perspective, the foundation for the creation of an adequate mental model of the situation is the orienting activity self-regulation (Bedny et al., 2014). Creation of an adequate dynamic mental model of the situation depends not only on cognitive but also on emotionally motivational components of activity. The factor of significance (emotionally evaluative mechanism of activity regulation) plays the main role in the selection of adequate information and evaluation of the situation. The relationship between cognitive and emotionally motivational components of activity also affects the timeliness of responses in such situations.

An operator can quickly create an adequate mental model of the emergency situation based on its isolated features due to automaticity of perceptual and thinking skills. However, incomplete information may in some cases lead to the formation of an inadequate mental model and inadequate subsequent executive actions. Such errors often occur under a time limit that provokes hasty responses. An inexperienced operator makes errors in emergency situations due to inadequate analysis of the situation and hasty response. An operator commits errors not because he or she acted too late but because he or she rushes to action. This explains why an experienced operator often responds slower than the inexperienced one in an emergency situation. An adequate reflection of the situation and creation of its mental model includes not only verbal but also imaginative and nonverbalized components. Thus, the concept of self-regulation of orienting activity is the basis for analyzing errors in emergency situations. SSAT considers situation awareness (SA) as just one of the mechanisms of activity regulation that is involved in the reflection of the situation

The other concept that is important in the analysis of errors is the *reserve of precision*. This term was introduced by Kotik (1978). Reserve of precision according to a specific parameter is the more allowable additional deviation compared to the minimum deviation that can be achieved by the operator in considered conditions. Evaluation of reserve of precision should be performed according to more critical parameters. Reserve of precision can be defined in the following way:

$$\delta_{res} = D_j - \delta_{Jmin}$$

where
   δ_{res} is the reserve of precision
   D_j is the maximum admissible deviation of parameter j, which is permissible for an operator
   δ_{Jmin} is minimal amount of deviation that can be achieved by an operator (maximum of precision that can be achieved by an operator)

Reserve of precision has objective and subjective meaning (Bedny and Karwowski, 2007). It can be objectively given and subjectively accepted by an operator. The subject can use different strategies to attain task precision depending on the subjectively accepted *reserve of precision*. The more risky strategies are used by the operator, the less reserve of precision he or she utilizes. Underestimation of objectively existing reserve of precision can result in errors or failure. A subjectively accepted reserve of precision depends on the significance of the considered parameter for the subject. For example, in Section 3.4 of Bedny et al., 2014, it was demonstrated that while tapping on a narrow target, the subject selected a risky strategy. It was explained by the fact that a slight increase in accuracy resulted in a significant increase in the time tapping on the target. Subjects prefer to sacrifice precision in sustaining the speed of performance. In any task, reserve of precision is evaluated according to a specific parameter. Therefore in any task, an operator has one reserve of time but different reserves of precision.

The complexity of a task can be increased by introducing more rigorous requirements to precision. According to the concept of self-regulation, this can be explained by the functioning of the following functional blocks or mechanisms of self-regulation: *assessment of task difficulty, assessment of sense of task, subjective standard of successful result,* and *subjective standards of admissible deviation*. For example, in simple tasks where achieving a required precision is not difficult for an operator, the significance of the task may be reduced. Hence, the emotionally evaluative mechanism and its associated motivational factor decreased in their intensity. This also influences the functioning of separate cognitive processes such as attention. Concentration during task performance is decreased. If a task has more rigorous requirements for precision, the difficulty of the task and its significance for an operator can increase. The significance of the task can also depend on the significance of the goal of the task for an operator. For example, the significance of a task can be very low for an operator. Increasing task precision requirements in such situations may be ignored by him or her. The operator may develop subjective criteria of precision, which can be lower than any objective requirements for precision. He or she in the considered situation can sacrifice accuracy, by reducing difficulty and, for example, decreasing the time of performance. Speed–accuracy trade-off depends on the strategies utilized by the operator, which in turn depends on the mechanisms of self-regulation. Error analysis and its associated precision of human performance cannot be considered outside of the analysis of strategies of performance and therefore the mechanisms of activity self-regulation.

Error taxonomy plays an important role in error analysis. Therefore, we consider the error classification principles. We distinguish two major groups of human errors. One group of errors is considered on a business organizational level (systemic level) and another one on the individual-psychological level (subsystem level). One of the important issues of error analysis on the system level is the relation between the formal structure of the business organization and the users' opportunity to perform their individual duty. An individually psychological level is an error analysis during performance of a specific task. A starting point in understanding human errors is the development of various error taxonomies that can be utilized during task analysis.

Any taxonomy can be helpful if it can be utilized for the error analysis of human performance. In cognitive psychology, the categories of human errors are derived from the information processing approach. In the former Soviet Union, the better known taxonomy was suggested by Zarakovsky and Medvedev (1979). This taxonomy utilized data obtained in applied activity theory. In SSAT, the classification of errors is derived from the analysis of activity self-regulation and the classification of human cognitive and behavioral actions. The concept of *strategy* is particularly important in the discussed taxonomy. Later we present error taxonomy that was developed in SSAT. This taxonomy includes five basic criteria and their parameters and dimensions, which corresponds to the hierarchical decomposition of criteria for error analysis (Table 9.1).

The proposed classification of errors allows to integrate into unified system a large number of different factors, accumulated in the activity theory and cognitive psychology. It suggests in each particular situation to choose a sequence of preferable steps in the analysis of the causes of errors.

As an example, let us consider a possibility to use error analysis based on SSAT principles. The use of these principles in some cases leads to the same

TABLE 9.1

Error Taxonomy

Criteria	Parameter	Dimension
General characteristics of errors at the system level	In what particular system or subsystem errors occurred and the time of their appearance.	In what system or component of the system, at what stage of system operation, and on what work shift.
	Working conditions of the system or subsystem in which errors occur.	Good, bad, overloaded, underloaded, etc.
	External manifestation (error consequences).	Consequential, nonconsequential.
General characteristics of errors at the individual task level	Relation of errors to the task in which they occur.	In what task do errors occur.
	Detectability of errors.	Obvious, hidden.
	Operator's awareness of the errors.	Operator is aware, unaware about the error.
	Existing causes.	Predictability
	Probabilistic characteristics	
	Typical	Usual, unusual.
	Expectedness	Expected, unexpected.
	Kind	Constant, variable.
	Quantitative	Frequency of occurrence and number of occurrences.
	Information workload	Excessive quantity or rate of information flow, lack of information, improper distribution of information over time.
Position of errors at the action performance level	To what particular cognitive actions do errors belong.	Errors resulting from sensory and perceptual actions, mnemonic actions, thinking actions, and decision-making actions.
	To what particular verbal and motor actions do errors belong.	Errors connected with discrete and continuous motor actions, verbal errors, or errors connected with undesired involuntary responses.
Cause of errors derived from inadequacy of activity self-regulation	Goal formation and evaluation of the meaning of the situation and task.	Incorrect understanding and formulation of goals, conceptual and dynamic mental model.
	Evaluation of strategies of task performance.	Wrong and/or untimely selected strategies of activity, or incorrect transition from one strategy to another. The relationship between strategies of task performance and errors. Analysis of objective and subjective criteria of success, existing feedback, etc.

(Continued)

TABLE 9.1 (*Continued*)

Error Taxonomy

Criteria	Parameter	Dimension
	Evaluation of specific task characteristics: the relationship between complexity and difficulty.	Under- or overestimation of task difficulty, self-evaluation of personal abilities, and task requirements.
	Sense of task and motivation. Personal attitudes.	Personal significance of task and level of motivation to follow safety requirements, etc.
	Inadequacy of activity regulation level.	Three levels 1. Level of stereotypy or automaticity of performance 2. Level of the conscious regulation of activity in terms of acquired rules and familiar strategies 3. Level of regulation of activity based on general knowledge, principles, and heuristics
	Team performance strategies.	Errors resulting from inadequate coordination of team members' performance strategies.
General causes of errors	Errors stem from idiosyncratic (personal) characteristics, equipment design, or interface characteristics.	Errors caused by human factors (human erroneous actions were discovered), or errors are derived from technical factors of the system. Errors caused by interaction of these factors.
	Errors stem from the functional state of users or operators.	Errors caused by fatigue, boredom, monotony, and decreased vigilance.
	Stress-producing factors	Time limitations, danger, and other external influences.
	Technical factors.	Incorrect distribution of functions between human and machine, inadequate information and instructions, inadequate equipment design.
	Organizational factors.	Errors occurred during integration or coordination of systems, subsystems, or professionals. Inadequate supervision and/or work/rest schedules.
	Operator's experience.	Insufficient training and experience, insufficient knowledge and skills, inability to select the required strategy of activity, and timely transfer from one strategy to another.
	Idiosyncratic (personal) features.	Operator is unsuited for work because of inadequate cognitive characteristics, physical fitness, or emotional stability.

(Continued)

TABLE 9.1 (*Continued*)

Error Taxonomy

Criteria	Parameter	Dimension
Analysis of consequences of errors from three points of view	Influence on efficiency of the system.	System malfunctioning or shutdown, failure to achieve goal in assigned time, accident.
	Influence on operator's activity.	Incorrect cognitive or behavioral actions, incorrect sequence of actions, untimely performance of actions or their further performance.
	Influence on operator's state.	Produce stress, fatigue, loss of attention, inability to continue activity, etc.

classification of errors in cognitive psychology and SSAT. For example, the cause of erroneous actions may be the incorrect formation of the goal of the task or actions, incorrect interpretation of the goal, and inadequate selection of strategies of goal attainment. The operator incorrectly formulates the goal of actions and as a result performs wrong actions, which according to Norman (1988) will be a mistake. But when the operator correctly formulates the goal of actions but accidently performs the wrong action, this is a slip. Thus, the concept of goal is important in the separation of mistakes from slips.

Let us consider another example. Errors more often can occur when the level of activity self-regulation is inadequate. For example, in the face of unpredictable changes of situation, the level of stereotypy or automaticity of activity regulation during task performance may result in errors. In general, the conclusion about stems of errors from a personal factor could be made only when the erroneous human actions are discovered. Hence, the concept of goal and action, as they are understood in activity theory, helps us to describe and classify errors and develop methods of their prevention.

Errors during task performance depend on the user's strategies, adequacy of the mental model of the task, evaluation of task difficulty, significance, subjective criteria of success, and feedback (Bedny and Karwowski, 2007; Sengupta et al., 2008).

In AAT and SSAT, three levels of activity regulation are described: (a) the level of stereotypy or automaticity of performance, (b) the level of conscious regulation of activity in terms of acquired rules and familiar strategies, and (c) the level of regulation of activity based on general knowledge, principles, and heuristic strategies. All of these levels have a hierarchical organization. We can see from Table 9.1 that inadequacy of the activity regulation level and conditions of performance may be the reasons for errors. For example, in the face of unpredictable changes of situation, stereotyped methods of performance may result in errors. In dynamic situations with low levels of predictable sequence of events, the ability to use flexible strategies of activity assumes greater significance. In contemporary working conditions and in computer-based tasks in particular, errors occur more often in difficult problem-solving situations, and therefore, the concepts of strategy, self-regulation, feedforward and feedback connections, etc., become particularly important. Errors may vary widely depending on the evaluation of task difficulty and its significance, selection of subjective criteria of success, etc. For example, it was discovered that the accuracy with which a pilot can read an aviation instrument often depends more on its significance (subjective importance) than on its visual features. The previously presented taxonomy of errors demonstrates that precision of performance and produced errors depends on the specificity of activity self-regulation and its derived strategies of performance.

9.3 Error Prevention and Training

In this section, we consider some aspects of error prevention that derive from the analysis of mechanisms of self-regulation. In all of the self-regulation models outside of activity theory, the role of feedback and corrections of errors is considered from homeostatic principles of activity regulation. According to this principle, there are three basic mechanisms: *input function, goal or standard*, and *comparator* and *output function* where the output has an impact on the environment (see, e.g., Carver and Scheier, 2005; Vancouver, 2005). Receiving information is provided by an input function mechanism. The comparator compares the input against the goal or standard, assessing whether they are the same. If they are not the same, there is a discrepancy or an error that can be corrected through output functions. Such models cannot explain how a person can not only correct but also prevent errors. In these models, corrections of actions or activity are based only on errors that have been already committed. According to the activity approach, errors can not only be corrected but also prevented. Humans can correct errors as well as anticipate them. In engineering psychology and ergonomics, prediction and anticipation of errors are considered in a narrow field that is known as tracking tasks (Wickens and Holland, 2000). In this book, the anticipation of errors is seen from a broader perspective as a problem of self-regulation of activity. In this chapter, anticipation of errors is analyzed in the process of learning and training.

Let us consider the correction of errors by using a simple example. Suppose a person pours a cup of tea. How does he or she adjust the position of his or her hands? When a person is pouring tea from the teapot, he or she perceives the information obtained from the tea's stream. If the stream approaches the edge of the cup, the person adjusts his or her position so that the stream flows closer to the center of the cup. If the teapot is very high above the cup, then splashes of tea come out. Therefore, the teapot should be lowered and brought closer to the cup. Can this task be performed by a child who has no relevant knowledge and skills to correct his or her actions as an adult does?

It is clear that the goal and subjective standard of success are not the same thing in this task. The goal of this task is the image of the desired result, which is *pour tea into the cup*. The position of the teapot and its height vary within certain limits. A person uses not just visual but also other types of information. If the person is at the banquet, in nice clothes, he or she will use a safer strategy not to spoil it. Thus, adults do not simply perceive information but also know how to evaluate their actions and correct them. Adjustment of actions and prevention of errors, even in such a simple situation, require not only some of the mechanisms that are involved in task formation and execution but also mechanisms for the evaluation of results and prevention of errors.

During the training or learning process, errors cannot be totally eliminated. Errors in the learning process have an informational purpose. They can be reduced, but totally eliminating them is impossible. Their significant reduction depends on the relationship between the conscious and unconscious levels of self-regulation. For example, in gymnastics when a trainer uses instructions such as *straighten arms upward* or *turn your head to the left*, he or she influences a gymnast's behavior on a conscious level of self-regulation and the gymnast consciously performs actions while at the same time exerting an influence on an unconscious level. When the gymnast turns his or her head, this immediately influences muscle tension of his or her back that by trial and error is corrected unconsciously to achieve the required muscle tension. A gymnast gradually transfers from a strategy of error elimination to a strategy of error prevention. This process involves a complex relationship between the conscious and unconscious levels of movement regulation.

For instance, Novikov (1986) discovered that some technical devices can totally prevent errors during the skill acquisition process. However, when trainees attempt to perform the same actions without technical devices, they commit errors again. He demonstrated it in an experiment where a trainee learned how to apply the exact effort to particular controls when complete visual information about efforts was presented on an oscilloscope. The trainees quickly performed the required motor actions without errors. However, when information has been removed, they committed errors again. When an instructor began using discreet feedback instead of an oscilloscope, which included turning of a bulb only during a period when trainees committed errors, it resulted in elimination of errors after trainees started performing a task without training devices. This demonstrates that elimination or prevention of errors is possible only after the required mechanisms of activity self-regulation are developed.

Regulation of motor action is achieved by introducing correction during performance. It is necessary to take into account that self-regulation is always done based on the analysis of the interaction/relationship of spatial, temporal, and force parameters of motor action. One also needs to take into account that regulation of motor actions is carried out based not only on the misalignment or deviation of a controlled parameter of an action (e.g., position of instrument) but also the time derivatives (velocity and acceleration), which greatly increase their speed of correction. Immediate feedback is usually utilized for the prevention of errors, but such feedback often does not allow preventing errors; therefore, it is necessary to use an advanced feedback.

An immediate feedback occurs as soon as a current deviation of a parameter exceeds a critical value and a simulator produces a signal about an error. A worker can quickly eliminate an error. However, in some cases, a worker has to prevent the error from occurring and an advanced feedback has to be utilized, which works differently.

The simulator can detect deviations of controlled instrumental movements not only based on values of some movement parameters at a given time t_1 but also with an aid of a derivative to calculate a possible deviation in time $t_1 + t_2$. Given that motor actions are governed not only by a current value of a controlled parameter deviation but also by the speed of deviation and acceleration of deviation of such parameters, a special calculating device can be introduced to the simulator that would calculate these parameters. Thanks to this, the simulator can provide a worker with the necessary information about the possible deviations of movements' controlled parameters in the near future based on the analysis of present parameters of motor actions. This makes it possible not only to correct errors but also to prevent them.

Using a combination of deviations of derivatives in time helps capture not only the magnitude of deviations of motor actions at any given time but also *predict* the possible deviations of movement and correct motor actions in advance. Hence, the analysis of principles of motor action regulation includes predictive mechanisms that can prevent errors. This principle can be used in the design of simulators. Correction of motor actions is possible with a time delay of 0.06–0.12 s (Chkhaidze, 1970; Novikov, 1986). For example, it has been revealed that experienced workers regulate their movements based first of all on such indexes as speed and acceleration. Students regulate their movement at the early stages of skill acquisition based on the position of a tool. During the training, they gradually start regulating their movements also based on such parameters as speed and acceleration. Hence, the relationship between these two methods of movement regulation is changing and a possibility of error prevention increases.

Procedures that facilitate the development of an adequate dynamic mental model of a training situation play an important role in the training process. Such model provides a dynamic orientation of students during a training process. At the first stage of a training process, a student formulates a goal or interprets and accepts a goal of task, which is given through an instruction. At the next stage, a student creates and clarifies a mental dynamic model of activity during the skill acquisition process. However, skills can be sufficiently complex and a student cannot create an adequate mental model of all components of an acquired skill. An adequate dynamic mental model plays a critical role in shifting student's attention to those components of a skill that are more important in a particular phase of training. An instructor can change his or her explanation of the task over time and that in turn would change the students' focus of attention on different components of a task while neglecting the other components to some degree. Transformation from one stage of training to another is accompanied by changes in students' attention focusing on various parts of the same activity.

The dynamic features of a mental model depend on the stages of the training process. Correspondence of a student's dynamic mental model to stages of skill acquisition is a critical factor in the prevention of errors at the executive stage of task performance.

According to our models of self-regulation, a student's dynamic orientation should be provided in accordance with the student's dynamic mental model that changes and evolves over time during training. Adequate changes and development of a student's mental model is facilitated by dynamic instructions.

The dynamic orientation of students is closely connected with the dynamic evaluative stage of activity. Therefore, feedback about errors should be introduced based on the dynamic criteria of success. These criteria should be changed depending on the stage of training and student's success. For example, a student should not be informed about his or her errors that are not important at an ongoing stage of skill acquisition. In SSAT, such method of students' orientation is a combination of individual parts and a whole method of training because students perform a whole task and at the same time concentrate on particular aspects of tasks that are more important at each specific stage of training.

In summary, we can say that by changing the system of verbal and written instructions and by bringing the student's attention to diverse aspects of activity depending on the various stages of a training process, an instructor can help students develop an adequate dynamic mental model of a situation and adequate strategies of performance in general.

Humans are capable of performing mental actions with the images that are analogous to the operations they carry out with real objects. In SSAT, mental manipulation of images according to a formulated goal is an example of imaginative actions. Hence, one can perform motor actions in an imaginative form that is relevant to studying ideomotor training. This type of training demonstrates a possibility of improving motor action performance by manipulating images of objects. Thanks to ideomotor actions, we can prevent or reduce the number of actual errors during performance of real actions. This method has been demonstrated very clearly in the training of gymnasts. It has been discovered that highly skilled gymnasts perform their routine in an imaginative form at about the same time as they perform a real one. Dimersky (1965) experimentally demonstrated the effectiveness of an imaginary action to restore and maintain the skills of pilots. The author showed that the mental performance of all phases of a critical flight's tasks is an important method for improving the effectiveness of a pilot's task performance. Andrianov and Dubrovitskij (1971) showed that at the final stage of preparation for a flight, a pilot should mentally reproduce the most important stages of a flight, as it is particularly important. Ideomotor training prevents real potential errors.

It is important for the process of motor and mental skill acquisition to develop strategies of self-control for error prevention (Bedny, 1987). Self-control is always in compliance with the goal of the activity. A student has to learn to identify all the necessary features of a situation that should be taken into account by self-control. These features may change depending on the stage of a training process. Self-control plays an important role not only in elimination but also in prevention of errors.

Students learn to use a variety of sensory-perceptual features of their movements, or technological parameters of task for correcting their actions and preventing errors.

Interpreting a cognitive process not just as a process but also as a system of cognitive actions can facilitate a better understanding of how people use feedback in cognitive actions. This idea is best illustrated using the game of chess. A chess player performs a complex mental activity before making a simple motor action of moving a figure into a new position. A chess player promotes various hypotheses, evaluates them, corrects, and selects a new one if necessary. In the course of such activity, a chess player performs perceptual, mnemonic, and thinking actions. In order to develop various hypotheses and analyze them, a chess player perceives figures' position (perceptual actions), analyzes their interaction (thinking action based on visual information), extracts possible similar situations from memory (mnemonic action), evaluates possible actions–responses of a competitor (thinking action), and so on. Every action has a goal and can be evaluated based on the analysis of its goal, criteria for evaluation, and expected consequences of actions. The concept of feedback is also important for the regulation of cognitive actions. Intellectual mental operations are also important for self-control. They allow comparing mentally selected features of a situation and based on it anticipate further changes in a controlled process. Self-control includes a variety of cognitive actions such as sensory-perceptual, mnemonic, thinking, decision-making, and imaginative actions. Self-control can be considered as self-examination by a subject of his or her own activity. It is important for self-control to distinguish between such terms as deviations and errors. Deviations occur within a range of tolerance. They serve as identification characteristics of a possibility of errors. Self-control of actions or a whole activity can be ongoing and final. Ongoing self-control is used during performance and final self-control is used after completion of a certain part of a task.

Formation of self-control is an essential component of the training process. It is primarily associated with conscious components of evaluating an ongoing activity. It can be changed during training; some parts of it can be automated and become subconscious. Other components of self-control are always conscious. Self-control includes not only cognitive but also emotional and motivational components. The latter have a significant impact on the selection criteria and strategies for self-control. Self-control is dynamic and changes during the learning process. In the beginning of learning, self-control strategy largely depends on an instructor. Individual features of personality also influence the specificity of self-control strategies.

Self-control is an essential part of self-regulation of activity in general. Written and oral instructions should always be developed based on discovering effective strategies of self-control. Self-control can also be tailored to the individual characteristics of a student. Self-control includes itself forecasting strategies. Previously described models of self-regulation (Bedny et al., 2014) are the theoretical basis for developing efficient strategies of self-control. An instructor should develop pedagogical methods that facilitate the formation of efficient self-control. Some strategies of self-control are more task-specific; some of them are general and can be used during the performance of various tasks. Self-control is important not only in performing the motor components of activity but also in performing its cognitive components. Task analysis should always include identification of effective strategies of self-control for error prevention.

Techniques of self-control can be often explained verbally. Students acquire such techniques by trial and error, which takes a long time. The process can be improved by giving special instructions and usage of simulators and training devices. Before introducing such means, it is necessary to perform a psychological analysis of a task at hand and of individual motor and cognitive actions.

Mastering self-control can be very difficult and a student cannot control his or her actions based on all required parameters simultaneously. In such cases, the most effective are methods of dynamic orientation in the process of self-control. At various stages of skill acquisition, a student's attention is focused on the most important aspect of task performance for this stage. After acquiring self-control skills, students start regulating their actions mainly based on the prediction of errors.

The relationship between various sense organs changes during the acquisition of self-control and development of motor skills. In the early stages of mastering motor actions, information from exteroreceptors plays a leading role in the skill acquisition process. The visual sense organ is important in particular. In the process of motor skill development, the interoreceptors become important. When, for example, instead of visual feedback the subject utilizes only kinesthetic information for motor action regulation, this means that regulation of motor activity shifts from an external to an internal control loop.

An external control loop takes 0.1–0.2 s of the regulation cycle in average. When an inner control loop is used, a regulation cycle takes in an average of 0.04–0.05 s. When a worker regulates his or her motor actions based on the internal control loop, he or she can shift attention from his or her own actions to the results of an action with a tool. During transition to the internal control loop, the role of the conscious level of self-regulation is reduced. Not all information that is perceived by various sense organs is related to feedback but only information that is used to regulate actions and is necessary to achieve an action goal. In some cases, information from various sense organs that is not used for the regulation of an action could be a source of interference in their regulation.

Self-control plays an important role in the self-regulation concept of learning, developed in the framework of SSAT (Bedny and Karwowski, 2007; Bedny and Meister, 1997).

According to the self-regulation concept of learning, there are conscious and unconscious levels of actions or activity regulation. The main type of human learning involves the conscious level of self-regulation that is associated with the achievement of a conscious goal of activity or actions. The unconscious level of self-regulation includes an explorative behavior with motor operations (associative level of learning). This level of learning should be distinguished from the conscious explorative activity with motor actions (cognitive level of learning). These two levels of self-regulation are interdependent and can be transferred into each other. From an SSAT perspective, self-regulation is the main mechanism for the creation of associations during learning. An instructor, at least partly, can monitor the subconscious level of self-regulation during the training process. One method that permits to introduce consciousness in action regulation is reorientation of attention based on the dynamic situation of a specific instruction. Introducing different instructions, prompts, reformulating a goal, and so on result in the shift of a student's attention to different components of activity. Such modification of conscious self-control can influence the unconscious level of self-regulation of behavioral and cognitive actions. It is particularly obvious in gymnastics. A gymnast can acquire extremely complex movements without awareness of their significant parts.

Analysis of the presented material brings us to the following conclusion. Most models of self-regulation outside of activity theory assume that a person can regulate motor actions and activity in general based on error correction as a result of motor responses, but this is not the only way of regulating motor actions and activity in general. Regulation based on error correction is specific for a training phase. Thanks to the improvement of teaching methods, the number of errors and their magnitude can be reduced. Gross errors can also be eliminated in a training process through development of special skills of self-control. When motor action skills are developed, their regulation can be achieved by predicting errors. Cognitive actions play a critical role in error prevention. At various stages of skill acquisition, a person uses different types of information about deviations in action performance. These deviations are not always errors but might be just information about the possibility of errors. A person begins utilizing mental operations that allow him or her to change a strategy of a motor action performance, making him or her appropriate for specific conditions and prevent errors.

9.4 Application of Queuing Theory to Human Error Analysis

It is quite difficult to apply mathematical tools in psychology. We will demonstrate the first example that illustrates the use of morphological analysis that was developed in SSAT for further quantitative analysis of an operator's performance. As proposed in this book, the morphological analysis of an activity allows to significantly expand the possibility to use mathematical tools for the assessment of human performance. In this chapter and in the following sections of the book, we will focus on the coverage of methods of studying human work on the basis of not only qualitative but also quantitative methods. Here we consider the possibility of the use of a mathematical tool that can be helpful in cases where the main source of error is the lack of time. We describe the possibility of using the queuing theory in studying some type of human errors. In Sections 9.1 through 9.3, we discussed the qualitative aspects of error analysis.

We often have to deal with queuing systems. For example, we wait in line to get airline tickets and to check in at the gate where the flight boards are and the plane joins the line of planes waiting to use the runway for takeoff. In telecommunication systems, there is a great number of call requests that should be served in a particular period of time. If the intensity of calling requests would be higher than the intensity of service, then a lot of call requests can be lost. When multiple processes run on the same computer processor, they wait in a queue and should be completed in a certain time frame. The queuing theory helps not only in coordinating the intensity of requests and intensity of service but also in directing diverted calls via different paths. In a queuing system, calling units (requests) such as people in line, waiting planes, waiting computer processes, and service mechanisms such as ticket agents, dispatchers, and computer processors who serve possible requests are critical factors in these systems. The queuing theory allows to develop such mechanisms of service where some requests are kept in the system for a specific period of time until the system resources are freed up. The queuing theory can develop various orders of service and helps to determine the queuing discipline or the manner in which the exchange between possible requests and services is coordinated.

The queuing theory is the mathematical study of waiting lines or queues. It is used for the analysis of the resources needed to provide service for queue, predicting its length and waiting time. It found its application in military, engineering, computer science, economics, transportation system, telecommunication, etc. Scientists who use the queuing theory try to estimate the expected waiting time, queuing length, percentage of idle time for cerevice facilities, and so on.

Calling units can be classified as either patient or impatient. A patient calling request stays in line, regardless of the state of the system, while an impatient one can leave the line before receiving the service during a particular time. For example, the customer can leave the line if he or she waits for more than 1 h. A device or tool that can at any time serve only one calling request called in the queuing system is a channel. If we have multiple channels capable of simultaneously serving calling requests, we are talking about a multichannel system. Channels can perform either one type of service or various kinds of services. A set of rules used for the selection of various orders of service from the queue is called service discipline. These rules can vary. It can be the service in accordance with the order of the queue (first come, first served), service in accordance with the existing scale of priorities (last in the queue should be serviced first), etc. Another important factor of the system is the mechanism of service. Its main characteristics are the service time and the capacity of the system to serve all requests. Service time can be considered as a random variable in certain situations. Calculations are carried out for the evaluation of a system's ability to perform the required service processes within a specified time period. The capacity of a system is its ability to adequately serve all calling requests for which a system is designed.

All of these ideas can be used in ergonomic studies. In the man–machine system, signals presented to the operator can be seen as calling requests, and an operator who processes information as a service mechanism. The service time of some signals (requests) is strictly limited. Failure to service such signals in accordance with the time limit can result in serious consequences and is considered an error of the system. This brings us to the conclusion that an operator often deals with impatient calling requests. In all of these cases, the queuing theory can be a suitable method to calculate the reliability or precision of an operator's task performance. Thus, in all cases when the main source of human errors is the time limit, this theory can be used. It becomes especially important when stress is added to the time limit. Such systems can be encountered in the military or in any situations where the consequences of human errors are very dangerous and the system's resources that are used for serving the considered queue are limited.

Usage of the queuing theory in ergonomics is associated with considerable difficulties due to the fact that an operator receives information from many devices, making it difficult to describe the arrival process. Service mechanisms are associated with interpretation of information from various displays and performance of required human responses. In these circumstances, it is difficult to determine what constitutes calling requests, a queue, waiting lines, etc. The *Handbook of Engineering Psychology* provides some recommendations on the use of the queuing theory (Lomov, 1982). These recommendations are very ambiguous and cannot be easily applied. For instance, the authors recommend combining methods from the queuing theory with methods in the information theory to gather the required initial data. It is recommended to use Shannon's method for calculating the time for receiving and processing information (Shannon and Weaver, 1949). However, as we have demonstrated in our studies (Bedny and Meister, 1997), information theory cannot be applied for this purpose. For instance, it is very difficult to calculate correctly the amount of information received by an operator from numerous visual displays. In this case, it is impossible to determine the length of the alphabet of the signals from which the signal is selected. We have shown that a person is constantly changing the length of the alphabet of possible states of the signal when he or she is using long-term memory in a choice situation (Bedny and Karwowski, 2011). Konopkin (1980) showed that the choice reaction time may depend on the current subjective expectations of the signals in a particular situation and not on the actual informational value of the signal. Similarly, Neumenn and Time (1975) provided data about the reaction time in relation to significant and insignificant displays. Differences in the reaction time between significant and insignificant displays have been discovered. Informational theory ignores the pace of work performance (Bedny and Karwowski, 2007). Networks of queues are very complex when an operator receives information from multiple devices. In most cases an operator does not respond to the readings of separate displays but performs a system of tasks in which the readings of displays are interrelated. Therefore, the number of arriving calls (requests) depends first of all on the number of tasks performed by the operator and not on the signals from the individual displays. Scientists who attempt to use the queuing theory did not suggest an adequate method of estimating the service time for tasks that have a very flexible sequence of action performance. Such method has been developed only in SSAT (Bedny and Karwowski, 2007). It is very important to understand what should be considered as possible requests and queues' serving mechanisms during evaluation of human performance in critical situations and to determine the most critical tasks, their priority, task performance time, reserve of time for service, etc. Ergonomists working together with specialists in the queuing theory can make all the required calculations to determine the safety or reliability of a man–machine system. Similarly, such calculation can be performed for computerized systems. In order to do that, ergonomists should be familiar with the general principles of the queuing theory.

Let us consider some basic mathematics used by the queuing theory. The traffic of calling requests can be regular or random. The number of calling requests arriving at a unit of time is called *intensity of the mean arrival rate* λ. If λ is a constant or is a function of time, the flow is regular, but more often the mean arrival rate λ is random because it depends on the number of arriving requests per unit of time, which is a random variable. Random arrival rates are characterized by probability distribution functions. A queuing model usually is based on the Poisson distribution process.

The Poisson distribution of calling requests has the following characteristics:

1. *Stationary (permanent) characteristics*—the probability of appearance of a particular number of requests at the particular time interval *t* depends only on the length of the time interval and does not depend on where on the time axis this time interval is located. Stationary conditions can satisfy traffic or a waiting line of calling requests, the probabilistic characteristics of which are independent of time and, in particular, when the average number of requests per unit of time is constant.

2. *Independence of events*—the property of traffic or waiting line when in any nonoverlapping intervals of time the number of calling requests that fall into one of the interval is independent of the number of requests that fall on other intervals of time.

3. *Ordinary traffic*—the property of traffic when the chance of events that two or more calling requests can fall into a particular interval of time Δ*t* is disparately small compared to the probability of one event. This means that the calling requests come singly and not in pairs, triplets, and so on. If the arrival process includes calling requests that come only in pairs, triplets, etc., then such not ordinary traffic can be reduced to the ordinary.

This is the basic feature of the Poisson distribution. Analysis of the Poisson distribution demonstrates that specialists who are using the queuing theory should know the pattern of calling requests or arrivals. Calling units arrive to the queuing system either according to a predetermined schedule or in a random fashion. If the arrival is scheduled, analytical queuing models are irrelevant. If they are random, however, it is necessary to determine the probability of time distribution between arrivals. It is proven in the probability theory that the conditional probability of any future event depends only on the present state of the system. So we can predict the probability of calling requests *k* in the period of time *t*. It can be calculated based on the Poisson probability function. Under the previously described conditions, the probability of calling requests that have arrived $p_k(t)$ in the time interval from 0 to *t* can be calculated using the following formula (Vencel and Ovcharenko, 1969):

$$P_k(t) = \frac{(\lambda t)^k}{\kappa!} e^{-\lambda t}, \tag{9.1}$$

where
  k is the number of calling requests that have arrived
  t is size of the time interval
  λ is the mean arrival rate per unit of time

This is an exponential distribution for an interval of time for arriving calling requests. This formula is called a Poisson distribution.

Let us consider how we can determine experimentally signal arrival distribution. The mean arrival rate λ can be determined experimentally by chronometric measuring of the number of calling requests for a certain period of time. Suppose we need to calculate the signals' mean arrival rate (λ) coming to the operator from a specific device. Sometimes the rate can be very high but the timely response to the reading is important for safety. The device has several indexes. Depending on the combination of the indexes on the instrument, it is possible to give several responses. The possible response time depends on the combination of indexes. It is necessary to estimate the mean arrival rate λ and determine its law of distribution. To calculate the mean arrival rate λ, we have to divide work time into chosen intervals. For example, we choose the length of time interval to be 5 min. Then count the number of time periods when there was no signal or an operator received two signals, three signals, and so on. The total number of signals is determined by the following formula:

$$N = A_1 X_1 + A_2 X_2 + \cdots + A_N X_N \tag{9.2}$$

where $A_1, A_2, \ldots, A_N$ is the number of time periods when signals are coming in the quantity of $X_1 + X_2 + \cdots + X_N$. Then the mean arrival rate per unit of time (average number of requests per unit of time) (λ) can be calculated as follows:

$$\lambda = \frac{N}{t} \tag{9.3}$$

where
  N is general number of signals
  t is the general period of time when the chronometrical study was performed

The experimental distribution of the arrival of different numbers of signals for a particular time interval can be obtained from the next formula:

$$I_k = \frac{A_k}{A} 100 \tag{9.4}$$

where
  A is the general quantity of the signals that have arrived
  A_k is the observed number of arrivals of k signals in a considered period of time

We can make the assumption that, in our example, we have a Poisson distribution. At the next step, we determine the theoretical distribution $p_k(t)$ for various k according to Formula (9.1) and multiply obtained data by 100 to convert the results into percent. Obtained data can be presented in a table form (see Table 9.2).

TABLE 9.2

Signals' Arrival Distribution

Number of Signals during 5 min	Quantity of Cases for the Arrival of Signals for Various k	Observed Frequency of Signal Arriving I_k for Various k (%)	Theoretical Frequency of Signal Arriving $P_k(t)$ (%)
k_1	A_{k_1}	I_{k_1}	P_{k_1}
k_2	A_{k_2}	I_{k_2}	P_{k_2}
⋮	⋮	⋮	⋮
K	A_{kn}	I_{kn}	P_{kn}
Sum	A	I	P

Then, using the criterion χ^2, we can check whether in our example the hypothesis about the Poisson distribution is valid. Coefficient χ^2 is calculated utilizing the well known in the statistics formula. In our case it is:

$$\chi^2 = \Sigma \left\{ \frac{\left[I_k - p_k(t) \right]^2}{p_k(t)} \right\} \tag{9.5}$$

where

I_k is the observed frequency of signals arriving for various k
$p_k(t)$ is the theoretical frequency of signals arriving for various k

The next important characteristic of the queuing model is service time. Service time can be constant or random in nature. Usually service time is a random variable for an operator not only because an operator's behavior varies. In our example, the same device can present different information that would lead to a response that depends on such information. Ergonomists should determine how to describe this random response. In many cases, service time is exponentially distributed. The probability that service time τ will be less than t can be determined from the following expression:

$$F(t) = p(\tau < t) = 1 - e^{-\mu t} \tag{9.6}$$

where

μ is the intensity of the service rate
t is the considered interval of time
τ is the service time

$$\mu = \frac{1}{t_{serv}} \tag{9.7}$$

For the purpose of serving signals on time, λ must be less than μ ($\lambda < \mu$). However, even if $\lambda > \mu$, there can be a continuous accumulation of signals for some time periods that require adequate responses and an operator does not have enough time to respond because of their irregular input (the queue will continue to grow). Such signals can remain in line and wait for an operator's service. If a signal remains in a waiting line for more than t_{res}, it can be lost by an operator and result in errors or failures. The probability that service time τ will be more than t can be determined using the following expression:

$$F(t) = p(\tau > t) = e^{-\mu t} \tag{9.8}$$

If a system is able to collect signals and keep them in a waiting line for some period of time, then the probability of locating k signals in the queue can be determined using the following expression:

$$p_k = \rho^k (1-\rho) \tag{9.9}$$

The probability of absence of signals in a waiting line can be calculated as follows:

$$P_0 = 1 - \rho \tag{9.10}$$

where

$$\rho = \frac{\lambda}{\mu} \tag{9.11}$$

The average waiting time in the queue (before service of a signal) is

$$t_{queue} = \frac{1}{\mu} \times \frac{\rho}{1-\rho} \tag{9.12}$$

Other distributions and methods of calculation from the queuing theory can also be used.

We described some basic principles of using the queuing theory when there is only one source of possible signals (calling requests or arrivals) that come from a single device. Therefore, the number of traffics or waiting lines considered in the queuing system in this example is equal to 1; there is one device and an operator should timely and adequately respond to several possible versions of a signal. In reality, an operator interacts with multiple displays and manipulates groups of controls. An operator is often using a computer during task performance. So an ergonomist should take into account that readings from various displays might be interrelated and an operator does not just perform a single reaction in response to readings from a display but rather performs a logical system of interrelated actions. In other words, an operator performs a number of tasks that are spread in time irregularly.

Typically, one task involves the use of a number of devices and various controls.

SSAT suggests principles of utilizing the queuing theory for such purposes. The proposed approach reduces the number of calculated waiting lines and uses a more efficient method of determining service time.

It is obvious that the number of tasks performed by an operator is much smaller than the number of individual responses to the readings of single devices. The mean arrival rate in the queuing system should not be calculated based on the signals presented on an individual display, requiring an appropriate response, but should be calculated based on the analysis of quantity of the more significant tasks arriving per units of time that should be performed by an operator. The task performance time should be used for calculating the mean service rate when we model a queuing system. Indeed, an operator can quickly respond to data presented on one device but give a slow response to data presented on other devices. As a result, an operator can exceed the reserved time allocated to a specific task.

Therefore, Table 9.2 should be modified as follows (see Table 9.3).

In some cases, a signal from an individual device can be critical and relatively independent of the data presented on other devices. Moreover, not one simple action but a certain system of actions is required in response to such a signal. Such situation can be also observed as an independent task. Thus, instead of calculating the signal arrival distribution, we calculate the task arrival distribution. According to SSAT, modeling of a queuing system requires the following steps (Bedny, 1979):

A. *Modeling random arriving events (tasks) of a queuing system*

    1. Identify the most important tasks performed by an operator and the tasks that must be performed in restricted time conditions. Each task is considered as a calling request or a calling unit.

    2. Determine the importance of the tasks and the possible consequences of their late or improper performance.

    3. Determine the distribution of time between arrivals (tasks) that are important for the analysis of the queuing system.

TABLE 9.3

Task Arrival Distribution

Number of *Tasks* during t (min)	Quantity of Cases for the Arrival of *Tasks* for Various k	Observed Frequency of *Task* Arriving I_k for Various k (%)	Theoretical Frequency of *Task* Arriving $P_k(t)$ (%)
k_1	A_{k1}	I_{k1}	P_{k1}
k_2	A_{k2}	I_{k2}	P_{k2}
$\vdots$	$\vdots$	$\vdots$	$\vdots$
K	A_{kn}	I_{kn}	P_{kn}
Sum	A	I	P

B. *Steps needed for modeling service mechanisms*

The mean service time for selected tasks and other temporal characteristics of task performance are defined in accordance with the principles described in Section 7.4. In this chapter, we use the proposed approach for modeling a queuing system. We want to stress the fact that based on the suggested method it is possible to determine not only mean service time, but also task performance time for any version of task, including performance with maximum or minimum required time.

In accordance with these principles, the following are necessary:

1. Describe algorithmically each of the selected tasks; select the most representative version of the algorithms of performance for each selected task that would allow to determine an average, maximum, and minimum execution time of tasks with the probability of occurrence of such variant of the algorithm to determine the mean service time to perform the selected task.

2. Determine the reserve time and its distribution for each selected task.

3. Determine priorities of task performance based on each task's importance for the system and characteristics of its reserve time.

4. If tasks do not overlap, the possibility of an operator serving each task can be determined independently. If tasks do overlap, the possibility of serving each task when an operator is switching his or her attention from one task to another should be determined.

Let us consider a simplified example utilizing data collected when subjects worked on an experimental control board presented in Figure 7.1. On one side of the board, there was a panel for participants, and on the other side, the panel for the experimenter. The performance time has been measured automatically with 0.01 s precision. There were five indicators and controls on the participants' panel. A system of signals that utilized five indicators and that allowed creating various combinations of signals and 110 versions of an algorithm performance was presented to participants. An experimenter could present different signals to participants with varying probabilities, which affected the average execution time of the algorithm. As we have shown in previous studies, chronometric analysis requires choosing not all but only the most representative versions of an algorithm. For the considered physical model of an operator's control board, it was sufficient to choose three of the most representative versions of an algorithm. This significantly simplified the analysis of service mechanisms of our queuing system. For simplification of the calculation and further discussion, we suppose that only four indicators and four controls are used (digital indicator 3 and

TABLE 9.4

Algorithmic Description of Task on Experimental Control Board with Four Indicators and Four Controls (the General Algorithm of Task Performance)

Members of Algorithm	Description of Member of Algorithm
O_1^α	Look at the first digital indicator.
$l_1 \overset{1}{\uparrow}$	If the number 1 is lit, perform $_1O_2^\varepsilon$; if the number 2 is lit, perform $_2O_2^\varepsilon$
$_1O_2^\varepsilon$	Move the two-positioned switch (6) to the right.
$\overset{1}{\downarrow}_2 O_2^\varepsilon$	Move the two-positioned switch (6) to the left.
O_3^α	Determine whether to turn on signal bulb 4 or 5.
$L_2 \overset{2(1-3)}{\uparrow}$	If neither bulb 4 nor 5 is turned on ($L_2 = 0$), perform O_4^ε; if bulb 4 or 5 is turned on, perform l_5.
$\overset{2(1)}{\downarrow} O_4^\varepsilon$	Move the right arm to the fourth position's hinged lever 7, grasp the handle, and press button with the thumb.
$O_5^{\alpha w}$	Determine the pointer's position on pointer indicator 2.
O_6^α	If the pointer position is 1, perform $_1O_6^\varepsilon$; if 2, perform $_2O_6^\varepsilon$...; if 4, perform $_4O_6^\varepsilon$.
$L_3 \overset{3(1-4)}{\uparrow}$	Move hinged lever 7 to the position that corresponds to number 1.
$\overset{3(1)}{\underset{1}{\downarrow}} O_7^\varepsilon$	Determine whether signal bulb 4 (red) or 5 (green) is turned on.
$\overset{3(4)}{\downarrow}_4 O_7^\varepsilon$	If the red bulb (4) is turned on ($l_5 = 0$), perform O_9^ε; if the green bulb (5) is turned on, perform O_{10}^ε.
O_8^α	Move the arm to the red button (9) and press it.
$\overset{4(1)}{\underset{1}{\downarrow}} O_9^\varepsilon$	Move the arm to the green button (10) and press it.

its associated multipositioning switch 9 are eliminated). For simplification of our discussion, we also eliminate from consideration the waiting period of time, which was equal to 3 s. An algorithmic description of the considered task is presented in Table 9.4. The actions of the operator in an emergency situation can be seen as an example.

If we know (with some approximation) the probabilistic structure of signals, we can define the average, maximum, and minimum execution time of tasks. It also becomes possible to determine the mean service time, reserve time, etc., based on the methods that were described in Section 7.3. In this example, there is only one waiting line.

Utilization of the queuing theory presupposes that preliminary data can be obtained from observations and experiments. Therefore, we pose some preconditions to demonstrate the ability to apply this theory. Emergency tasks (calling requests or arrivals) are distributed randomly. Observation and records of a task during a specific time interval demonstrate that their distribution with some approximation can be described by the Poisson probability distribution. Based on chronometrical study and Formulas (9.2) and (9.3), we calculate λ—the mean arrival rate per unit of time (average number of tasks-requests per unit of time). Based on these formulas we obtained the following data:

$$\lambda = 0.167$$

The task in question had various methods of performance and has been described using a probabilistic algorithm. Hence, the mathematical mean of task performance time or service time ($\bar{t}_{serv}$) should be calculated according to the following formula:

$$\bar{t}_{serv} = \Sigma P_i\, t_{serv.i} \qquad (9.13)$$

where
$t_{serv.i}$ is the service time of the I_{th} version of the algorithm
P_i is the probability of the algorithmic I_{th} version of the task

Suppose that for the considered task according to Formula (9.13), $\bar{t}_{serv}$ is 2.5 s (2.5 s is the mathematical mean of task performance).
  The intensity of service rate μ according to Formula (9.7) is determined as $\mu = 1/2.5$.
  Assume that the maximum time that is given to perform any version of the task algorithm is 7 s.
Then the reserve time is

$$\bar{t}_{res} = T - \bar{t}_{serv} \qquad (9.14)$$

where T is the maximum allotted time for task performance.
  In an emergency situation, any queuing has a waiting time. In our example, it will be a waiting time for task performance (calling request) in the system (t_{wait}). If $t_{wait} > \bar{t}_{res}$, the task is not served (calling requests or arrivals) and the case is considered to be an error.
  In our further discussions, we use the term *task algorithm*. Under the task algorithm, we understand a version of task performance that is described algorithmically. The work cycle time should not exceed this value, that is, T, in any case. Suppose, according to the system requirements, T is 7 s. If an operator cannot perform any version of task algorithm during 7 s, a system failure occurs. Then the average reserve time ($\bar{t}_{res}$) can be defined by formula (9.14) as:

$$\bar{t}_{res} = T - \bar{t}_{serv} = 7 - 2.5 = 4.5 \text{ s}$$

In fact, the time of service of the i_{th} version of the algorithm can be more or less than $\bar{t}_{serv}$. As a result, $t_{res.\,i}$ may be less or greater than average $\bar{t}_{res}$. Hence, reserve time is a random quantity. Reserve time can be shortened by the fact that the I_{th+1} version of the algorithm can come at a time when the previous version of the algorithm is not completed. As a result, subsequent algorithms will be idle for as long as the previous one is served.

The mean leaving rate of the queue by some versions (algorithms) of task performance is defined as the inverse value of the mean reserve time:

$$V = \frac{1}{\bar{t}_{res}} \tag{9.15}$$

Analysis of the previously presented material shows that we have the following data for calculation:

1. Mean arrival rate per unit of time (average number of tasks/requests per unit of time) $\lambda = 0.167$
2. Maximum time for emergency task performance $T = 7$ s
3. Service time for emergency task performance $\bar{t}_{serv} = 2.5$ s
4. Average reserve time $\bar{t}_{res} = 4.5$ s

Using these data, we can determine the average business of an operator and the reliability of task performance in critical situations. In order to do that, we need first to calculate μ—the intensity of service rate; V—mean rate of a version of task leaving the queue; and n—number of serving mechanisms (Vencel and Ovcharenko, 1969).

The inverse value of an average duration of service (μ the intensity of service rate) is calculated as follows:

$$\mu = \frac{1}{\bar{t}_{serv}} = \frac{1}{2.5}$$

The mean rate of a version of a task leaving the queue (V) is determined as follows:

$$V = \frac{1}{\bar{t}_{res}} = \frac{1}{7 - 2.5} = \frac{1}{4.5}$$

At the next stage, it is necessary to introduce the following dimensionless coefficients for determining the serving queuing dynamic when an operator is working on the analyzed control board (Vencel and Ooncharenko, 1969):

$$\alpha = \frac{\lambda}{\mu} = 0.42$$

$$\beta = \frac{V}{\mu} = 0.56$$

$$\gamma = \frac{\lambda}{V} = 0.75$$

$$\delta = \frac{n\mu}{V} = 1.8$$

where n, the number of serving mechanisms, is equal to 1 (one operator).

Then it is required to calculate the percentage of the shift time when an operator is busy working. For this purpose, the following formula is used in the queuing theory:

$$\bar{k} = \frac{\alpha R(n-1, \alpha) + nP(n, \alpha)\dfrac{R(m+\delta, \gamma) - R(\delta, \gamma)}{P(\delta, \gamma)}}{R(n, \alpha) + P(n, \alpha)\dfrac{R(m+\delta, \gamma) - R(\delta, \gamma)}{R(\delta, \gamma)}}$$

To simplify the calculations, we assume that δ is equal to 2. The maximum number of available positions in the queue m is also equal to 2 (where m is the number of algorithms in the queue waiting for the service).

For the calculation of k, additional data need to be obtained using tabulated functions of the Poisson probability distribution:

$$R(m, a) = \sum_{k=0}^{k=m} \frac{a^k}{k!} e^{-a}$$

$$P(m, a) = \frac{a^m}{m!} e^{-a}$$

According to the earlier formula, values of R and P depend on their parameters (m and a). For example, for $R(n - 1, \alpha)$,

$$R(n - 1, \alpha) = R(0; 42) = 0.67$$

Similarly, we can calculate the values of R and P with different parameters. Based on the obtained data, the value of k can be calculated as follows:

$$\bar{k} = \frac{0.42 \times 0.67 + 1 \times 0.27 \times \dfrac{1 - 0.95}{0.15}}{0.94 + 0.27 \times \dfrac{1 - 0.95}{0.15}} = 0.39$$

Therefore, an operator will be busy only 39% of his or her shift and can be assigned some additional functions.

To evaluate the reliability of task performance, we need to determine the average number of calling requests (various versions of a task algorithm) that can be served at a unit of time:

$$\lambda_0 = \bar{k}\mu = 0.39 \times \frac{1}{2.5} = 0.156$$

And the reliability of task performance is

$$P_{serv} = \frac{\lambda_0}{\lambda} = \frac{0.157}{0.167} = 0.94$$

Studies demonstrated that SSAT has developed a method for the algorithmic description of a task and new principles of the time study of an operator's work. This allows using the queuing theory for the assessment of reliability and accuracy of an operator's work in conditions of a possible time limit when it is important to coordinate a possible request from various arriving tasks and the possibility of the operator to respond to them adequately. This method can also be used when an operator performs computer-based tasks.

Application of the queuing theory can also provide valuable information for the design of indicators, which can be used in an emergency situation. For example, based on the queuing theory, it becomes possible to design indicators that can inform an operator about the more important emergency tasks, their reserve time, the order in which they should be performed, and so on.

Application of the queuing theory to ergonomic studies suggests joint efforts of ergonomists and specialists in the corresponding field of mathematics. An ergonomist would select the critical tasks to analyze and gather all necessary data for the following analysis. The queuing theory specialist would be involved in the application of the queuing theory methods. In this chapter, we demonstrated the possibility of applying the queuing theory in laboratory conditions based on analysis of hypothetical example. This method can also be checked by computer simulation.

Section III

Task Complexity Evaluation

10

Complexity, Difficulty, and Intensity of Work

10.1 Job Evaluation and Complexity of Work

The concept of complexity is used in many disciplines. Various aspects of this issue have been studied in mathematics, economics, biology, engineering, etc. For example, in economics, some scientists considered complexity from the standpoint of the transdisciplinary approach (Rosser, 2010a,b). According to this approach, the term complexity allows to study from a single position such disciplines as economics, physics, and biology. Assessment of the complexity of human labor is not considered by these scientists. However, evaluation of individual work, an assessment of its productivity, and adequate payment for more complex work are important aspects of economics. Complexity measurements can be utilized for assessment of efficiency of human performance and therefore is critically important not only for economics but also for ergonomics and work psychology. Here we study the psychological aspects of complexity. Increasing task complexity correlates with decreasing productivity, increasing errors and time performance of various tasks, and increasing cognitive efforts. However, before we start to consider the psychological aspects of complexity, it is necessary to examine some general aspects of this issue.

According to Simon (1999), complexity is increasingly acknowledged as a key characteristic of any system. Human work activity can also be considered as a system. The term *complexity* originates from Latin, which means *twisted together*. According to Edmonds (1999), the term *complexity* can be interpreted as a feature of an entity that consists of a combination of several interdependent and difficult to separate components. Simon (1999) postulates that complexity is a basic property of a system. The study of complex systems stimulates the development of an interdisciplinary field of science called a *science of complexity*. The major purpose of studies in this field is the development of formalized and quantitative evaluation methods of complex systems structure. Hence, it is necessary to describe the structure of the system first and only after that to perform the evaluation of this systems' complexity.

In different fields, scientists emphasize different aspects of complexity. For instance, in mathematics, there is the computational complexity theory. This theory is used to analyze what can be accomplished by using a computer during solving computational problems (Du et al., 2000). Computational complexity can be approached from various perspectives such as time, memory, and other resources used by the system to solve the problem. In engineering, there is the concept of design complexity under which one can assume either an artifact complexity or a design process complexity (Braha and Maimon, 1998). The development of a technologically complex product makes it expensive; therefore, this factor becomes critical for business survival (Singh, 1997).

The concept of complexity is very important in economics for the evaluation of business organizations (Veikher, 1978). Complexity is considered a major factor in the study of organizational–environmental interaction. A very dynamic and complex environment condition yields a high degree of uncertainty. The latter, in turn, is a driving force that influences organizational decisions (Thaompson, 1968). When some events occur in the environment with a high level of uncertainty, the consequences of the events could not be predicted with reasonable precision and it makes organizational decisions less efficient. Such decisions have a certain negative impact on business development. The complexity theory has been used in the field of strategic management and organizational studies (Levinthal and Gavetti, 2004). This theory treats organizations and companies as collections of business strategies and structures. When an organization consists of a small number of relatively simple and efficiently connected structures and strategies, it is more adaptive and has a better chance of surviving. As we will show further, human behavior and activity might also be considered as a structured system that can be evaluated based on its complexity. This is the psychological aspect of complexity. The increase in psychological complexity correlates with the increase in the time of task performance, increase in the number of errors, and decrease in human productivity in general (Bedny, 1987; Thomas and Richards, 2012). The increase in task complexity is accompanied not only by the reduction of productivity but also by the increase of emotionally motivational tensions of human performance. The quantitative evaluation of complexity of human performance presents an important approach to the enhancement of work efficiency, improvement of design solutions, etc. This, in turn, can bring us closer to optimal performance. It is well known that the more complex the task is, the lower the reliability of the task performance (Bedny et al., 2010). For example, the more complex the task is, the greater the likelihood of errors. Therefore, the task complexity evaluation is an important factor for the prediction of human reliability (Miller and Swain, 1987). Evaluation of task complexity during human interaction with equipment is essential in the assessment of efficiency of a design solution (Bedny and Karwowski, 2007, 2011). However, oversimplification of a task can lead to monotony and boredom. Therefore, in our further discussions, we speak not about simplification of tasks but about their optimization. Further, it is important to note that the concept of complexity is important not only for increasing productivity but also in assessing the usability of products. For example, if a consumer product is too complex to set up or to use, it negatively affects such product's sale. The complexity factor is also important for entertainment-related tasks. If, for example, a videogame is too easy, it is not exciting enough for the players to choose. Thus, complexity is an important concept in psychology, ergonomics, and economics. There are multiple factors that determine task complexity. However, this does not provide a rationale for introducing various types of complexity in psychology. Depending on the specific task, the conditions of its performance, and the degree of preparedness of a person to perform a task, there may be different cognitive demands for task performance. Thus, complexity emerges as a multidimensional system. An increase in the complexity of the task is accompanied by an increase in the probability of more errors, increase in the time of its performance, increase in mental fatigue, and reduction of productivity.

We would like to differentiate between the concept of task complexity (evaluation of activity complexity during task performance) and the concept of cognitive complexity or difficulty in cognitive psychology (Wickens and Hollands, 2000). According to the concept of difficulty in cognitive psychology, the more difficult the task is, the more mental resources it requires. However, the solution of this issue is reduced to the analysis of the *performance resource function* (PRF), which depends on the relationship between automatic and controlled processing. In practice, the relationship between automatic and controlled processing is constantly changing. The more automatic the task performance becomes, the less resources it requires. This theoretical approach is useful for the analysis of time-sharing tasks. However, it cannot allow evaluating the complexity of activity during task performance because this approach ignores the structure of activity as a system. This is a parametric method of study because it is focused on the separate aspects of complexity evaluation.

During the training process, we not only observe the changes in relationship between automatic and controlled processing but also can see how strategies of activity are being utilized as the training gradually changes. The trainee attempts to transfer to a new strategy. The new strategy is tested by the trainee, and if evaluated to be more efficient, it is used, and if inefficient, it is discarded. The trainee's activity strategies are transformed from a less efficient to a more efficient one (Bedny and Meister, 1997). Some elements of activity can be eliminated and new elements can be introduced. Some elements of activity are performed in parallel due to the performance of actions with less conscious control, while some others are performed in sequence. The more complex the acquired activity is, the more intermediate strategies are utilized by the learner. The changes between the automatic and controlled processes are only one aspect of activity acquisition. The process of learning leads to the systemic changes in the structure of activity. Finally, in cognitive psychology, the terms *complexity* and *difficulty* are used synonymously. In systemic-structural activity theory (SSAT), they have different meanings.

Human activity is a process that unfolds in time as a structure. Therefore, all measurement procedures for complexity evaluation should be adapted for quantitative evaluation of the time structure. Quantitative analysis of complex structures is not possible without a preliminary morphological analysis of the system. Thus, morphological analysis of activity is critically an important step of task complexity evaluation. In our work, the purpose of morphological analysis is the description of the architecture of the activity structure. Activity as a complex structural system consists of logically organized units, which in turn is organized hierarchically. The structure of activity depends on the strategies of performance that is considered as a dynamic interaction with equipment or machinery. This dynamic interaction has a probabilistic relationship between the structure of activity and configuration of equipment. Therefore, if the configuration of equipment changes, it changes the strategies of performance in a probabilistic manner and it leads to the changes in the activity structure. At the same time, one can change the strategies of performance when a worker utilizes the same equipment. The concepts of complexity and difficulty were considered earlier when we studied complexity of computer based tasks (Bedny et al., 2014). Particular attention was paid to the complexity from the standpoint of functional analysis or activity self-regulation. In SSAT there are two models of activity self-regulation. These models of activity self-regulation contain the function block *assessment of task difficulty*. This block has a purpose to evaluate the objective complexity of the task. Evaluation of task difficulty is connected with the subject's past experience and his or her beliefs that he or she does or does not possess the necessary abilities to perform the tasks. Such evaluative process influences on strategies of task performance. This is a functional analysis of complexity and difficulty. Further in this chapter, we consider complexity not from the self-regulation activity perspective. In this section, we consider the concept of complexity and relate it to the concept of difficulty from a morphological analysis perspective. The major units of analysis in such situations are not function blocks but human cognitive and behavioral actions. This is the morphological approach to task complexity evaluation. The approach considered in this chapter can be utilized in improving the efficiency of performance and increasing productivity, in equipment design and safety, in job evaluation and time study, in reliability analysis and increasing efficiency of training, etc. Morphological and functional analysis of complexity are two interdependent approaches.

Contemporary ergonomics, economics, and work psychology are tightly interconnected because the major purpose of these fields is to study human work and productivity. In this section, we will briefly consider the significance of the concept of complexity for the study of work in economics and ergonomics. The reason is that both ergonomists and economists are concerned about the efficiency of labor and its evaluation. One important aspect of job evaluation is its use for the establishment of workers' compensation rate. It is based on the assumption that each job possesses an inherent worth that is independent of market forces of supply and demand (Arnaut et al., 2001). Economists recognize that inaccurate job evaluation may result in a highly unequal distribution of income between workers. Job evaluation is a procedure that is used for determining the relative value of jobs in the organization. This information in turn is useful for determining the level of compensation paid (Muchinsky, 1990). Therefore, job evaluation requires job classification according to the developed criteria. Presently, various techniques of job evaluation have been developed. They can be divided into two groups: nonanalytical and analytical. Nonanalytical job evaluation is based on the comparison of one job considered as a holistic system with another existing job. Such method is based on the integrative evaluation of a job without consideration of its separate factors and quantitative evaluation of their complexity. We can mention here such methods as job ranking, paired comparison, job classification, etc. These methods have limited precision because many aspects of the job are not being considered. Analytical methods are more precise because by these methods, each job is broken down into elements based on a number of factors. Such factors as education, skills, responsibilities, decision making, dexterity, etc., are being taken into consideration. The basis for the analytical job evaluation procedures lies in the choice of factors and in their weights. When workers are paid with piece rates, the inequality factor arises naturally (Neilson and Stowe, 2010). Attention also should be paid to the fact that work can be of various levels of complexity. So analytical job evaluation procedures should be applied to analyze and determine the complexity of work. Therefore, the theoretical basis of any method for job evaluation is the concept of simple and complex work (Armstrong et al., 2003).

The optimization of work performance according to these characteristics determines how efficiently humans interact with technical components of the system. Excessive simplification of human work can lead to monotony. On the other hand, inadequate design of hardware or software with respect to complexity characteristics may impair the cognitive mechanisms of activity regulation. Further in our study, we will put more emphases on the optimization of human performance according to the complexity characteristics, rather than on the simplification of work. It is important to understand that from a technological point of view, the system can be excessively complex but human interaction with it can be simple and vice versa. Hence, we will be interested only in the psychological aspects of complexity, that is, in those aspects of complexity that are associated with human labor. In fact, any method of job ranking is performed based on such factor as complexity of work, which is considered as one of the most important criterion for evaluation of job worth. It should be pointed out that during a long period of time, one of the more important criteria in evaluating work complexity was the level of the worker qualification for a particular job. As a result, other factors were not being taken into consideration sufficiently. Overemphasis on the qualification factor can lead to the situation that workers with a high level of qualification will be paid in accordance with what they potentially can do. At the same time, workers with a lower level of qualification will be paid based on what they really do (Veikher, 1978). Presently, most economists are in agreement that the concepts of complex and simple work have an important meaning for job evaluation (Moshensky, 1971). Complex work is a function of a qualified workforce that requires more mental efforts than the simple work does. According to many economists, complex work requires more concentration of attention, more precise motions, and more mental efforts in general. Complex work results in changes of the value of time units of work. Such units become more significant components of work in comparison with time units in simple work. All of the previously discussed is evidence that in any method of job evaluation, the factor of work complexity must be taken into account. Job evaluation that is based on the work complexity assessment permits to eliminate job evaluation where the dominant principle is the expenditures associated with training and education. Without understanding the differences between complex and simple work, it is difficult to explain why expenditures of mental efforts for these two kinds of work during the same time period are different. The development of principles of work complexity evaluation is a theoretical basis of job evaluation. Economics as a science cannot evaluate the complexity of work with sufficient precision without utilizing the psychological methods of work evaluation. As a result, due to the insufficient consideration of psychological factors, there are no direct methods for the measurements of work complexity in economics.

10.2 Complexity and Difficulty as Basic Characteristics of a Task

Simon (1999) postulates that complexity is the basic property of a system. Human activity can also be considered as a system. However, there are a lot of contradictions in the study of complexity of human cognition and behavior. In this section, we will consider the psychological aspects of complexity and its relation to the concept of complex work in greater details. Complex work requires greater concentration of attention, more precise motions, and mental efforts in general (Veikher, 1978). Such understanding of work complexity brings psychologists and ergonomists to viewing the complexity of work from psychological perspectives. Task complexity is the key aspect of this area of study. The more complex the task is, the more mental or cognitive efforts are required for its performance. Therefore, task complexity is a psychological characteristic of task that determines the cognitive demands for task performance (Bedny, 1987; Bedny and Karwowski, 2007; Bedny and Meister, 1997). The psychological aspects of task complexity are also important for specialists that study the complexity of human labor.

There are multiple factors that determine the complexity of a task (Bedny, 1987; Bedny and Karwowski, 2008; Bedny and Meister, 1997; Galwey and Drury, 1986; Payne, 1997). The complexity of task depends on the quantity of the task elements and the specificity of their interaction and on the number of static and dynamic elements. The degree of uncertainty or unpredictability of the task is also an important component of task complexity. An increase in the information processing speed under time restrained conditions is one possible factor of task complexity. According to cognitive psychology, the main causes of the complexity of a problem-solving task are the complexity of the rules for its solving, ease of developing such rules and their application, and memory workload during utilization of these rules. Reducing duration of keeping information in working memory during task performance is also an important factor in reducing the difficulty of a problem-solving task (Kotovsky and Simon, 1990). Demands that are imposed on the short-term memory by the task is an important factor for the task complexity (Jacko, 1997). Task complexity influences users' mental workload and affects their performance in general (Jacko and Ward, 1996). According to Arend et al. (2003) and Jonassen (2000), when task complexity increases, the amount of cognitive resources that is required for task performance also increases. Therefore, increase in task complexity is usually accompanied by deterioration in performance, making task complexity one of the most general characteristics of a task.

Let us consider some cognitive characteristics of task complexity from the activity theory perspective. The specificity of memory workload is one important source of task complexity. Keeping more than three intermediate data about dynamic objects in memory is evaluated subjectively as a difficult situation, which produces errors (Zarakovsky and Magazannik, 1981). The duration of keeping various pieces of information in working memory is also a critical factor. The specificity of extracting information from the long-term memory also influences task complexity. The level of familiarity of reproducible information is also a factor influencing the complexity of retrieving information. If the retrieved information has similarities with task-irrelevant information, then the complexity of the task increases. The sensory-perceptual characteristics of information also influence the complexity of task performance. If perceived information is in a threshold area of our senses, it results in an increase of complexity. The stimulus difference in the discrimination process is another sensory-perceptual factor that contributes to work complexity. These sensory-perceptual factors are tightly interconnected with decision making at the sensory-perceptual level; therefore, sensory and nonsensory factors are interdependent. The factor of similarity between different features of the task influences the complexity. For example, the more similar signals appear, which require different responses, the more complicated the perceptual components of the task are.

Various characteristics of the decision-making process at the verbal-thinking level are also important factors of task complexity. For example, the number of contradicting solutions influences the complexity of the task. The decision-making process is more complicated when it involves extracting information from memory. On the other hand, decision-making processes become easier to perform when they are predominantly determined by external stimuli or information provided by external sources. Complexity also increases when the subject is required to alter stereotypical actions. Finally, the level of concentration of attention is also a critically important characteristic of complexity. The higher the level of concentration is required, the more complex the task is.

Complexity can also be employed to evaluate the motor components of activity (Bedny and Meister, 1997). For example, the more precise a motor action is, the more concentration of attention it requires, and the more complex the action becomes. However, when one considers the amount of physical efforts during the performance of motor actions, it is viewed as an evaluation of the physical characteristics of a task. Therefore, the concept of complexity can be applied to the motor aspects of a task or—more specifically—to the mental regulation of movements. A subject cannot perform a complex motor task without significant mental effort and concentration. The concept of complexity is also associated with the emotional-motivational component of activity. Emotional tension and motivational forces increase as the task complexity increases. Therefore, it is important to distinguish the cognitive aspects of complexity, which depend on the specificity of information processing, and emotional-motivational aspects of complexity that reflect the energetic aspects of cognitive activity. These two aspects of complexity are interdependent and influence each other. The relationship between the different components of a task (cognitive, motor, and emotional-motivational) is critical in evaluating the complexity of a task performance. The specificity of the combination of elements of activity is another factor of complexity.

Complexity is an objective characteristic of a task. Task difficulty is another characteristic that depends on task complexity. If complexity is an objective characteristic of a task, then difficulty is the worker's subjective evaluation of the effects of task complexity. Therefore, complexity and difficulty cannot be considered as synonymous. The same task complexity can be subjectively perceived as a task with a different level of difficulty for workers with different past experiences and individual features. An increase in the complexity of a task increases the probability of the performer's requirement to exert more cognitive effort. Complexity itself does not have a subjective component. A worker cannot directly experience complexity of the task by itself but rather perceives its subjective difficulty.

The concept of difficulty can be approached from two different perspectives. It can be considered as a subjective characteristic of task complexity. It also can be studied as a functional mechanism of activity regulation that is in a dynamic relationship with other mechanisms of activity self-regulation. The second approach is related to the functional analysis of activity (Bedny and Karwowski, 2007; Bedny and Meister, 1997). From the functional analysis perspective, it is important not only to evaluate the objective characteristics of the task but also to evaluate a person's belief that he or she possesses the necessary abilities and experience to accomplish the goal of the task. Therefore, the self-concept of ability, self-efficacy, self-esteem, etc., becomes important in the task difficulty evaluation from the functional analysis perspective. The distinguishing concepts of complexity and difficulty permit us to collect and use a significant amount of interesting psychological data that used to be excluded from work analysis. For example, the evaluation of task difficulty influences the formation of the subjective criteria of success. A worker can subjectively evaluate the task as very difficult and not a personally significant one. In such situations, the worker can decrease his or her quality standard or even reject to perform a task (Bedny and Meister, 2007). Kim (2008) even introduced such terms as *pretask* and *posttask difficulty*, which are confusing terms. According to SSAT, there is a subjectively perceived task difficulty before execution of a task and subjective evaluation of a task after its execution. From self-regulation of activity point of view (see models of self-regulation in Sections 3.2 and 3.3 of Bedny et al., 2014) there are three most important functional mechanisms or function blocks that determine the complexity-motivation relationship. The first mechanism is responsible for the subjective evaluation of the task significance, the second one is responsible for the subjective evaluation of the task difficulty, and the third one is involved in the formation of the level of motivation during task performance. If, for example, a subject evaluates the task as a personally significant and difficult one, the level of activity motivation that is required for task performance would increase. At the same time if a subject evaluates the task as a difficult but not a personally significant one, the level of motivation can drop or a subject can even reject the task altogether. Therefore, resource allocation is a complicated self-regulative process. When one compares the concepts of complexity and difficulty, it is useful to utilize the term *idealized worker*. Such worker can be seen as an *average* subject who has the required past experience in the considered field and the appropriate average level of abilities to perform the considered tasks. For such an idealized subject, the difficulty and complexity of the task are the same. For the real subject, complexity and difficulty are not the same. The more complex the task is, the greater is the probability that the task will be more difficult for him or her. Difficulty depends not only on the objective complexity of tasks but also on the idiosyncratic features of the individual.

Let us examine a task complexity review that was performed by Liu and Li (2012). They did a useful job of gathering a lot of information on the subject. However, we cannot agree with some of their interpretations of the basic concepts of task complexity. Summarizing the analysis of task complexity, Liu and Li (2012) identified three viewpoints: structuralist, resource requirement, and interaction. According to the structuralist vision, complexity is defined by the structure of task. For example, it can be defined as a function of the number of elements of which the task is composed and by the relationship between these elements (Liu and Li, 2012, p. 554). A number of distinct acts, multiple paths, multiple outcomes, the nature of the relationship between task input and task products, etc., are examples of such characteristics. From the resource requirement viewpoint, task complexity is defined as the physical and mental demands and cognitive efforts. The interaction viewpoint reflects the subjective task complexity from the task performers' standpoint. The performer can interpret the complexity of the same task differently. We presented only some characteristics of complexity identified by these authors. In our opinion, such classification is very ambiguous. For example, how did the authors define acts, how did they evaluate them, and how can one can calculate their number in a task? Similarly, how can the input and product relationship be utilized for the evaluation of task complexity?

When Liu and Li (2012, p. 555) considered the relationship between complexity and resource requirements, they noted that Bedny et al. (2012), Robinson (2001), Wood (1986), etc., have different perspectives on the relationship between complexity and resource requirements. For example, according to Liu and Li, Bedny et al. thought that task complexity is one aspect of resource requirement. We cannot understand what the authors are trying to say in this sentence. We cannot also agree with Liu and Li's statements that Bedny and the other mentioned authors consider complexity and task load or task demands as synonymous. Liu and Li confuse the cause and effect relationship. The heavier the tasks are, the more physical efforts are needed for task performance. Similarly, the more complex the tasks are, the more cognitive efforts are required. Excessive mental or physical efforts for task performance are not desirable. Investment of cognitive or physical efforts for task performance suggests that there are some resources for such investment. We cannot agree that there are three separate viewpoints on task complexity listed earlier. For example, the more complex the structure of the task is, the more likely that it would require more resources for performance. However, past experience and individual differences also influence the relationship between the structure of the task and the required resources for performance. The authors confuse the factors that affect complexity with the various viewpoints on complexity.

The concept of resources is important for the analysis of task complexity, consideration of attention, etc. (Bedny and Karwowski, 2011; Kahneman, 1973; Wickens and McCarley, 2008). Attention is a mechanism that integrates all cognitive processes to achieve a task goal. In Bedny et al., 2014, Chapter 5, we considered the concept of resources when we described the model of attention. Wickens and McCarley (2008, p. 179) described the neurophysiological study of Just et al. (2003) and wrote:

> … brain activation constitutes a form of attentional resources. Resources supply is the total available amount of brain activation, and resources demand for a given task is the amount of activation needed for task performance. Resting state brain activation levels thus provides a measure of resources availability, and activation levels during task performance provide a measure of the task's resources demands.

Thus as a consequence of the limited cognitive resources and increased task complexity, the possibility of committing errors can increase, the probability of achieving a task's goal may reduce, and the level of mental fatigue can elevate.

Liu and Li (2012) erroneously interpret data from psychology when they considered such concepts as task complexity, resources, task demands, etc. Their comparative analysis of such terminology is difficult to comprehend and interpret because the authors use incorrect criteria for such discussion.

Based on their analysis, Liu and Li (2012, p. 559) recommend using the following basic terms: objective task complexity, subjective task complexity, and task difficulty, which can be classified further as pretask difficulty and posttask difficulty. Task difficulty according to these authors is defined as "…extent to which task performers feel difficulty in performing a task." At the same time, these authors define subjective task complexity as "a perceived complexity by the task performer." The authors state that "the objective perspective considers task complexity to be related directly to task characteristics and is independent of task performers." However, task complexity is a psychological concept that cannot be totally separated from a performer. Task complexity should be considered as the source of cognitive demands imposed on a performer by a task. For some, these requirements will be excessive, while for others, they are moderate. The more complex the task is, the more is the probability that this task will be difficult for subjects. When we evaluate task complexity, we can predict the difficulty of this task for future performers. Therefore, objective complexity and subjective difficulty are interdependent and objective complexity of the task reflects the possible difficulty for a performer.

Liu and Li try to separate such terms as perceived complexity by a task performer and a feeling of task difficulty. Words like *feeling* and *perceiving* are related to a subjective opinion. It is not so easy to separate these two subjective judgments about a task. It is particularly confusing when we consider a situation of *pretask feeling of difficulty, posttask feeling of difficulty,* and *perceiving of task complexity.* Posttask feeling of difficulty is practically not distinguishable from perceived or experienced task complexity. Both terms reflect a subjective judgment or experience about a task after its execution.

Liu and Li argue that pretask difficulty is a measure of self-efficacy. However, pretask difficulty may be overestimated or underestimated due to unfamiliarity of the subject with the task ahead. The estimation process can be affected by past experience that is not related to self-efficacy. Further, according to Bandura (1989), all motivational factors are a result of self-efficacy. However, as we have discussed in Bedny et al. (2014), motivation is a complex self-regulative mechanism of activity that cannot be reduced to a self-efficacy mechanism. As discussed earlier, the goal can be perceived as very difficult due to low self-efficacy. However, if this goal can be very significant for a subject, he or she will be motivated to perform this task. Self-efficacy is only one possible mechanism of motivation. Thus, pretask difficulty cannot be considered as a measure of self-efficacy. According to SSAT, complexity is an objective characteristic of task that determines task difficulty in a probabilistic manner. Complexity is a relatively stable characteristic of the task. Difficulty of a task is a more varied characteristic than complexity, which depends on the individual characteristics of a performer and in particular on his or her past experience and individual feature of personality. Estimation of task difficulty by a subject who is not familiar with it and is going to perform it for the first time depends on the past experience and his or her ability to correctly evaluate the task to be performed, subjective estimation of task difficulty, significance of the task, etc. In SSAT, task complexity is considered as the most important psychological characteristic of human work that determines the cognitive demands of task performance. From the functional analysis perspective, evaluation of task difficulty can be considered as an important mechanism of activity self-regulation that can influence the selection of adequate strategies of performance, motivation, acceptance or rejection of the task, etc. (see Sections 3.2 and 3.3 in Bedny et al., 2014). This is another aspect of task difficulty analysis. This analysis is important for discovering possible strategies of task performance, which is a qualitative systemic analysis that precedes the quantitative assessment of task complexity. A quantitative assessment of task complexity cannot be executed without discovering the most preferable strategies of task performance. Liu and Li (2012, p. 558) stated that according to Bedny et al. (2012), difficulty is a functional mechanism of activity and behavior regulation involves motivational and emotional activities. As presented in Sections 3.2 and 3.3 (see Bedny et al., 2014), the models of activity self-regulation such as *assessment of task difficulty, assessment of sense of task* (emotionally evaluative mechanism), and *formation of the level of motivation* are interacting mechanisms of activity regulation. There are functional and morphological analyses of activity in SSAT that are not the same. The previously mentioned authors do not see the differences between the quantitative assessment of task complexity and the analysis of the mechanism of activity self-regulation depicted by the function block *assessment of task difficulty in the model of activity self-regulation.*

11

Theoretical Principles of Complexity Measurement

11.1 Informational and Energetic Components of Complexity

Analysis of psychological, ergonomics, and economics literature demonstrates that there are four basic characteristics of task or, more precisely, work activity during task performance: physical demands (heavy work), intensity of work, complexity, and difficulty of work. It should be noted that, at present, these terms are not precisely defined and very often are being used synonymously. For example, heavy work is considered synonymous with difficult work (Rodgers and Eggleton, 1986, vol. 2). We consider these characteristics to be totally different. In this section, we will describe physical demands and intensity of work and will perform comparative analysis of these concepts.

Sometimes in order to determine the efficiency of the utilized terminology, it is useful to compare it with similar terminology in other languages. For example, Russian work physiologists Kandror and Demina (1978) describe the state of work to be heavy as the heaviness of work. This is not a recognized term in English. However, this term can be useful for analysis of physical demands of the job. Some economists who consider *heaviness of work* are concerned with the effect of the energetic components of work during task performance. Physical demands of work can be extremely heavy, moderate, or light. If physical efforts are minimal, we can neglect these characteristics of work. Usually though, the degree of physical effort is an important characteristic for physically manual tasks. Physical efforts have subjective components such as feeling of physical stress that is affected by the individual's physical conditions. Physical and mental efforts can influence each other. For example, coordination of physical actions is much more difficult for a subject when physical efforts are increased. Physical efforts or *heaviness of work* can be measured by utilizing such indexes as energy expenditure and heart rate estimation.

Another important characteristic of task in ergonomics and psychology is the intensity of work. This is also important when studying human work in economics (Boisard et al., 2003; Gal'sev, 1973; Veikher, 1978). The work intensity depends on the speed or pace of the performance. Studies demonstrate that the workers reported an increase in work intensity when they were exposed to higher pace of work or had insufficient time to complete the job. In recent years, it has become increasingly evident that work becomes more and more intense. Economists sometimes utilize the term *work tension* instead of the term *work intensity*. Tension of work includes such characteristics as pace of performance and amount of mental and physical efforts per unit of time (Kholodnaya, 1978). Russian psychologist Nayenko (1976) made a distinction between what he called operational and emotional tensions. Operational tension is determined by a combination of a task difficulty and the lack of available task performance time. Emotional tension is determined by the personal significance of an activity to the worker. These two types of tensions are closely interrelated, and under certain conditions, one type of tension causes the other.

Analysis of such concepts as intensity and tension demonstrates that these two concepts are very similar and are contaminated with such concepts as heaviness, complexity, and difficulty. In the next section, we will consider the latter two described concepts. According to our analysis, heaviness of task (degree of physical effort), task complexity, and task difficulty are the most productive characteristics of task. Intensity of work can be used as an additional characteristic of work, which we associate with the lack of available time for task performance. If the mental aspect of work is a dominating factor, then the intensity of such work can emerge as a component of task complexity. However, if the work requires physical efforts, its intensity should be characterized as the aspect of the work heaviness (heavy work).

A major part of further discussion will be the analysis and quantitative evaluation of task complexity. The task can be simple or complex. The range between simplicity and complexity of a task can be very broad. The complexity of a task affects human performance in various ways. Complex tasks are less reliable and accompanied by a more significant quantity of errors. Simple tasks are learned more quickly than complex ones. Complex tasks have increased cognitive requirements for work and are the cause of mental fatigue. The more complex the task is, the more mental efforts are required for its performance. It says that task complexity is the basic psychological characteristic determining the demands of the task. The level of concentration of attention is an important aspect of the mental effort evaluation. The model of attention developed by Kahneman (1973) considers attention as mental efforts and can be helpful in our analysis of the issue. According to Kahneman's model, any subject has limited mental resources. The quantity of these resources is relatively constant and can be slightly increased as a result of the increased activation of the nervous system. One drawback of this model is that it does not take into consideration informational aspects of human activity sufficiently enough. In the model of attention described in Bedny et al. (2011), informational and energetic mechanisms of attention were considered. In this model, we describe the mechanism responsible for coordination of energetic and informational components of activity. According to this model, when task complexity increases, in addition to the increase in nervous system resources, the intensification of the informational processes can be observed as well. This, in turn, results in increased complexity of coordination of informational and energetic aspects of activity. Energetic components of activity include activation of the nervous system and motivation, among others. It was demonstrated that the more complex human information processing, the more energy is demanded in this process (Bedny and Karwowski, 2007).

Bloch (1966) characterized attention as the intensity of neuropsychological energy or the level of activation involved in task performance. The higher the level of activation, the higher the level of wakefulness (attention). According to Bloch, performance is a linear function of these two variables. Attention is considered to be one of the levels of wakefulness that depends on the level of activation of the neural centers. An increase in the level of wakefulness raises the level of attention at the same time. There are specific and nonspecific levels of activation. Nonspecific or global level of activation is closely connected with the functioning of the reticular activity system of the brain. It involves general changes in the functional state of the brain. Specific activation is related to the regional changes in different brain subsystems (sensory, motor, and associative). As has been demonstrated by Aladjanova et al. (1979) and Lazareva et al. (1979), difficulty of performance is connected to nonspecific forms of activation. This kind of activation is not directly associated with the content of activity and can be considered as a continuum. Specific activations should not be considered during complexity evaluation. Hence, the more difficult the task is for the performer, the higher is the level of activation of the brain and the level of wakefulness. In our discussion, we consider only the ranges of wakefulness and the nonspecific activation related to work activity.

Based on an analysis of the theoretical data, it was shown that, at the present time, five levels of complexity can be identified with high accuracy (Bedny and Karwowski, 2007). Complexity and the difficulty of activity associated with it can be viewed as a continuum. Actions that require a minimal level of wakefulness (attention) are the simplest, whereas actions requiring maximum level of wakefulness are the most complex ones. This continuum of complexity can be depicted as a straight line. This continuum can be divided into a five-level ordinal scale for different categories of complexity. Any category of complexity can be considered as an interval. Differences in activity complexity within each category can be ignored. This principle of categorization is similar to the measurement procedures developed for the *accuracy, interchangeability, and measurement* used for mass manufacturing processes. At present, it is possible to develop a five-point scale of complexity. The scale could be developed based on theoretical analysis of data obtained by the use of Methods–Time Measurement (MTM-1) system and data derived from cognitive psychology and activity theory. In the MTM-1 system, the elements of activity are clustered into three groups based on different levels of concentration of attention (UK MTMA, 2001). The simplest group requires a low level of attention. For example, the element of activity *RA* (reach to object in fixed location) requires minimal concentration of attention. We relate it to the first category of complexity. The element *RB* is more complicated (reach to a single object in the location that may slightly vary from cycle to cycle), and according to the MTM-1 system, it requires an average level of concentration of attention. According to systemic-structural activity theory (SSAT) rules, such elements of task correspond to the second category of complexity. The element *RC* is even more complicated (reach to an object jumbled with other objects in a group) so a search and select occur, which requires a higher level of attention. This element, according to the SSAT rules, corresponds to the third category of complexity. There is a cognitive element of activity in the MTM-1 system, which is used when an operator is required to recognize an object and make an *if-then* or *yes–no* type of decision. According to the MTM-1 system, this simplest cognitive element of work activity requires the highest level of concentration of attention similarly to the element *RC*. Hence, it is also related to the third category of complexity according to SSAT rules.

This data enabled us to introduce a three-point scale for behavioral (motor) elements of activity. The simplest behavioral elements that require a minimum level of attention corresponds to the first category of complexity. The second group of elements that requires an average level of concentration is related to the second category of complexity. The elements that are associated with the highest level of concentration correspond to the third category of complexity. The simplest cognitive elements of activity, which are based on our analysis of the MTM-1 system, can also be related to the third category of complexity. Each category represents a range of complexity, which means that complexity varies inside an interval and such variation can be ignored. This evaluation principle is used for determining interchangeability of parts in mass production operations. Parts varying within a given tolerance are considered to be the same.

Analysis of attention, wakefulness, and activation combined with the MTM-1 system allows the definition of three categories of complexity for motor components of activity depending on the level of concentration of attention during performance. Figure 11.1b depicts the scale that has three categories of complexity for motor activity.

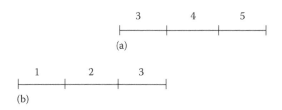

FIGURE 11.1
Five-point ordinal scale for evaluation of complexity. (a) Order scale for cognitive activity; (b) order scale for motor activity.

Let us consider an example. The MTM-1 element *Reach (RA)* (reach object in fixed location) requires a minimum level of control and attention. The element *Reach (RB)* is more complex. It involves reaching an object in a location that can vary and requires an average level of attention. The element *Reach (RC)* is the most complicated. Its purpose might be to reach an object mixed with other objects. This last element requires a high level of control or concentration of attention.

The MTM-1 system has one microelement that describes cognitive component of activity. It is an eye focus time. This is the time required for recognition of a relatively simple object, or simple decision-making action, at sensory-perceptual level (performing decision *yes–no* or logical decision *if-then*, etc.). Performance time *EF* is approximately 0.30 s. In the MTM-1 system, this microelement requires a high level of control, which is equivalent to the third category of complexity. Hence, even the simplest cognitive operations and actions should be related to the third category of complexity. All cognitive actions and operations performed with a high level of automaticity are related to the third category of complexity.

However, activity may have more complicated components, for example, in cases of the decision-making action. When the required responses are unknown in advance, it is a more complicated decision-making action than the decision-making where the required responses are already known. Very often, an operator has to perform some actions or make decisions in an ambiguous situation, for instance, when a signal on a screen moves forward but the operator is required to move a lever backwards to react to this signal. This kind of situations requires remembering instructions, which in turn requires a greater level of concentration of attention. This analysis allows us to conclude that mental actions accompanied by overloaded working memory, perceptual actions involving perceiving unclear signals, and decision-making actions where the required responses are not known in advance and decision requires analysis of contradicted information should be placed in the fourth category of complexity. In a similar way, motor actions that comprised the third category of complexity becomes more complex (Bedny and Karwowski, 2007). For example, operator performs a motor action such as a move control forward into the exact position when a controlled object moves backward, and therefore, this motor action should be related to the fourth category of complexity (see Figure 11.1a).

In some cases, a worker performs some task in a stressful situation (emergency conditions). Hence, actions performed in such stressful situations can be elevated to an even higher level of complexity. Thus, it is necessary to develop a five-point ordinal scale for motor and cognitive activity elements (see Figure 11.1a). As discussed earlier, the simplest cognitive component of activity belongs to the third category and the most complicated cognitive elements cannot exceed the fifth category. Hence, the five categories of complexity can be presented as depicted in Figure 11.1.

From this figure, one can see that the motor and cognitive scales partly overlap each other, with the complexity of the most complicated motor components of activity corresponding to that of the simpler cognitive components. Based on the presented material, a formalized system of rules and procedures was developed for translation of qualitative concepts into quantitative indices. These enable the identification of a strictly monosemantically determined complexity category of activity elements associated with time intervals for various elements of activity. In this system, the concept of *complexity of a time interval* is used to describe the complexity of various elements of activity performed in a given time interval.

The categories of complexity demonstrate mental efforts that are required for the performance of particular elements of activity. Some additional requirements should also be taken into consideration during evaluation of complexity, which will be discussed in the following section. Introduction of additional categories of complexity is not possible at this time because there are currently no scientifically grounded criteria for it.

11.2 Review of Some Methods of Task Complexity Measurement

Activity is a process that is unfolding in time. At the same time, this process is a composition of qualitatively different interdependent units of activity. Some elements of activity can be performed simultaneously and others in sequence. Activity elements have logical and hierarchical organization. Thus, activity is a system with complex structure, which unfolds over time. According to Simon (1999), a *complex system* is made up of a large number of parts that have many interactions. Hence, activity can also be considered as a complex system. This makes the development of procedures for quantitative measurement of activity complexity important. The first and most important aspect of complexity evaluation is selection of units of measurements. The second important aspect of this problem is development of measurement procedures that permit comparison of different elements of activity that unfold in time. This is an important issue that has not yet been resolved. For a long time, scientists and practitioners have attempted to develop a quantitative method of task complexity evaluation. Any measurement procedure requires a selection of adequate units of measure. Some scientists suggest to utilize different units of measure, such as the number of controls and indicators and the number of actions and alternatives in multiple-choice tasks (Galwey and Drury, 1986; Payne, 1976). For the complexity evaluation of computer-based tasks, such measures as task solving time, the number of different transitions, and the total number of states of the system that describes the task solving process were suggested by Rauterberg (1996). However, the units of measure suggested by Rauterberg and others are incorrect from a mathematical point of view because they are incommensurable units of measures. For example, it is impossible to say which is greater, 2 in. or 2 lb. The suggested measures do not always correlate with complexity. Sometimes a more complicated task can be performed at the same time as a simpler one. The subject can spend more mental effort with a task that is performed in a short time or with fewer transitions during performance. Manipulation with one control can be more complex than manipulation with several controls. Similarly, it would not be accurate to calculate the amount of actions performed by an operator during task performance for task complexity evaluation. For example, one motor action or cognitive decision-making action can be more complicated than several simple ones.

Recently, the quantitative evaluation of task complexity from reliability engineering perspectives has been described by Park (2009, 2011). For example, Park and Jung (2007) considered a type of task that they described as emergency operating procedures (EOP). He suggested the following units of measure of task complexity.

Park and Jung wrote (2007, p. 1104):

> … each task can be decomposed into one or more procedural steps, it is expected that the complexity of a task is the sum of the complexity of procedural steps that belong to the task being considered. The second question related to the complexity of EOPs is "how can the complexity of the basic unit be properly quantified?" Unfortunately, the answer to this question seems to be somewhat tricky, because there is no plausible framework to evaluate the complexity of procedural steps.
>
> For this reason, software complexity measures were applied to quantify the complexity of procedural steps because similar factors (such as lots of data included in a source code, a complex logic, and lengthy source codes) have been considered to quantify the complexity of software. Traditionally, in software engineering, graph entropy concepts have been widely adopted to evaluate the complexity of software. Typical graph entropy measures are the first-order and the second-order entropies.

The author does not specify how to select procedural steps of task performance, to determine their start and end points, psychological characteristics, and principles of steps' classification. Subjects are compared with a computer. Each step of the task, in the author's opinion, has approximately the same complexity and includes cognitive and behavioral components that can be *quantified* as the steps of a computer program. In the following, we present as an example two *arbitrary procedural steps*, S_1 and S_2, listed in Park and Jung's work (Park and Jung, 2007, p. 1104):

Step 1. *If* pressurizer pressure is less than 123.9 kg/cm²A, *then* verify if *both* SIAS and CIAS are actuated.

Step 2. *If* pressurizer pressure is less than 121 kg/cm²A *and* SIAS is actuated, *then* perform *both* of the following: (a) Stop *one* RCP in each loop. (b) *If* RCS subcooling margin is less than 15°C, *then* stop *all* RCPs.

As can be seen from the description of these two steps, cognitive and behavioral components in each step are not separated. The equivalence of these steps is in doubt due to the fact that the number of substeps in the second step is greater than in the first one and we cannot evaluate the complexity of these substeps from this description because it is carried out only in technological terms, not in standardized psychological terms. This leads to an ambiguous description of an operator's activity and ambiguity of interpretation of the earlier presented records. The suggested method of describing human activity is not adequate for quantitative analysis of task complexity from a psychological perspective. Approximately, similar steps are not commensurable units of measurement. Park (2011) evaluates each step of any task based on five criteria: step information complexity, step size complexity, step logic complexity, step abstraction hierarchy complexity, and step engineering decision complexity.

Let us consider step size complexity analysis (Park, 2011, p. 2526). The author wrote "…it is anticipated that human operators may be faced with a higher level of task complexity when they have to carry out a lot of actions to accomplish a required task." It is obvious that the quantity of actions can influence task complexity. However, in Park's method, the term *action* has a totally different meaning in comparison to the psychological meaning of this term. The term *action* has different meanings in different fields of psychology. For example, in activity theory there are various types of cognitive and behavioral actions. In view of the fact that such concept as *action* is critical for assessment of complexity, we consider this notion in a detailed manner. Park utilizes the recommendation of the Department of Energy (1998), which gives suggestions for action description during the development of instructions (see Park, 2009, p. 66). However, such recommendation does not have any relationship for selection of units of measurement. From physics, we know such units of measurement as minutes, seconds, kilograms, and pounds. Park is trying to use objective measurement procedures, and measurement units must comply with this method. We present some examples of Park's action description method (Park, 2009, p. 67) (see Table 11.1).

This example clearly demonstrates that the term *action* is used to describe instructions given to operators in technological terms. The meaning of *action* in psychology and particularly in activity theory is totally different. For a detailed description of the content of actions, Park recommends to decompose them into action specification steps. In his example of preparing

TABLE 11.1

Action Verbs and Their Meaning (Fragment)

Action Verb	Action Meaning
Align	Arrange equipment in a specific configuration to permit a specific operation
Close	Manipulate a device to allow the flow of electricity or to prevent the flow of fluids, other materials, or light
Cool (down)	Lower the temperature of equipment or environment

TABLE 11.2

Comparing Key Contents of Two Arbitrary Actions

Action Description	Content	Corresponding Description
Cream together the butter and the brown sugar until smooth	Action verb	Cream
	Object	Batter (mixture of butter and sugar)
	Action specification	Until smooth
Using a mixer fitted with paddle attachment, cream the butter and sugar together until very light, about 5 min	Action verb	Cream
	Object	Batter
	Action specification	Until very light
		A mixer with a paddle (a dedicated means)
		Operation time (5 min)

chocolate chip cookies with his daughter, he described actions as follows (Park, 2009, pp. 3, 67–68): (A1) Cream together the butter and the brown sugar until smooth; (A3) Using a mixer fitted with paddle attachment, cream butter and sugar together until very light, about 5 min. These two actions are further decomposed into three parts as shown in Table 11.2 (Park, 2009, p. 68).

These are not behavioral or cognitive action descriptions but rather a description of technological process involved in task performance. What actions are performed by a person is not clear from this description. Also, what kind of human actions can take 5 min? Actions as elements of human activity or behavior cannot last that long. As an example, we present duration of some actions in SSAT. A simultaneous perceptual action involved in identifying a relatively simple symbol takes 0.3 s; a perceptual action involved in perceiving a seven-digit number takes 1.2 s; a thinking action that includes a syllogistic solution (deductive reasoning) when the number of premises is 2 requires 7 s; a motor action moving a hand at 30 cm distance and pressing a button requires 0.3 + 0.16 = 0.46 s. We can only assume that 5 min is the duration of the whole task. We cannot quantitatively evaluate the complexity of human activity based on task description in technological terms. For this purpose, technological units of analysis should be transferred into psychological units of analysis. Moreover, for quantitative evaluation of task complexity, it is necessary to describe human actions in a standardized form.

All human actions should be described and classified according to psychological principles. The author ignores the fact that activity unfolds over time as a process. Each cognitive and motor action has a beginning point, an end point, and a duration, which means that we cannot evaluate quantitatively the complexity of actions without knowing their duration, distribution in time, possibility of being executed sequentially or simultaneously, logic of transition from one action to another, and so on. The complexity of actions depends on the level of concentration of attention during their execution. The activity has logical probabilistic structure and its components have hierarchical organization. Without analyzing the temporal structure of activity and choosing appropriate units of measures, the complexity of activity cannot be assessed. Quantitative evaluation of task complexity is first of all a psychological problem, because we evaluate the complexity of human behavior or activity. Let us consider as an example another hypothetical task, *controlling the water level in the tank* (Park, 2009, 2011, pp. 2525–2526). Park's diagram of the water-level controlling system is presented in Figure 11.2.

As depicted in Figure 11.2, there are level indicators that demonstrate water level and four valves that are used to change the water level in the tank. CV-1 regulates the rate of outflow from the tank by adjusting its open position from 0% to 100%. There are two bypass valves (BV-1 and BV-2). In normal conditions, they are closed. When the water level becomes too high, these valves can be opened to provide an additional flow path of water from Tank A. According to the authors, a series of actions that should be followed by human operators can be properly identified from the task analysis results (see Table 11.3).

Presumably, on the basis of the presented list of actions, we can quantitatively evaluate the task complexity. From the presented figure, we can see two actions that are required to perform this task: (1) control CV-1 if the water level in Tank A is less than 5.5 mm; (2) open BV-1 and BV-2 if the water level in Tank A is greater than 5.5 mm. Table 11.3 depicts symbolically indicator L, Tank A, valve controller CV-1, and two bypass valves BV-1. From

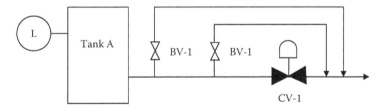

FIGURE 11.2
Diagram of the water-level controlling system. (From Park, J., *The Complexity of Proceduralized Tasks*, Springer-Verlag, London, U.K., 2009.)

TABLE 11.3

Controlling the Water Level in the Tank

Task Goal	Controlling the Water Level in Tank A Less Than 5 mm
Action	1. Control CV-1 if the water level in Tank A is less than 5.5 mm
	2. Open BV-1 and BV-2 if the water level in Tank A is greater than 5.5 mm
Indicator	1. Level indicator: L-1 (range: 0–7)
Controller	1. Valve controller: CV-FIK-1 (jog control, 0%–100%)
	2. Bypass valve switchers
	BV-HS-1 (selecting mode: Open, Close)
	BV-HS-1 (selecting mode: Open, Close)

Source: Park, J., *The Complexity of Proceduralized Tasks*, Springer-Verlag, London, U.K., 2009.

the figure, we cannot understand the basic physical characteristics of equipment that affects task complexity. For example, we do not know what type of indicator presents information; we also do not know the physical characteristics of controllers that would be important for a cognitive psychologist. According to the activity approach without adequate description of physical characteristics of equipment, one cannot identify cognitive and behavioral actions used by an operator during task performance and therefore cannot evaluate cognitive demands for this task performance. It is well known that the types of display and their locations are important factors for perceiving information. Similarly, characteristics of controls, their resistance, relative position, relationship between indicators and controls, etc., are important factors that should be taken into consideration during analysis of cognitive demands imposed on an operator by a task. The author wrote "a task complexity will become high if human operators have to carry out many actions" (Park, 2011, p. 2526). However, the quantity of actions cannot be utilized as a measure of complexity. One action can be more complex than several other ones. A smaller number of cognitive and behavioral actions can be more complex than a bigger number of simple actions.

We have taken into consideration that a decision-making action has various numbers of outputs with various probability. An operator can perform decision-making based on externally presented information or information extracted from memory. Decision-making can be performed in a limited time, in stressful conditions with dangerous consequences, etc. The duration of decision-making actions in such conditions would be different. Motor actions can be described similarly. Physical characteristics of controls, their resistance, distance, relationship between their movement, and movement of controlled objects influence the complexity and duration of motor actions.

From the action description presented in Figure 11.2, we can see that the task contains two actions. In further discussion (see Figure 11.2), *the amount of actions with the associated actions sequence* (Park, 2011, p. 2526), the author describes four actions. They are as follows: (1) verify that water level in Tank A is higher than 5 m; (2) open BV-1; (3) open BV-2; and (4) control CV-1.

However, the author considers only three actions: (1) open BV-1; (2) open BV-2; and (3) control CV-1. And then the author wrote (Park, 2011, p. 2526, paragraph 3.2):

> In addition, human operators may be exposed to a higher level of task complexity when they are faced with a task containing a lot of possible paths. In other words, since the number of possible paths is proportional to the number of decision points, human operators need to use additional cognitive resources to follow the correct sequence of actions with respect to the situation at hand.

Surprisingly, on the same page in paragraph 3.3, decision-making is used as a criterion for determining the amount of domain knowledge. The number of decisions, number of actions, and amount of knowledge are not commensurable units of measures for quantitative evaluation of task complexity. They are qualitative criteria that cannot be directly applied for quantitative evaluation of task complexity. Activity is a process, and therefore, a time structure of activity should be described at the first critically important stage of morphological analysis of activity. Without a clear classification of cognitive and behavior actions in a standardized manner, it is also impossible to quantitatively evaluate task complexity.

Parker described actions such as open BV-1, open BV-2, and control CV-1 to explain *step size complexity factor*, which are technological units of analysis. An accurate description of units of analysis and units of measure of activity are necessary for quantitative analysis of task complexity.

From activity theory perspectives, *open BV-1*, *open BV-2*, and *control CV-1* are motor actions. In order to assess behavioral actions quantitatively, one needs to know their type, specifics of their regulation, distance of movement, their performance time, level of concentration of attention during their execution, ability to perform them simultaneously or in sequence, motions that are included into the content of motor actions, and so on. There are also perceptual and decision-making actions. Without a clear understanding of the types of controls and displays, their arrangement, etc., we cannot understand what type of cognitive actions is used by an operator. In the previous example, the only factor that attracts the author's attention is the decision-making process preceding motor action execution. However, a motor action can be very complex or very simple independently of preceding decision-making.

The complexity of human activity is not a sum of the submeasures of complexities of the independently executed sequential steps. The complexity of one stage of activity may have an impact on the complexity of its subsequent steps because activity is a structure, not a series of independent and equally complex steps. Activity is a multidimensional system and its complexity assessment cannot be reduced to one numerical measure as suggested by Park. Optimization of activity based on its complexity should be performed utilizing a set of measures that characterize it as a system with complex structure and multiple interdependent characteristics, which requires conducting morphological analysis of activity as a system. Human activity varies even when performing the same task. As shown in the previous section, human activity should be described utilizing an algorithmic analysis developed for this purpose. A graph theory cannot be applied for quantitative evaluation of task complexity. Activity elements unfold in time as continuous processes with a complex structure. A graph theory is more suitable for the description of discontinuous events, which can be presented as a set of points (events) and arcs connecting them. In order to quantitatively assess task complexity, one has to develop a time structure of activity because activity is a process. The suggested method (Park, 2009) of quantitative evaluation of task complexity totally ignores the time parameters of activity because the duration of separate elements of activity and the possibility to perform them simultaneously or sequentially are critical factors.

Park utilizes software complexity measures and the concept of graph entropies to evaluate cognitive demands for task performance (Park, 2009). However, a human is not a computer and this method clearly demonstrates a computer-reductionist approach to studying extremely complex psychological processes. This approach considers complex human behavior or activity, including internal psychological functions, as an aggregation of computer operations. This leads to the postulate of additive organization of mental operations in the human brain. Activity is a multidimensional structure that should be described by a set of measures, and optimization of activity, equipment, and software should be performed according to these measures. Approaches that consider human beings as working computers ignore psychological aspects of human activity.

11.3 Units of Measurement and Formalized Procedures of Complexity Evaluation

The quantitative evaluation of task complexity is referred to choosing adequate units of measure that would permit a comparison of different elements of activity. Activity is a structure that unfolds in time, which suggests using time intervals for qualitatively different elements of activity as units of complexity measurement (Bedny and Karwowski, 2007). If the time structure of activity during task performance, and the time for qualitatively different elements of activity and time of performance of a holistic task are known, one can calculate a mathematical mean to evaluate the fraction of qualitatively different elements in task performance. Then, each fraction can also be evaluated in accordance to the five-point ordinal scale of complexity. In other words, anything that can be qualitatively described and measured in time precisely can be evaluated quantitatively, because all elements of activity can be transferred into one surface of measurement.

The following criteria are used to classify qualitatively different units of measure of task complexity: (1) qualitative content of activity elements during a particular interval of time; (2) complexity of these elements of activity according to a five-point scale; (3) possibility of their performance simultaneously or sequentially; (4) probability of appearance of a particular element of activity during task performance. For example, according to the first criterion, one should distinguish between the time interval devoted to cognitive activity and the time interval devoted to motor activity. The time interval devoted to mental activity is classified based on the dominant cognitive process, such as the time interval when a worker perceives weak signals in the sensory threshold area, the time for perception of various signals, the time for keeping information in the working memory, and decision-making. Each qualitatively different interval of time is related to a particular category of complexity according to the five-point ordinal scale of complexity. At the next step, one should determine the possibility of performing elements of activity simultaneously or sequentially. This is due to the fact that simultaneous performance of elements of activity changes the complexity of the considered time interval. Then, one should evaluate the probability of appearance of these time intervals during task performance. Finally, a specialist would calculate the mathematical mean and fraction of time for every qualitatively different element of activity. In this way, one could estimate the amount of cognitive efforts during task performance. It should be noted that there are special rules that determine the possibility of combining activity elements in time (see Section 7.3).

Formalized procedures of complexity evaluation require considering two interdependent aspects. The first one is associated with the possibility to perform two elements of activity simultaneously. The second one is associated with evaluating the complexity of time interval when activity elements are performed sequentially, or when they are performed simultaneously. What elements of activity could be performed either sequentially or simultaneously depends on the complexity of each element, on the strategies of performance, on the significance of the situation, etc. For example, in a dangerous situation, when each element of activity has a high level of significance, an operator performs them sequentially, even if they are simple. However, in normal situations, when an operator has well-developed skills, these elements of activity could be performed simultaneously. Therefore, one needs to know preferable strategies of task performance. It also needs to be taken into account what kind of elements of activity can be performed simultaneously. The basic feature of elements of activity that determines a possibility of their simultaneous performance is the complexity of each element or the level of concentration during their performance (see Section 7.3). In the following, we present an example of formalized rules, which is based on the theoretical data described earlier and on the data obtained in cognitive psychology, systemic-structural activity theory, and the MTM-1 system.

Rule 1. Time intervals for motions requiring either a low (A), average (B), or high (C) level of concentration of attention (see MTM-1 system) can be related according to our rules to the first, second, or third category of complexity, respectively.

Rule 2. If an activity is performed in a stressful situation, then time intervals related to the third and fourth categories of complexity should be elevated as the fifth category of complexity. Time intervals related to the first and second categories of complexity should be elevated as the third category. For instance, if an operator performs a simple decision-making action (the third level of complexity) but this action is taking place under stressful conditions, the action should be considered as being of the fifth level of complexity.

Rule 3. If a worker recognizes an object and makes a simple *yes–no/if–then* decision for the next selection out of two possible actions, this is considered as a decision-making action at sensory-perceptual level, which according to the data obtained in psychophysics, for example, Green and Swets (1966), should be divided into two separate cognitive operations. One of them is related to a sensory-perceptual operation and the other one to a decision-making operation. During algorithmic description of task performance, $1/2EF$ should be related to afferent operators (O^a) and another $1/2EF$ should also be related to the simplest decision about *what should be done based on obtained information* (logical condition *l*). Hence, when applying activity element *EF* during algorithmic description of activity and the design of its time structures, it is important to pay attention to the relationship between the detection or recognition stage and the decision-making stage. According to MTM-1, this simplest cognitive element of work activity requires a high level of concentration of attention and thus can also be related to the third category of complexity.

Rule 4. If cognitive activity coincides with motor activity, then complexity of such time interval depends on the specificity of motor and cognitive elements. The same defined complexity should be assigned separately to cognitive and motor elements of activity.

There are two types of work time. One period of time occurs during an actual performance of the task. The other period of time is associated with active waiting period, when an operator is observing the ongoing production process and is not directly involved in performance. In such situation, we use the following rule:

Rule 5. Active waiting period includes the continuous generation of expectations and hypotheses about the nature of the ongoing events. Such active waiting periods should be evaluated according to their complexity, which depends on the level of concentration of required attention and the presence or absence of any emotional stress. Additionally, we present some other rules with their graphical interpretation in Figure 11.3.

In Figure 11.3a, elements of activity *A* and *B* are presented as horizontal rectangles. Their durations designate the duration of activity elements.

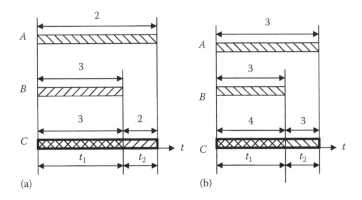

FIGURE 11.3
Graphical interpretation of complexity: (a) according to rule 6 and (b) according to rule 7.

Element of activity A has the second category of complexity and the element B has the third category of complexity. Here, we apply rule 6.

Rule 6. The period of time when two elements of activity that have different categories of complexity are performed simultaneously should be evaluated according to the complexity of the more difficult element (see Figure 11.3a, bold line rectangle C). The time interval t_1 has a third category of complexity and the remaining time interval t_2 has the second category of complexity.

Figure 11.3b demonstrates a situation when two elements of activity belong to the third category of complexity and begin at the same time. However, in the considered example, elements of activity are not completed simultaneously (see Figure 11.3b, elements A and B) and element A has longer duration than element B, so we utilize rule 7. Two elements of activity that have the third category of complexity can seldom be performed simultaneously. Usually only identical motor elements of activity that take place in optimal visual field can be performed simultaneously.

Rule 7. The period of time when two elements of activity that require high level of concentration of attention (the third category of complexity) are performed simultaneously should be associated with the fourth category of complexity. The combination of these kinds of elements of activity requires the highest level of resource mobilization and often leads to increase in performance time.

From Figure 11.3b, we can see two elements of activity that have different durations and the third category of complexity (see Figure 11.3b, elements A and B). The period of time when two elements of activity are performed simultaneously (period of time t_1) is related to the fourth category of complexity. The period of time t_2 remains unchanged and is related to the third category of complexity (see Figure 11.3b).

Rule 8. The period of time when two elements of activity that require low or average level of concentration of attention are performed simultaneously (motor elements of activity of the first or the second category of complexity) and the complexity of the overlapping time interval remains unchanged.

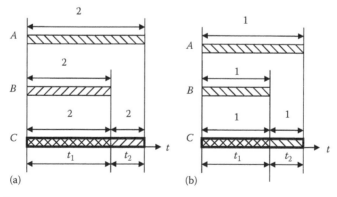

(a)   (b)

FIGURE 11.4
Graphical interpretation of complexity according to rule 8: (a) motor elements of activity of the second category of complexity and (b) motor elements of activity of the first category of complexity.

From Figure 11.4a, we can see two elements of activity that have different durations and the second category of complexity (see Figure 11.4a, elements *A* and *B*). The period of time when two elements of activity are performed simultaneously is unchanged and all period of time *t* is related to the second category of complexity (see Figure 11.4a). A similar rule is applied for the first category of complexity (see Figure 11.4b).

Rule 9. Time interval related to simple decision-making (when operator knows in advance how to react to a particular situation) can be related to the third category of complexity. This is the simplest decision-making action that related to the third level of complexity. If an operator must make a more complex decision, that is, one where he or she does not know in advance how to react to varying situations or decisions, which requires extraction of information from memory, then this interval of time is related to the fourth category of complexity. Similarly, decision-making in a contradicted situation is more complicated than choosing from known alternatives. Such decision-making also belongs to the fourth category of complexity.

Rule 10. In a simple situation when the operator has to recognize an object (O^α) and make *if–then* or *yes–no* decisions (logical condition *l*), this is the simplest cognitive element of work activity that requires a high level of concentration of attention and thus can also be related to the third category of complexity (see rule 3). However, when the characteristics of the object to be perceived are not easily distinguishable (i.e., when the characteristics of an object are in the threshold area), it is necessary to introduce two elements of *EF* during the design of a time structure: the first element associated with recognition of the object (operator O^α) and the second with the decision (logical condition *l*) as to what should be done based on the data obtained. In some cases, the object recognition in the threshold region and its following categorization may be so difficult for the subject that the members of the algorithm (the first *EF*, and the operator O^α and logical condition *l* related to it) may be referred to as the fourth category of complexity.

There are some additional rules in complexity evaluation (Bedny and Karwowski, 2007; Bedny and Meister, 1997) that we do not consider here. Complexity measures and their application will be discussed further as we will consider some examples. It should also be noted that in SSAT there is the system of description and classification of cognitive and motor actions. For example, there are sensory actions, simultaneous perceptual actions, imaginative actions, decision-making actions at a sensory-perceptual level, decision-making actions at a verbal-thinking level, thinking actions, etc. (see Section 6.3).

In Section 7.3 (see Figure 7.4), we described the time structure of one stage of cylindrical pin installation into the halls when two pins have different shapes. We consider a situation when both pins grasped by right and left hands have a flute. In such situation, there is a complicated logical structure and therefore a complicate time structure of activity. We remained that when the fluted pin is picked up by the left hand, it must be placed so that the flute

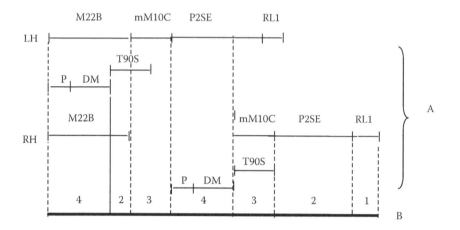

FIGURE 11.5
Time structure of activity and category of complexity of elements of activity: A, time structure of activity; B, category of complexity of elements of activity.

is in the hole. When the fluted pin is picked up by the right hand, it must be placed so that the flute is above the hole. Figure 11.5 demonstrates not only the time structure of this element of activity during task performance but also the complexity of various time intervals during installation of pins, when both have a flute.

All lines in Figure 11.5 (A) depict the time structure of activity. The bottom line in Figure 11.5 (B) describes the complexity of the time intervals related to the various steps of activity. The time structure of activity at this stage of task performance has been described before (see Section 7.3, Figure 7.4). A mental action (PDM) includes a simultaneous perceptual operation that involves recognition of clearly distinguished stimuli (fluted pin), operation (P), and mental operation decision-making (DM) (how to turn a pin into a vertical position). PDM has the fourth category of complexity because decision is made based on information extracted from memory when a subject decides *how to turn a pin*. Therefore, according to rule 9, mental action PDM is a cognitive action that belongs to the fourth category of complexity.

Let us consider the complexity of work activity when its elements are combined in time (see line B on the bottom of Figure 11.5). We assign complexity to various time intervals when elements of activity are performed simultaneously. Two *M22B* motions belong to the second category of complexity and their combination in time does not change the time interval complexity. However, combining them with a mental action (PDM) that has the fourth category of complexity elevates this time interval into the fourth category of complexity (see bottom line, the first time interval). The small time interval that is second from the left has the second category of complexity because this period of time includes the second and the first categories of complexity (two elements *M22B* for the left and right hands [second category of complexity] and *T90S* [first category of complexity]). Therefore, the second interval (line B) has the second category of complexity. The third interval from the left (see bold line B) is a combination of elements of activity with the third (*mM10C*) and the first (*T90S*) categories of complexity. Therefore, this interval has the third category of complexity. The fourth interval includes cognitive action with the fourth (PDM) and second (*P2SE*) categories of complexity. Hence, this time interval belongs to the fourth category of complexity. The fifth interval from the right has the third level of complexity because element *mM10C* according to the level of attention concentration belongs to the third level of complexity and *T90S* to the first. PSE (*PSE*-installation of cylindrical pin into the hole, easy to handle requires average level of attention concentration) belongs to the second level of complexity and *T90S* to the first. Similarly, we determine the complexity of the two time intervals on the right side of line B on the bottom. This example demonstrates how we can evaluate the complexity of the elements of activity when they are performed not only in sequence but also simultaneously. Depending on the combined complexity of the described intervals, it is possible to assess the overall complexity of the stage of activity depicted in Figure 11.5. Although we do not consider the overall activity complexity in this example, we would like to indicate that, according to the ratio between intervals with various complexities, the complexity of the considered stage of activity can also be evaluated (Bedny and Karwowski, 2007). The time structure of activity depends on the constructive features of equipment and method of performance. Hence, it can be utilized not only for time study and assessment of complexity of task performance but also for assessment of equipment design solution.

12

Quantitative Assessment of Task Complexity

12.1 Measures of Task Complexity

In Bedny et al. (2014), we described a new method of reliability assessment of human performance that derives from the SSAT approach. The application of this method has been demonstrated using human–computer interaction. However, this method can be utilized for the reliability assessment of any system including ones where an operator does not directly interact with a computer. Hence, in this chapter, we will not discuss reliability assessment.

In this chapter, we consider general principles of quantitative evaluation of task complexity. Task complexity evaluation is the final stage of systemic-structural analysis of activity. It can be performed only after qualitative and then morphological analysis of activity (algorithmic and time structure analysis). The activity structure is a multidimensional system. Therefore, not one but multiple measures should be used for task complexity evaluation. Such measures can be used for optimization of equipment design solutions and optimization of human performance, reducing dangerous points of the work process, and increasing of human reliability in the system. For the development of measures of complexity, we can define the general performance time for a considered member of the algorithm and then determine the fraction of time for afferent components of activity (receiving information), logical conditions (decision making), motor activity, different complexity elements, etc. Let us consider examples of calculating some of such measures.

In a timed study during a long period, production operations are considered as a sequence of behavioral elements and the time of their performance is determined by a simple summation of execution time of its individual elements. However, contemporary production operations or tasks have a complex combination of behavioral and cognitive components with complex logical organization. Hence, analytical methods of task execution time calculation cannot be reduced to a simple summation of execution time of separate elements of a task. Task execution time can be determined only after an algorithmic description of task and analysis of time structure of activity during task execution. Hence, the algorithm (task) execution time in the general form can be calculated according to the following formula:

$$T = \sum_{j=1}^{n} P_i t_i$$

where
P_i is the probability of occurrence the *i*th member of the algorithm
t_i is the duration of the *i*th member of the algorithm

This is the first measure that is needed for the calculation of other measures of task complexity. The formula takes into account the probability of transition from one member of an algorithm to another. In addition, calculations are carried out taking into account the overlapping elements of performance over time. This principle will be maintained in calculating all other measures. Let us consider as an example some other measures.

The performance time of all logical conditions (decision-making process that determines the logic of the transition from one member of an algorithm to another) can be evaluated according to the following formula:

$$L_g = \sum_{i=1}^{k} P_i^l t_i^l$$

where
P_i^l is the probability of occurrence the *i*th logical condition
t_i^l is the duration of the *i*th logical condition

The next step would be to determine the relationship between the time spent on logical conditions and the time spent on execution of the whole task (fraction of time for logical components of work):

$$N_l = \frac{L_g}{T}$$

where
L_g is the time for performance of logical conditions
T is the time for the entire task performance

This measure characterizes the complexity of the decision-making process during task performance. There are several other measures that characterize the decision-making process from various aspects of activity performance.

The time taken for afferent operators and the executive (response) components of activity can be evaluated in the following way:

$$T_\alpha = \Sigma P_r^\alpha t_r^\alpha; \quad T_{ex} = \Sigma P_j t_j^0$$

where
P_r^α and P_j are the probability of rth afferent and jth efferent operators
t_r^α, t_j^0 are the performance time of the rth afferent and jth efferent operators

The time related to recognizing and identifying weak (i.e., approaching to threshold range) signals can be determined utilizing the following formula:

$$'T_\alpha = \Sigma' P_{r'}, t_{r'}$$

where
$'P_{r'}$ and $t_{r'}$ $P_{r'}$ is the occurrence probability
$t_{r'}$ is the occurrence time of r'th afferent operators, characteristics of which approach a threshold value

The proportion of time for afferent operators to time for executive activity is evaluated in the following way:

$$N_\alpha = \frac{T_\alpha}{T_{ex}}$$

The proportion of time for afferent operators related to recognizing and identifying weak (i.e., approaching to threshold range) task signals to the time for executive activity is evaluated in the following way:

$$Q_\alpha = \frac{'T_\alpha}{T_{ex}}$$

The proportion of time for logical components of work activity depending largely on information selected from long-term memory rather than externally presented information is evaluated as follows:

$$L_{ltm} = \frac{l_{ltm}}{L_g}$$

where l_{ltm} is the time for logical components of activity whose operational nature is predominantly governed by information retrieved from long-term memory.

The measure that characterizes the workload of the working memory can be determined according to the following formula:

$$N_{wm} = \frac{t_{wm}}{T}$$

where

N_{wm} is the fraction of time for retaining current information in working memory during a period of time for the entire task performance

t_{wm} is the time for storing current information in working memory during task performance

T is the task performance time

Activity may be either stereotyped (repetitive) or changeable (variable). The performance of a stereotyped activity is normally easier; if procedures always take place in a set order, or a given procedure always follows some particular member of an algorithm, these logical components of activity are stereotyped. Where procedures and the transition from one action to another have probabilistic features, these procedures are considered variable. Those members of the algorithm that always follow in the same sequence can be considered as stereotyped components of activity. Their sequence is subjectively perceived by the operator as the habitual performance of the same order of actions. If the habitual performance of a stereotyped efferent operator is always followed by the same afferent operator and its associated logical condition, then the afferent operator and its logical condition are also related to the stereotyped activity. It can be hypothesized that the more time in a process that is devoted to variable procedures, the more complex this process is. It is possible to calculate measures of stereotyped and variable (changeable) components for executed activity and logical conditions. The time devoted to stereotyped and variable operators and logical conditions during activity performance can be determined according to the following:

$$t_{st} = \Sigma P^o_{jst} t^o_{jst}; \quad t_{ch} = \Sigma P^o_{jch} t^o_{jch}$$

$$l_{st} = \Sigma P^l_{ist} t^l_{ist}; \quad l_{ch} = \Sigma P^l_{ich} t^l_{ich}$$

where

P^o_{jst}, P^o_{jch} are the probability of the appearance of jth stereotyped and variable operators

t^o_{jst}, t^o_{jch} are the performance times of jth stereotyped and variable operators

P^l_{ist}, P^l_{ich} are probability of the appearance of jth stereotyped and variable logical conditions

t^l_{ist}, t^l_{ich} are the performance times of ith stereotyped and variable logical conditions

Accordingly, the measure of stereotyped and variable logical components of activity can be determined from the expressions

$$L_{st} = \frac{l_{st}}{L_g}$$

$$L_{ch} = \frac{l_{ch}}{L_g}$$

where l_{st} and l_{ch} are the mathematical means of performance time of stereotyped and variable logical activity.

In the same way, we can determine the stereotyped and variable executive components of activity.

Let us now consider methods for evaluating the complexity of a time interval connected with an active waiting period. Sometimes an operator may perform tasks that periodically do not require any active involvement in performance; such intervals of waiting time may be encountered both within and between tasks. In spite of the absence of externally observable behavior, such waiting times require concentration of attention. The operator must be ready to become immediately involved in performance as the situation requires, for example, if an emergency arises. The complexity of an active waiting period can be evaluated with respect to the level of concentration of attention required during the waiting period, in accordance with the following rules:

1. If waiting periods require a low, average, or high level of concentration of attention, they are described by the first, second, and third categories of complexity, respectively.

2. When waiting periods convey emotional stress (i.e., there is danger of trauma or accident), they are described by the fourth category of complexity.

3. When waiting periods of any level of complexity require that information be kept continuously in the working memory, their complexity category should be increased by one.

The existence of an active waiting period in a task requires the introduction of additional measures of task complexity. One such measure is the *fraction of active waiting period in the entire task execution time* that is calculated according to the following formula:

$$\Delta T_w = \frac{t_w}{T}$$

where
t_w is the entire time for the active waiting period in the work process
T is the total task execution time

If an active waiting period consistently occurs following a particular element of the activity or task, it is considered to be a stereotypy active period, measured by the following formula:

$$W^{st} = \frac{t_{wst}}{t_w}$$

where
t_{wst} is the time for stereotypy waiting components
t_w is the total duration of waiting period

A task can include several active waiting periods of time. If the internal psychological content of waiting periods of time is identical, they are repetitive. The fraction of time for repetitive waiting periods of work activity in the entire time for an active waiting period can be calculated in a similar fashion. We considered some measures as examples. All measures of task complexity evaluation and their psychological interpretation are presented in Table 12.1.

Measures presented in Table 12.1 can be helpful for evaluating the complexity and efficiency of performance, time study, analysis of the training process, identifying the critical point of the tasks, assessment of human reliability during task performance, and evaluation of usability of equipment.

TABLE 12.1

List of Complexity Measures and Their Psychological Interpretation

No.	Name of Measure	Formula for Calculation	Variables	Psychological Meaning
1	Algorithm (task) execution time	$T = \Sigma P_i t_i$	P_i—probability of occurrence t_i—duration of the ith member of the algorithm	Duration of task performance
2	Sum of the performance time of all afferent operators	$T_\alpha = \Sigma P^\alpha t^\alpha$	P^α—probability of occurrence t^α—duration of the rth afferent operator	Duration of perceptual components of activity
3	Sum of the performance time of all thinking operators	$T_{th} = \Sigma P^{th} t^{th}$	P^{th}—probability of occurrence t^{th}—duration of the rth thinking operator	Duration of thinking components of activity
4	Sum of the performance time of all operators that requires keeping information in working memory	$T_{wm} = \Sigma P_{wm} t_{wm}$	P_{wm}—probability of occurrence t_{wm}—duration of the wmth operator associated with keeping information in working memory	Time for retaining current information in working memory

(Continued)

TABLE 12.1 (*Continued*)

List of Complexity Measures and Their Psychological Interpretation

No.	Name of Measure	Formula for Calculation	Variables	Psychological Meaning
5	Sum of the performance time of all logical conditions	$L_g = \Sigma P_i t_i$	P_i—probability of occurrence t_i—duration of the ith logical condition	Duration of decision-making components of activity
6	Sum of the performance time of all efferent operators	$T_{ex} = \Sigma P_j t_j$	P_j—probability of occurrence t_j—duration of jth efferent operators	Duration of executive components of activity
7	Sum of the performance time of all cognitive components of the task (including perceptual activity)	$T_{cog} = T_\alpha + T_{th} + L_g + T_{wm}$	$T_\alpha; T_{th}; L_g;$ (see previous text) T_μ—duration of keeping information in working memory (some cognitive components can be equal to zero in a particular task)	Total duration of cognitive components of activity
8	Sum of the time spent on discrimination and recognition of perceptual distinctive characteristics that are approaching threshold of sensory receptors	$'T_\alpha = \Sigma P_{r'} t_{r'}$	$P_{r'}$—occurrence probability $t_{r'}$—duration of the r'th afferent operators, characteristics of which approach the threshold value (required additional EF)	Duration of the perceptual process connected with weak stimuli (approaches threshold characteristics)
9	Fraction of time for afferent operators in the time for the entire task performance (N_α)	$N_\alpha = T_\alpha / T$	T_α—performance time of afferent operator T—time of the entire task performance	Fraction of perceptual components of activity in the performance of the entire task
10	Fraction of time for thinking operators in the time for the entire task performance (N_{th})	$N_{th} = T_{th} / T$	T_{th}—performance time of thinking operators (see previous text)	Fraction of thinking components of activity in the performance of the entire task
11	Fraction of time for logical conditions in the time for the entire task performance	$N_l = L_g / T$	L_g—time for performance of logical conditions T—time for the entire task performance	Fraction of decision-making components of activity in the performance of the entire task
12	Proportion of time for cognitive components of task (including perceptual activity) to the performance time for all efferent operators	$N_{cog} = T_{cog} / T_{ex}$	T_{cog}—time for performance of cognitive components T_{ex}—performance time for all efferent operators	Relationship between cognitive and external behavioral (executive) components of task

(Continued)

TABLE 12.1 (*Continued*)

List of Complexity Measures and Their Psychological Interpretation

No.	Name of Measure	Formula for Calculation	Variables	Psychological Meaning
13	Fraction of time for logical components of work activity, which largely depends on information selected from long-term memory rather than from external sources of information, in the entire time for logical condition performance	$L_{ltm} = l_{ltm}/L_g$	l_{ltm}—time for logical components of activity whose operational nature is predominantly governed by information retrieved from long-term memory	Level of memory workload and complexity of the decision-making process
14	Fraction of time for retaining current information in working memory in the time for the entire task performance	$N_{wm} = T_{wm}/T$	T_{wm}—time for storing information related to task performance in working memory	Level of working memory workload during task performance
15	Fraction of time for performance of all efferent operators in the time for the entire task performance	$'N_{beh} = T_{ex}/T$	T_{ex}—time for external behavioral (executive) components	Fraction of external behavioral (executive) components of activity in the entire task
16	Fraction of time for cognitive activity in the time for the entire task performance	$'N_{cog} = T_{cog}/T$	T_{cog}—time for cognitive components	Fraction of cognitive components of activity in the performance of the entire task
17	Fraction of time spent on discrimination and recognition of perceptual distinctive characteristics that are approaching threshold of sensory receptors in the time for the entire task performance	$Q = T_\alpha/T$	T_α—time for discrimination and recognition of various perceptual distinctive characteristics that are approaching threshold of sensory receptors	Characteristics of complexity, sensory, and perceptual processes
18	Measure of stereotypy of logical processing of information	$L_{st} = l_{st}/L_g$	l_{st}—time for stereotypy of logical processing of information L_g—time for performance of logical conditions	Characteristic of inflexibility or rigidity of the decision-making process

(Continued)

TABLE 12.1 (*Continued*)

List of Complexity Measures and Their Psychological Interpretation

No.	Name of Measure	Formula for Calculation	Variables	Psychological Meaning
19	Measure of changeability of logical processing of information	$L_{ch} = l_{ch}/L_g$	l_{ch}—time for changeable logical processing of information	Characteristic of irregularity or flexibility of the decision-making process
20	Measure of stereotypy of executive components of work activity	$N_{st} = t_{st}/T_{ex}$	t_{st}—time for stereotypy of executive components of activity T_{ex}—time for executive components of activity	Characteristic of inflexibility or rigidity of executive components of activity
21	Measure of changeability of executive components of work activity	$N_{ch} = t_{ch}/T_{ex}$	t_{ch}—time for changeable executive components of activity	Characteristic of irregularity or flexibility of executive components of activity
22	Scale of complexity (a) Algorithm (b) Member of the algorithm	X_r—level of complexity $(1, 2, ..., 5)$	Level of concentration of attention during task performance (1, minimum concentration; 5, maximum)	Level of mental effort during task performance and performance of various elements. Unevenness of mental effort and critical points of task performance
23	Fraction of time for repetitive logical components of work activity in the performance time of all logical conditions	$Z^l = t_{rep}/L_g$	t_{rep}—time for performance of identical logical conditions	Characteristic of habitualness of information processing
24	Fraction of time for repetitive afferent components of work activity in the performance time of all afferent operators	$Z^\alpha = t_{rep}/T_\alpha$	t_{rep}—time for performance of identical afferent components	Characteristic of habitualness of the perceiving process
25	Fraction of time for repetitive efferent components of work activity in the performance time of all efferent operators	$Z^{ef} = t_{exrep}/T_{ex}$	t_{exrep}—time for performance of identical efferent components	Characteristic of habitualness of executive components of activity

(*Continued*)

TABLE 12.1 (*Continued*)

List of Complexity Measures and Their Psychological Interpretation

No.	Name of Measure	Formula for Calculation	Variables	Psychological Meaning
26	Fraction of active waiting period in the entire work process	$\Delta T_w = t_w/T$	t_w—entire time for active waiting period in the work process	Relationship between active waiting period and performance
27	Category of complexity of active waiting periods	X_w—1 ... 4	Concentration of attention during waiting period (1, minimum; 4, maximum)	Level of mental efforts during active waiting period
28	Fraction of time for repetitive waiting periods of work activity in the entire time for active waiting period in the work process	$Z^w = t_{wrep}/t_w$	t_{wrep}—time for repetitive waiting periods	Characteristic of habitualness of waiting periods
29	Measure of changeability of waiting periods	$W^{ch} = t_{wch}/t_w$	t_{wch}—time for changeable waiting periods	Characteristic of irregularity of waiting periods
30	Measure of stereotypy of waiting periods	$W^{st} = t_{wst}/t_w$	t_{wst}—time for stereotypy waiting periods	Stereotypy of waiting periods

12.2 Complexity Evaluation of Production Operations

In this section, we consider a quantitative method of complexity evaluation of an assembly operation (task) in the real manufacturing assembly lane process. Such tasks are not physically demanding. They are repetitive with a short performance cycle and energy demands for these tasks are usually quite low. Such tasks often require precise motor actions under perceptual or muscle control and coordination of actions. Moreover, some tasks can include relatively simple thinking operations and decision-making actions. In view of the high frequency of these tasks, workers have to decide in what order to perform specified actions varying in a restricted range numerous times during every shift. Correct performance of such actions requires constant attention that leads not only to physical but also mental fatigue by the end of the shift resulting in degradation of the quality of work, errors, and even injuries. General principles of analyzing such operations are not significantly different from studying tasks in semiautomatic, automated man–machine, and highly computerized systems where motor components of activity are significantly reduced and mental components dominate. It is clear that even simple mental actions repeated thousands of times per shift present a problem for workers making complexity the main characteristic of such tasks.

Evaluation of task complexity is important not only in ergonomics or work psychology studies but also for economics. Complex work can be executed in the same time frame as a simple one. For example, in some cases, due to innovative technology, manufacturing operations can be performed in a shorter period of time. However, this may be accompanied by increasing the complexity of human labor. Thus, higher productivity can lead to increasing complexity of the human work. Complex work is often characterized by a higher mental effort and requires more training. This means that not only duration but also the complexity of the work can influence work compensation. However, there are no methods in economics to measure the complexity of work. The systemic-structural activity theory (SSAT) offers a method of assessment of work complexity. The complexity of a task is the main cause of mental workload during task performance (Hancock and Caird, 1993). Recently, a special issue on transdisciplinary perspectives on economic complexity presents several papers related to complexity (Barkley Rosser, 2010). However, complexity of human work is not discussed in this issue. Complexity of human work in economics can be evaluated only based on the analysis of psychological aspects of complexity. There are various factors that impact psychological complexity. Let us consider some examples.

We encountered an interesting case study when two manufacturing operations were performed in sequence, one after the other (Bedny et al., 2001). They consisted of the same movements and have the same performance time. Compensation for these two operations was also the same. However, as our studies have shown, the first operation was more complex than the second one. This led to a violation of compensation principles. Moreover, a more complex production operation has been performed with violation of product quality. The objective of the study was to determine whether our method of estimating the complexity of work can capture a difference in the complexity of the considered production operations. The second objective was to analyze the possibility of simplifying a more complex production operation and improving the quality of its performance.

Later, we consider these assembly production operations. Analysis of these tasks includes qualitative analysis (objectively logical analysis), morphological analysis (algorithmic description and time structure description), and quantitative assessment of task complexity. Morphological analysis is a prerequisite for quantitative assessment of task complexity. Analysis of these two manufacturing production operations (tasks) performed at an assembly line presents a significant theoretical interest. This is due to the fact that they are almost identical in terms of the existing methods of analysis of the production operations. In this chapter, we discuss some additional theoretical aspects of the analysis of these operations.

All physical actions and their constituent movements are identical in these two production operations. Motor actions overlap cognitive components of activity. In the analysis of manufacturing operations, this type of cognitive components, which are considered as components of motor actions or movements, is not analyzed. Moreover, such concepts as cognitive or motor actions are not used at all. So, for example, in Barnes' (1980) seminal book, the concept of *action* as the unit of analysis is not utilized. The title of the book *Motion and Time Study...* clearly reflects this ideology. Moreover, SSAT also utilize such units of analysis as a member of the algorithm, which can describe cognitive and motor components of work. We recall that a member of the algorithm can integrate several cognitive or motor actions by a high-ordered goal. Since we considered these operations earlier, here only fragments of their algorithmic analysis will be presented. After that we present a quantitative analysis of the complexity of these operations. Later, we present fragments of tables that algorithmically describe these two production operations that are performed in sequence (see Tables 12.2 and 12.3).

> The task requires the following major steps: O_1^ε, take a neck from the bin on the left by the left hand; put the jig into neck by the right hand and put the neck (with the jig) into a working position on the welding machine; O_2^α, determine the type of fixing arm; l_1^μ, if jig's arm is wide, take wide brackets from the front bin; l_4^μ, if the fixating arm is narrow, take the bracket from the rear bin; O_4^α, determine the quality of the bracket; O_5^α, check the position of the bracket on the hand; l_2, if the bracket is not suitable, take another one from the front bin; l_3^μ, decide if turning of brackets into 180° is required; O_6^ε, turn the bracket 180°; O_7^ε, set up the bracket from the front bin and weld it on; $'O_7^\varepsilon$, turn the neck with the jig 180°; O_8^ε, take the bracket from the rear bin; O_9^α, determine whether the bracket is suitable; O_{10}^α, determine simultaneously the position of the straps; l_5, if the bracket is rejected, repeat O_8^ε and perform l_6^μ; l_6^μ, if the three bracket is not in the required position, turn it in the correct position; O_{12}^ε, set up the bracket from the rear bin and weld it on; and O_{13}^ε, pass on the neck with welding brackets for the next production operation to another worker.

TABLE 12.2

Algorithmic Description of the Bracket-Welding Operation (Fragment)

Member of the Algorithm	Description of the Algorithm Member
O_1^ε	To take a neck from the bin, to put a jig into the neck, and to install it on the welding machine.
O_2^α	To discriminate the type of fixing arm.
$l_1 \overset{1}{\uparrow}$	If the fixing arm is broad ($l_1 = 0$), perform O_3^μ; if the fixing arm is narrow ($l_1 = 1$), perform O_8^ε.
$\overset{7\ 4\ 2}{\downarrow\downarrow\downarrow} O_3^\varepsilon$	While holding the neck and the jig with the left hand, take the bracket from the front bin and bring the neck with the bracket to the major working area.
....	....
O_{13}^ε	Pass on the neck to the next operation.

TABLE 12.3

Algorithmic Description of the Handle-Welding Operation (Fragment)

Member of the Algorithm	Description of the Algorithm Member
O_1^ε	Take a neck from the bin and put the jig into the neck and install it on the welding machine.
O_2^ε	Install it in the welding machine.
$\overset{1}{\downarrow} O_3^\varepsilon$	While holding the neck and the jig with the left hand, take the handle from the bin and bring the handle to the major working area.
O_4^α	Decide if the handle is acceptable.
……….	……………… …………..
O_{11}^ε	Pass on the neck to the next operation.

Note: Here we briefly present symbols utilized in morphological analysis to assist in reading the following text without returning to previous chapters.

After an algorithmic description of these tasks, we developed the time structure of activity for considered tasks. It is essential to remind that only after the technological units of analysis are transferred into standardized elements of activity or psychological units of analysis the time structure of activity can be developed. However, in our abbreviated description, we have not presented this stage of analysis. Only after temporal analysis of production operations it becomes possible to evaluate the complexity of considered tasks. For evaluation of production operation, we did not utilize all measures from Table 12.1 but only some of the most informative ones for analysis of this type of production operations (see Table 12.4).

Order numbers of described measures are presented in the left column. The names of measures are presented in the second column and their quantitative value is presented in the last three columns. Selection of measures depends not only on the specificity of the task under consideration but also on the specificity of other production operations on the considered production lane. This is an important factor in the comparative analysis of production operations.

Let us examine some quantitative measures obtained during the evaluation of complexity of the assembly operation *bracket welding* before intervention (see Table 12.4). These measures uncover critical points of this operation. The performance time for assembly operation is 13.7 s. The performance time of all efferent operators is equivalent to the time of task performance ($T = T_{ex}$), which means that all cognitive components of activity are overlapped in time by the motor components.

Logical conditions (decision making) l_1^μ, l_3^μ, l_4^μ, l_6^μ before improvement were performed based on information extracted from memory. Unacceptable brackets are encountered very seldom. The time for logical conditions l_2 and l_5 for rejection of such brackets is very short and can be neglected. Hence, we do not consider further any measures associated with these logical conditions. From this follows that the fraction of time for logical components of work

TABLE 12.4

Quantitative Evaluation of Complexity of Welding Operations

No.	Measures (Time Measured in Seconds)	Operation of Welding Brackets		Operation of Welding Handles
		Before Intervention	After Intervention	
1	Algorithm (task) execution time (T)	13.70	13.70	12
2	Sum of the performance time of all afferent operators (T_a)	0.39	0.52	0.004
5	Sum of the performance time of all logical conditions (L_g)	0.65	0.42	0.004
6	Sum of the performance time of all efferent operators (T_{ex})	13.40	13.40	12.0
8	Sum of the time spent on discrimination and recognition of perceptual distinctive characteristics that are approaching threshold of sensory receptors ($'T_a$)	0	0	0
11	Fraction of time for logical conditions in the time for the entire task performance (N_l)	0.05	0.03	0.003
13	Fraction of time for logical components of work activity, which largely depends on information selected from long-term memory rather than from external sources of information, in the entire time for logical condition performance (L_{ltm})	1.00	0	0
14	Fraction of time for retaining current information in working memory in the time for the entire task performance (N_{wm})	0.08	0	0
16	Fraction of time for cognitive activity in the time for the entire task performance ($'N_{cog}$)	0.75	0.70	0
17	Fraction of time spent on discrimination and recognition of perceptual distinctive characteristics that are approaching threshold of sensory receptors in the time for the entire task performance (Q)	0	0	0
18	Measure of stereotypy of logical processing of information (L_{st})	1.00	1.00	0
19	Measure of changeability of logical processing of information (L_{ch})	0	0	1

(Continued)

TABLE 12.4 (*Continued*)

Quantitative Evaluation of Complexity of Welding Operations

No.	Measures (Time Measured in Seconds)	Operation of Welding Brackets Before Intervention	Operation of Welding Brackets After Intervention	Operation of Welding Handles
21	Measure of changeability of executive components of work activity (N_{ch})	0.85	0.82	0.003
22	Scale of complexity			
	(a) Algorithm	2	2	2
	(b) Members of the algorithm	2;3;3;2; 3;3;3;3; 3;1;3;2; 3;3;3;3 3;2;1	2;3;3;2; 3;3;3;3; 3;1;3;2 3;3;3;3; 3;2;1	2;2;3;3; 1;2;3;3; 2;1
23	Fraction of time for repetitive logical components of work activity in the performance time of all logical conditions (Z^l)	0.41	0.25	0.50
24	Fraction of time for repetitive afferent components of work activity in the performance time of all afferent operators (Z^a)	0.003	0.25	0.50
25	Fraction of time for repetitive efferent components of work activity in the performance time of all efferent operators (Z^{ef})	0.38	0.38	0.20

Note: If the logical condition $l_1 = 1$, additional turning of the neck with the jig of 180° is required. During performance of the production operation, l_4 or l_7 is used because they are mutually exclusive logical conditions (if $l_4 = 1$, then $l_7 = 0$; or if $l_4 = 0$, then $l_7 = 1$).

activity, which largely depends on the information selected from long-term memory rather than perceived from external sources of information, in the entire time for logical condition performance is $L_{ltm} = 1$. This means that the operator performs decisions that also require a higher level of concentration.

The time for performing all logical conditions (decision making) for production operation L_g was 0.65 s. The fraction of time for logical conditions in the time for the entire task performance is $N_l = 0.05$.

The measure that characterizes utilization of working memory during performance of considered assembly operation (fraction of time for retaining current information in working memory) was $N_{wm} = 0.08$. It implies that during the execution of each operation, 1.1 s is used to maintain the information in working memory. Considering that this operation is repeated up to 2000 times per shift, the total time of working memory load per shift equals 33 min, and the total time for decision-making process is 19 min per shift.

Decision making or the logical processing of information is characterized by a high stereotypical level ($L_{st} = 1$).

It is also interesting to consider the measure "Fraction of time spent on discrimination and recognition of perceptual distinctive characteristics of task that are approaching to threshold of sensory receptors in the time for the entire task performance (Q)." The value of measure is 0 for this operation. This means that the production operation does not require high visual activity, or there are no special requirements to perceptual process.

A new method of task performance for the operation *welding the brackets* has been developed. The redesign procedure includes the following main steps:

1. Redesign the body of the jig ring in such a way that it consists of two halves of unequal width. By increasing the width of one half of the jig ring by 3 mm and decreasing the other half by 3 mm, produce a 6 mm difference between the halves that still preserves the overall weight of the jig. The wider section is associated with the wider arm, and the narrower section with the narrower arm. This facilitates recognition of the required arm working position. Such recognition can be conducted not only through vision but also based on the sense of touch.

2. Cover the wider section of the jig ring with yellow plastic, and the narrower section with a dark green plastic. These color cues enhance the discriminative properties of the two jig ring sections: the yellow color tends to enhance the impression of largeness, while the darker color tends to create an impression of diminished size. Thus, these two features intensify the intrinsic relationship between the appropriate jig arm and its associated bracket.

3. Cover the bins that contain the rear, wider brackets with yellow plastic, and the bin with the front, narrower brackets with dark green plastic. Correspondingly, colored coverings should also be applied to the table surfaces on which the bins rest.

4. Reconfigure the shape of the jigs' arms to produce a notch on one and an aperture on the other, providing an additional clue as to the relationship between each of the arms and the orientation of the corresponding brackets.

5. Reduce the overall weight of the jig by perforating the metal part of the jig's ring.

The following is the complexity evaluation of this task after the innovation.

The first three redesign recommendations simplify decision making and selection of appropriate bin with brackets (l_1^μ, l_4^μ). Decision making is performed based on externally presented information, not based on information extracted from working memory. Therefore, logical conditions l_1^μ and l_4^μ were transferred into l_1 and l_4. Redesign recommendation four simplifies decision making associated with correct orientation of brackets before welding (l_3^μ and l_6^μ). After innovations, the considered logical conditions are performed based on externally presented information and can be designated symbolically as l_3 and l_6. Thus, after improvement, all logical conditions were performed based on externally presented information or l_1^μ, l_3^μ, l_4^μ, l_6^μ were transferred into l_1, l_3, l_4, l_6. All previously discussed innovations changed the quantitative measures of complexity for the considered operation.

After implementation of the innovation for the operation *welding brackets*, the performance time for logical conditions (decision making) L_g is reduced from 0.65 up to 0.42 s. The fraction of time for logical conditions in the time for the entire task performance N_l reduced from 0.05 to 0.03. The fraction of time for logical components of work activity, which largely depends on the information selected from long-term memory rather than perceived from external sources of information, in the entire time for logical condition performance (L_{ltm}) reduced from 1 to 0, entirely eliminating the memory workload during the decision-making process. So the introduced improvement did not just reduce the decision-making component of work from 19 min up to 12 min per shift but also entirely eliminated the memory workload during the decision-making process while performing the operation. Before the innovation, N_{wm} was equal to 0.8, and after the innovation, it became equal to 0. Hence, a necessity to maintain information in the working memory was eliminated. It is worth noticing that qualitative psychological data were transferred into precise quantitative measures. Revised task complexity measures for welding brackets after innovation are presented in Table 12.4. Previously considered measures helped us to introduce some interventions and compare the complexity of production operation before and after improvement.

It should be underlined that the motions and motor actions of the worker before and after the innovation of the manufacturing operation *bracket welding* remained the same. Thus, we were able to evaluate the complexity of the production operation being performed before and after its improvement based on quantitative assessment of cognitive strategies of task performance. An interesting aspect of the analysis of the operation bracket welding should also be noted. For welding of the brackets, workers utilized a special jig, which has two arms. These arms provide alignment of brackets along the axis, guaranteeing that the brackets will be positioned precisely opposite one another. Since the brackets are of two different widths, there were two different width arms. Each calls for the corresponding bracket. The specifics of the carried out operation was in fact that after the completion of each operation, the initial operating position of the wide and narrow arms of the jig is alternating. It required from workers to constantly change their decisions made *to take a narrow* or *to take a wide* bracket from the adequate bin. The strong repeatability of production operations (up to 2000) is quite difficult for the worker. To ease the work, the workers rasp off a wide arm of the jig. This led to the disruption of technological requirements and quality of the product. Only after innovations workers have the opportunity to perform this operation without violation of the technological requirements. If we compare the quantitative measures of complexity of the operation *welding brackets* with the operation of *welded handles*, the following fact will become obvious. Two operations have the same movement but the first operation is much more complex than the second one.

This can be explained by the fact that the second operation does not requires making a decision on how to select and correctly orient the handles. The handles are just placed laterally on the flask because they are symmetrical and there is no distinction in orienting them on the flask. Because the handles are also identical there is not requirement to make a decision what type of handle should be selected. For the second task the fraction of time for logical conditions (decision making) is close to 0 (see measure 11) and the fraction of time for retaining current information in working memory (measure 14) equals to 0. The fraction of time for cognitive activity to the entire task performance time (see measure 16) also equals to 0. At the same time motor actions and motions for both operations are practically the same. Thus the traditional motion and time study method is not adequate for studying production operations that contain separate cognitive actions. It should be noted that the motor actions and motions contain in in their content cognitive components that provide their regulation.

Presented data demonstrate that work complexity evaluation is tightly connected with the time study and work design. During complexity evaluation, one has to develop a time structure of activity during task performance. This aspect of complexity evaluation helps us not only to find the more efficient strategies of task performance but also to determine the time standards and allowances for the task and reduce the complexity of the task. Adequate time standards can be used for planning work in time, for labor cost control, or for a wage incentive plan. This study demonstrates that quantitative measures of complexity can capture the difference in cognitive regulation difficulty of human activity during the performance of manufacturing operations with nearly identical behavioral characteristics before and after innovation. That is why the consideration of these manufacturing operations has not only practical, but also theoretical interest. This example demonstrates that the proposed approach to assessing the task complexity allows to capture even subtle difference in cognitive activity regulation. The proposed method of task complexity assessment was used in the analysis of task performed by an operator in semiautomatic systems, in designing man–machine systems, in the analysis of computer-based tasks, etc. Some examples will be considered in the following section.

12.3 Quantitative Complexity Evaluation of Tasks at the Stage of Automation*

Before we begin considering new aspects of applying principles of SSAT to the design of equipment in automated system, we would like to briefly mention an interesting example of design covered in our prior publications (Bedny, 1987;

---

* Section 12.3 is prepared by G. Z. Bedny and G. M. Zarakovsky.

Bedny and Meister, 1997). This example involves the creation of the principle of a remote control of unmanned underwater vehicle (UUV). An operator who is sitting at the control board in remote-controlled UUV can not directly observe the UUV and its surrounding environment which makes remote control of UUV very complex and cognitively demanding. We have suggested the new principle of remote control of UUV. Justification of the principle of control of the UUV was done based not only on qualitative analysis but also utilizing quantitative analysis of the complexity of remote-controlled UUV. In existing conditions remote control of UUV movements on the seabed is a complex task that can cause operators' errors and fatigue due to the fact that during rotation of UUV there is a distortion of correlation between the axis of the operator's body and that of the vehicle. Visual information on the screen contradicts with the operator's motor manipulation of the controls. We will not discuss all of these issues but just mention the fact that when, for instance, an operator turns a UUV and his or her control board is to the right, an image on the panel's display is rotated to the same angle in the opposite direction.

Utilization of such design significantly reduces the complexity of the remote control of UUV and increases reliability of task performance. Thus, the principles of design and the complexity evaluation of a task can be very useful when solving complex ergonomic problems.

Below, we consider a task that is performed by a military operator, which is involved in testing equipment. The essence of this task was to check the equipment and prepare it for further use. Some components of the system may be off due to technological reasons. So the task is to turn on all equipment components and prepare them for further use. This task should be performed reliably in restricted time conditions. Hence, complexity evaluation of such type of tasks is important. Later, we consider only one possible version of task performance. For our analysis, we chose a version of an algorithm when an operator has to turn on two subsystems. This allows us to demonstrate the method of task complexity evaluation for an operator working with a complex man–machine system. In the analyzed scenario, the first two subsystems are in the active state, and the last two subsystems are disabled. They need to be quickly turned on. At the first stage, we describe the task algorithmically and develop its time structure. The task elements are performed sequentially and therefore we present task performance in a tabular form (see Table 12.5).

The described algorithm gives a clear picture of the activity elements involved in task performance. It can be seen from Table 12.5 that there can be a situation when all four subsystems are turned off. However, we consider only a situation when only the last two subsystems need activating. Therefore, the first two logical conditions have the same output that requires bypassing some members of the algorithm. Let us consider the outputs of all four logical conditions. Logical condition l_1 has two outputs. However, in our version of the task, the first subsystem is in active state. Then logical condition l_1 has one activated output (value 1) or one decision: *If subsystem 1 is on, check the second subsystem* (perform O_9^a). Another output *If subsystem 1 is off (blocked),*

TABLE 12.5

Algorithmic and Time Structure Description of the Military Task *Test Equipment State*

Symbols	Members of the Algorithm (Technological Units of Analysis)	Classification of Actions (Psychological Units of Analysis)	Time (s)
O_1^α	Visual perception of three position switches on the right-hand side of the panel.	Successive perceptual action.	0.55
O_2^ε	Turning on three position switches on the right-hand side of the panel (move an arm and grasp a switch).	(1) Reach a single object in a location that may vary slightly from cycle to cycle (method R50B), (2) grasp (G1A), (3) turn switch and release it (T40°S), and (4) release (RL1): ($R \approx 50 + 60B + G1A + T40°S + RL1$).	0.95
O_3^ε	Raise hand and move eye into the required area.	Motor movement of hand with simultaneous eye movement.	0.50
O_4^α	Receiving information from the first alphanumeric display on the left side of the panel (identification of the first subsystem state).	Simultaneous perceptual action.	0.35
$O_5^{\alpha th}$	Interpretation of information about the first subsystem.	The simplest thinking action (performed based on visual information).	0.30
$l_1\uparrow^1$	If subsystem 1 is blocked, decide to unblock the system (go to O_6^ε). If the subsystem is on decide, check the second subsystem (perform O_9^α).	The simplest decision making from two alternatives (performed based on visual information).	0.30
$*O_6^\varepsilon$	Simultaneous movements of the right hand and eyes to the area with switches.	Reach a single object in a location that may vary slightly from cycle to cycle (method $R \approx 40–60B$).	—
$*O_7^\alpha$	Identification of the required switch and its position.	Successive perceptual action.	—
$*O_8^\varepsilon$	Grasp and turn the switch (unblock the first subsystem by using two positioning switches).	(1) Grasp (G1A), (2) turn the switch and release it (T40°S), and (3) release (RL1): (G1A + T40°S + RL1).	—
$\downarrow^1 O_9^\alpha$	Receiving information from the second alphanumeric display on the left side of the panel (identification of the second subsystem state).	Simultaneous perceptual action.	0.35
$O_{10}^{\alpha th}$	Interpretation of information about the second subsystem.	The same as $O_4^{\alpha th}$.	0.30
$l_1\uparrow^2$	If subsystem 2 is blocked, decide to unblock the system (go to O_{11}^ε). If subsystem 2 is on decide, check the third subsystem (perform O_{14}^α).	The simplest decision making from two alternatives (performed based on visual information).	0.30

(Continued)

TABLE 12.5 (*Continued*)

Algorithmic and Time Structure Description of the Military Task *Test Equipment State*

Symbols	Members of the Algorithm (Technological Units of Analysis)	Classification of Actions (Psychological Units of Analysis)	Time (s)
$*O_{11}^{\varepsilon}$	Simultaneous movements of the right hand and eyes to the area with switches.	Reach a single object in a location that may vary slightly from cycle to cycle (method R ≈ 40–60B).	—
$*O_{12}^{\alpha}$	Identification of a required switch and its position.	Successive perceptual action.	—
$*O_{13}^{\varepsilon}$	Grasp and turn the switch (unblock the second subsystem by using two positioning switches).	(1) Grasp (G1A), (2) turn the switch and release it (T40°S), and (3) release (RL1): (G1A + T40°S + RL1).	—
$\downarrow^2 O_{14}^{\alpha}$	Receiving information from the third alphanumeric display on the left side of the panel (identification of the third subsystem state).	Simultaneous perceptual action.	0.40
$O_{15}^{\alpha th}$	Interpretation of information about the third subsystem.	The simplest thinking action (performed based on visual information).	0.40
$l_3\uparrow^3$	If subsystem 3 is blocked, decide to unblock the subsystem (go to O_{16}^{ε}). If subsystem 3 is on decide, check the fourth subsystem (go to O_{19}^{ε}).	The same as l_1.	0.30
O_{16}^{ε}	Simultaneous movements of the right hand and eyes to the area with switches.	The same as O_6^{ε}.	0.85
O_{17}^{α}	Identification of the required switch and its position.	The same as O_{12}^{α}.	0.80
O_{18}^{ε}	Grasp and turn the switch (unblock the second subsystem by using two positioning switches).	The same as O_{13}^{ε}.	0.70
$\downarrow^3 O_{19}^{\alpha}$	Receiving information from the fourth alphanumeric display on the left side of the panel (identification of the fourth subsystem state).	Simultaneous perceptual action.	0.40
$O_{19}^{\alpha th}$	Interpretation of information about the fourth subsystem.	The simplest thinking action (performed based on visual information).	0.40
$l_4\uparrow^4$	If subsystem 4 is blocked, decide to unblock the subsystem (go to O_{20}^{ε}). If subsystem 4 is on, go to O_{23}^{α}.	The simplest decision-making action at the sensory-perceptual level (based on visual information).	0.30
O_{20}^{ε}	Simultaneous movements of the right hand and eyes to the area with switches.	The same as O_{16}^{ε}.	0.85

(*Continued*)

TABLE 12.5 (*Continued*)

Algorithmic and Time Structure Description of the Military Task *Test Equipment State*

Symbols	Members of the Algorithm (Technological Units of Analysis)	Classification of Actions (Psychological Units of Analysis)	Time (s)
O_{21}^{α}	Receiving information from the fourth alphanumeric display on the left side of the panel (identification of the fourth subsystem).	The same as O_7^{α}.	0.80
O_{22}^{ε}	Grasp and turn the switch (unblock the fourth subsystem by using two positioning switches).	The same as O_{18}^{ε}.	0.70
$\downarrow^4 O_{23}^{\varepsilon}$	Raise head and adjust gaze.	Simultaneous turning of head and gaze.	0.50
O_{24}^{α}	Receiving information from the fifth alphanumeric display *all system is unblocked* (identification of equipment state).	Successive perceptual action (repeat the same perceptual action twice).	0.65
O_{25}^{ath}	Interpretation of the information *all system is unblocked* and the conclusion *the task is completed*.	Simple thinking actions: (1) interpreting information (2) making judgment—*task is completed*.	1.00

Note: *O_6^{ε}–*O_8^{ε} and *O_{12}^{α}–*O_{13}^{ε}—these members of the algorithm should be omitted because subsystems 1 and 2 were unblocked.
O_{16}^{ε}–O_{18}^{ε} and O_{20}^{ε}–O_{20}^{ε}—these members of the algorithm should be performed (they are designated by bold lines) because subsystems 3 and 4 were blocked.

decide to unblock the system (go to O_6^{ε}) has value zero. Logical condition l_2 has two similar outputs. However, only one output is activated (has value 1). This output or decision is *If subsystem 2 is on, check the third subsystem* (perform O_{14}). Thus, members of the algorithm *O_6^{ε}–*O_8^{ε} and *O_{12}^{α}–*O_{13}^{ε} are not performed and should be omitted from our analysis. Logical conditions l_3 and l_4 have a different activated output (have value 1). For logical condition l_3, it is *If subsystem 3 is off (blocked) decide, unblock this subsystem*. Logical condition l_4 has a similar activated output. Therefore, all members of the algorithm designated by a bold line in Table 12.5 should be performed. In order to evaluate the complexity of the task in general, it would be necessary to evaluate the probability of performing all versions of the algorithms: when only one, only two, only three, and all four subsystems are turned off. Based on such data, an overall complexity of this task can be identified. Here, as an example, we evaluate the complexity of the task when only subsystems 3 and 4 should be turned on. Our analysis of this version of the task shows that actions are always performed in the same sequence that is specific to the considered situation.

Simultaneous performance of the actions is limited to the movement of eyes and head because the activity elements for this task are too complex to be performed simultaneously and, according to SSAT rules, should be performed sequentially. According to the rules created for developing the activity time structure, cognitive elements cannot be executed simultaneously. Moreover, the logic of the task in most cases does not allow performing actions simultaneously. Thus, all members of the algorithm are performed in the same sequence for this version of the task. The described algorithm gives a clear picture of activity elements that are involved in task performance. A symbolic description of each member of the algorithm presents in a standardized manner units of analysis of the operator's activity.

We do not utilize all measures presented in Table 12.1 for evaluating this task's complexity selecting only the most informative ones for analysis of this type of task. In our example, all members of the algorithm are executed sequentially and have a probability of 1. It noticeably simplifies the calculation of complexity measures. Initial data on the temporal characteristics of motor and cognitive components of activity were obtained from handbooks (Lomove, 1982; Myasnikov and Petrov, 1976) and chronometrical analysis of task performance. MTM-1 symbols were used for depicting basic motions only as a standardized language for motor action description and were not used for calculating their performance time. In our example, we used the following symbols: R-B (reach a single object in location that may vary slightly from cycle to cycle); G1 A (easily grasp an object); T40°S (turn an object 40°; the weight factor is small—*resistance during turn performance*); and release (normal release or simple opening of fingers). Quantitative measures of task complexity for the considered version of the task are presented in Table 12.6.

Measure 15 demonstrates that a significant part of this task is associated with cognitive activity. All cognitive components have the third level of complexity that is the minimal level of complexity for cognitive elements of work (see measure 20). Measure 8 demonstrates that there are a significant number of work components that are involved in receiving information. Measure 7 shows that discriminating small stimuli that approach the threshold area was not required.

The fraction of time for thinking operations (measure 9) is 0.19 and for the decision-making process (measure 10) is 0.1. Measure 12 demonstrates that the decision-making process is performed based on externally presented information, but not on information that is extracted from memory, and measure 13 shows that there is no requirement to retain information in working memory. Measures 14 and 15 demonstrate that cognitive components of work (including the stage of receiving information) dominate in this task. However, the complexity of cognitive components does not exceed the third category of complexity (see measure 20), which is the minimal level of complexity for cognitive components of activity. Measures 9, 10, 12, and 13 are specifically important for military tasks because they evaluate components of work that are susceptible to adverse effects such as stressful situations and strict time limit. However, in this particular example, these measures identify that cognitive elements of the task do not present considerable complexity, which suggests that in stressful conditions, the task will be executed with sufficient reliability.

In the second example, we consider complexity evaluation of navigational–tactical task. Such tasks involve determining parameters of relative position

TABLE 12.6

Quantitative Measures of Task Complexity of the Military Task *Test Equipment State*

N of Measure	Measures of Complexity	Value[a]
1	Algorithm (task) execution time (T)	12.95
2	Sum of the performance time of all afferent operators (T_α)	4.3
3	Sum of the performance time of all thinking operators (T_{th})	2.4
4	Sum of the performance time of all logical conditions (L_g)	1.2
5	Sum of the performance time of all efferent operators (T_{ex})	5.05
6	Sum of the performance time of all cognitive components of the task (including perceptual activity) (T_{cog})	7.9
7	Sum of time spent on discrimination and recognition of perceptual distinctive characteristics that are approaching threshold of sensory receptors ($'T_\alpha$)	0
8	Fraction of time for afferent operators in the time for the entire task performance (N_α)	0.33
9	Fraction of time for thinking operators in the time for the entire task performance (N_{th})	0.19
10	Fraction of time for logical conditions in the time for the entire task performance	0.1
11	Proportion of time for cognitive components of task (including perceptual activity) to the performance time for all efferent operators (N_{cog})	1.56
12	Fraction of time for logical components of work activity, which largely depends on information selected from long-term memory rather than from external sources of information, in the entire time for logical condition performance (L_{ltm})	0
13	Fraction of time for retaining current information in working memory in the time for the entire task performance (N_{wm})	0
14	Fraction of time for performance of all efferent operators in the time for the entire task performance ($'N_{beh}$)	0.39
15	Fraction of time for cognitive activity in the time for the entire task performance ($'N_{cog}$)	0.61
20	Scale of complexity (N_{ch})	
	(a) Algorithm	(a) 3
	(b) Member of the algorithm for motor activity: N_1, O_2^ε; N_2, O_{16}^ε; N_3, O_{18}^ε; N_4, O_{20}^ε; N_5, O_{22}^ε	(b) $N_1 = 2$; $N_2 = 2$; $N_3 = 2$; $N_4 = 2$
	All cognitive components have the 3d level of complexity	

[a] The performance time of various elements of the task is given in seconds.

of moving ships, aircrafts, etc. Our study considers a task where an operator monitors the movement of aircrafts. We do not describe motor actions used at the final stage of task performance. We restrict our analysis to studying only the cognitive component of the work. Therefore, the basis of the considered task is various cognitive actions and specifically thinking action performed based on visual information or information extracted from memory.

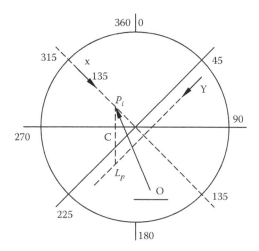

FIGURE 12.1
Graphical explanation of navigational–tactical task.

The following is the general description of the task: an object under the operator's control is designated in Figure 12.1 as O (aircraft), whereas objects X and Y (the other aircraft) are not under his or her control.

An operator has to determine the course by which object O, which is under the operator's control, could cross the trajectory of object X or coincide with this object. In such situation, both aircrafts can be at the same location and collision can happen. An operator also has to determine where aircraft Y will be at a particular time and determine the distance between aircraft Y and the intersection point of X and O designated by letter P_i. The operator can evaluate the distance and speed of moving objects, not only in natural units but also in conditional units. For example, an operator knows that 1 cm on the radar screen represents a speed of 600 km/h, and object O's speed is depicted by a horizontal line in Figure 12.1. It is not possible to determine the direction of object O's movement at the beginning of the process just based on its speed, which is depicted by a horizontal line in Figure 12.1. Figure 12.1 is used only for describing the task performed by an operator. Information about the parameters of the aircraft movement can be presented in various ways: the course and speed of movement can be given as numbers or vectors. Moreover, a computer can give a prognosis of the trajectory of an aircraft. A flight operator can perform multiple tasks simultaneously, but we are not going to discuss such situations. The main point of study is to evaluate the task complexity and, based on such data, develop recommendations for improvement of possible methods of information presentation to the operator.

We will evaluate the scenario when only numeric data are presented on the radar screen presenting positions of aircrafts by two numbers. One of them designates the speed and the other one reflects the direction of aircrafts. Later, we describe strategies of task performance, evaluate its complexity, and make conclusions about the efficiency of numeric presentation of information. To make formalized and quantitative analysis of the task clear, we describe what actions are taken by an operator when he or she interacts with a radar screen, utilizing for this purpose Figure 12.1. In Figure 12.1, information about moving objects is presented by vectors. In addition to the vector presentation, predicted trajectories of X and Y were given. However, we only use this figure for explanation of navigational task. We will consider further a different way of presenting information. In the considered version of the presentation of information, the speed and course of the objects will be presented only in numbers. Thus, Figure 12.1 is used only for explanation of navigation task only.

The strategy of task performance can be described as follows: An operator tries to predict the trajectory of objects X and Y. The length of the vector in Figure 12.1 depends on the speed of the object. Graphically, the unit of speed is presented by a horizontal line. This is the speed of object O. An operator needs to determine the course of object O under his or her control and predict the point of intersection P_i as well as the time object O reaches the point of intersection and also the location of Y at that instantaneous time (L_p). An operator mentally manipulates the line segments that represent the sides for the triangle OXP_i. The distances from two starting points to point P_i are different, but aircrafts O and X may reach P_i at the same time because they have different speeds. The operator has to determine the point of intersection and memorize this point. He or she also needs to determine the position of aircraft Y when objects O and X reach the intersection point. The operator attempts to determine the distance between P_i and L_p based on visual information.

Now we will consider the situation when an operator performs the same task problem when only numerical data are presented on the radar screen. Each aircraft position is presented only by two numbers. One number presents the speed, and the other direction. The goal of the task was to determine the position of object O when it can cross the course with object X that is moving at a 135° angle. Then the operator had to imagine the trajectory of the uncontrolled object Y that is moving at a 210° angle. The operator visually determines and verbally expresses the distance between P_i and L_p.

The objective of this study was to demonstrate that the algorithmic description of an operator's performance with a temporal analysis of his or her activity and the following quantitative evaluation of task complexity can predict the limitations of the numerical presentation of information and the benefits of the vector representation of information. The first step of analysis involves an algorithmic and temporal analysis of a considered task. Navigational–tactical task consists of four subtasks:

Subtask 1: Converting the numerical expression of the speed of moving objects and the course of objects X and Y into an imaginative form
Subtask 2: Determining of point of time when objects O and X meet

Subtask 3: Determining the position–location of object Y at the time of contact

Subtask 4: Determining the distance between objects O and Y when objects O and X reach the intersection point

Table 12.7 presents an algorithmic description and time structure of an operator's performance when a numerical presentation of information is used (subtask 1). An algorithmic description of the task was developed based on the observation of the operator's performance (see Table 12.7).

Temporal data were obtained from the *Aircraft Digital Monitoring and Control Systems* (Myasnikov and Petrov, 1976) and *Handbook of Engineering Psychology* (Lomov, 1982). Some data were obtained from chronometrical study of real performance. The left column of Table 12.7 contains a symbolic description of each member of the algorithm. These are psychological units of activity analysis that depict in a standardized form to which category of cognitive activity each member of the algorithm is related. The second column from the left presents a description of each member of the algorithm in technological units of analysis. The third column presents each member of an algorithm in terms of a standardized description of cognitive actions. These also are psychological units of analysis.

A combination of technological and psychological units of analysis allows us to clearly describe the content of each member of the algorithm. Sometimes it is difficult to extract single cognitive actions or their operations. If all cognitive actions in the content of each member of the algorithm are homogeneous and have similar characteristics, they can be considered without their specific separate description. The quantity of cognitive actions in each member of the algorithm is restricted by the capacity of short-term memory. Each member of the algorithm can include in its content only a similar type of cognitive actions. Therefore, each member of the algorithm can contain not only one but several cognitive actions. If an operator is not aware about the goal of the actions, then they are simply mental operations of a complex cognitive action. These are qualitative formalized rules. They cannot be directly utilized in the quantitative assessment of task complexity because actions are not commensurable units of measures. Only after developing the time structure of activity and utilizing corresponding units of measure (intervals of time for different elements of activity), one can perform a quantitative assessment of task complexity. The fourth column has a description of operative units of activity that are units of information (images, concepts, statements, etc.). They are semantically meaningful components of activity that are used for cognitive action performance. The capacity of working memory plays an important role in integrating such units into a meaningful whole. The last column shows the performance time of each member of the algorithm. This table demonstrates that utilizing technological and psychological units of analysis allows describing human activity in a precise manner. Different tasks can be approximately equivalent in terms of their complexity. The characteristics of cognitive actions depend on

TABLE 12.7

Algorithmic and Time Structural Description of Navigational–Tactical Task Solution (Subtask 1)

Member of the Algorithm	Technological Description of Subtasks	Actions (Psychological Units of Analysis)	Operative Units of Information	Time (s)
Subtask 1. Converting the numerical expression of the speed of moving objects and the course of objects X and Y into an imaginative form				
O_1^α	Receiving the numerical value of O's speed.	Simultaneous perceptual action	Digital numbers.	0.3
O_2^{rth}	Mental recoding of the numerical expression of the speed vector of O's movement into the line segment.	Recoding thinking action	Several digit number and imaginative templates in memory.	0.4
O_3^μ	Memorizing the standard units of length.	Mnemonic action	The same as in O_2^{rth}.	1.5
O_4^α	Receiving the numerical value of X's speed.	Simultaneous perceptual action	Digital numbers.	0.3
O_5^{rth}	Mental recoding of the numerical indication of trajectory of object X into an imaginative form as the vector of the movement trajectory.	Recoding thinking action	The digit number in memory and the image-based templates of angles.	1.5
O_6^α	Receiving the numerical value of X's speed.	Simultaneous perceptual action	Digital numbers.	0.3
$l_1 \overset{1(0.1;0.9)}{\uparrow}$	Decide to repeat mental recoding of the numerical value of object X into an imaginative form, or go to the next step of performance.	Simple decision-making action from two alternatives	If decision to repeat is *yes* ($p = 0.1$), go to O_7^{rth} and then perform O_8^{rth}. If no ($p = 0.9$), go directly to O_8^{rth}.	0.3
$\overset{1(0.1)}{\downarrow} O_7^{rth}$	Mental recoding of the numerical indication of trajectory of object X into an imaginative form as the vector of the movement trajectory.	Recoding thinking action	The digit number in memory and the image-based templates of angles.	1.5
$\overset{1(0.1+0.9)}{\downarrow} O_8^{rth}$	Vector extrapolation.	Recoding thinking action	Image of the vector in the form of a line segment.	0.4

(Continued)

TABLE 12.7 (*Continued*)

Algorithmic and Time Structural Description of Navigational–Tactical Task Solution (Subtask 1)

Member of the Algorithm	Technological Description of Subtasks	Actions (Psychological Units of Analysis)	Operative Units of Information	Time (s)
O_9^μ	Memorizing the speed vector and extrapolating the line.	Mnemonic action	Image of the vector in the form of a line segment.	1.5
O_{10}^α	Receiving the numerical value of Y's speed.	Simultaneous perceptual action	Digital numbers.	0.3
O_{11}^{rth}	Mental recoding of the numerical indication of trajectory of object Y into an imaginative form as the vector of the movement trajectory.	Recoding thinking action	The digit number in memory and the image-based templates of angles.	1.5
O_{12}^α	Receiving the numerical value of Y's speed.	Simultaneous perceptual action	Digital numbers.	0.3
O_{13}^{rth}	Mental recoding of the numerical indication of trajectory of object X into an imaginative form as the vector of the movement trajectory.	Recoding thinking action	The digit number in memory and the image-based templates of angles.	1.5
O_{14}^{rth}	Vector extrapolation.	Recoding thinking action	Image of vector in the form of a line segment.	0.4
O_{15}^μ	Memorizing the speed vector and extrapolating the line.	Mnemonic action	Image of vector in the form of a line segment.	1.5

Total performance time of the first subtask is 12.15 s

utilized operative units of information. For example, if an operator receives ambiguous information, simultaneous perceptual action can be transformed into successive perceptual actions. Moreover, the characteristics of cognitive actions can be classified according to the required level of attention concentration. This characteristic of cognitive actions also depends on operative units of information that are utilized by an operator. Therefore, constructive features of displays influence action description and their classification. Motor actions on the other hand depend on the configuration of controls.

Let us consider complexity measures that are specific to the first subtask (see Table 12.8).

Four members of the algorithm that belong to a considered subtask are perceptual activity (T_α). Their total execution time is 1.5 s. Three members of the

TABLE 12.8

Quantitative Evaluation of Complexity of the First Navigational–Tactical Subtask

N of Measure	Measures of Complexity	Value[a]
1	Algorithm (task) execution time (T)	12.15
2	Sum of the performance time of all afferent operators (T_a)	1.5
3[b]	Sum of the performance time of all thinking operators (T^{th})	5.85
4[b]	Sum of the performance time of all operators that require keeping information in working memory (T^μ)	4.5
5	Sum of the performance time of all logical conditions (L_g)	0.3
6	Sum of the performance time of all efferent operators (T_{ex})	0
7	Sum of the performance time of all cognitive components of the task (including perceptual activity) (T_{cog})	12.15
8	Sum of time spent on discrimination and recognition of perceptual distinctive characteristics that are approaching threshold of sensory receptors ($'T_a$)	0
9	Fraction of time for afferent operators in the time for the entire task performance (N_a)	0.12
10	Fraction of time for thinking operators in the time for the entire task performance (N_{th})	0.48
11	Fraction of time for logical conditions in the time for the entire task performance	0.02
13	Fraction of time for logical components of work activity, which largely depends on information selected from long-term memory rather than from external sources of information, in the entire time for logical condition performance (L_{ltm})	1
14	Fraction of time for retaining current information in working memory in the time for the entire task performance (N_{wm})	0.37
16	Fraction of time for cognitive activity in the time for the entire task performance (N_{cog})	1
21	Scale of complexity (N_{ch}) of the algorithm Almost 85% of cognitive components have the fourth level of complexity	4

[a] Time is given in seconds
[b] Instead of the symbol T_{wm}, we use here T^μ, and instead of the symbol T_{th}, we use here T^{th}.

algorithm are the mnemonic components of activity (T_{wm}) that take 4.5 s (instead of T_{wm}, we use the symbol T^μ). Other members of the algorithm are thinking components of activity, and more specifically, they are related to recoding actions (T^{th}). The performance time of these components of cognitive activity is 5.85 s. One action involves decision making that in SSAT is called logical condition.

According to SSAT rules, cognitive components of activity cannot be performed simultaneously. Therefore, none of the members of this algorithm can be performed simultaneously with other members of algorithm. There is a logical condition (l_1) that depicts the alternative to repeat the same member of the algorithm the second time (perform operator O_7^{rth} that is similar to O_5^{rth}). So most of the members of the algorithm are performed in the same sequence and only one member of the algorithm has a probability of being performed the second time. Logical condition l_1 is a checking action that emerges in critical situations. Such actions present self-control when performing some actions that are evaluated by an operator as subjectively significant. If an operator perceives a stage of task performance as being critical, he or she voluntarily or even involuntarily tends to repeat such stage mentally to check his or her own performance (Grimak, 1987).

Such actions periodically appear at certain stages of task performance. In our case, logical condition l_1, the duration of which is 0.3 s, reflects reexecution of O_5^{rth} (performing of O_7^{rth} that is similar to O_5^{rth}) or sanctioning to move further. Subjectively, such decision making is perceived as a dilemma of *repeating an action* or *moving on*. The need to re-execute a task element was accepted with a probability of 0.1, when an operator subjectively perceives the stage of execution as a critical one.

Each member of the algorithm can include in its content only the same type of cognitive actions (only thinking, only recoding thinking, and only mnemonic actions). The following are the possible actions in each member of an algorithm: simultaneous perceptual action, mnemonic actions of similar type, and recoding thinking actions. The sequential performance of cognitive actions significantly simplifies the calculation of complexity measures.

In SSAT, thinking actions can be divided into subgroups. For the task under consideration, recoding thinking actions should be separated because they constitute a significant part of task performance. Usually recoding actions should be eliminated in task performance if possible.

Let us calculate the performance time of recoding thinking actions. Members of the algorithm (operators) O_7^{rth} and O_5^{rth} are identical. Operator O_5^{rth} is performed with a probability of 1. The probability of repetition of O_5^{rth} (performing O_7^{rth}) is low and we assign it a probability of 0.1. Therefore, the performance time of O_7^{rth} is $0.1 \times 1.5 = 0.15$ s. Operator O_8^{rth} can be performed after repeating of the same operator with a probability of 0.1 (perform O_7^{rth}) or without repeating the same operator (directly) with a probability of 0.9. Hence, the probability of performance of operator O_8^{rth} is $P = 0.1 + 0.9 = 1$. Therefore, all performing operators O^{rth} except O_7^{rth} have a probability of 1. Then $\sum O^{rth}$ can be determined as

$$\sum T^{rth} = 0.4 + 1.5 + 0.1 \times 1.5 + 0.4 + 1.5 + 1.5 + 0.4 = 5.85 \text{ s}$$

Recoding thinking actions are similar to thinking actions that are performed based on information extracted from memory. These are cognitively demanding actions.

The performance time of the first subtask (T) is equal to 12.15 s because all members of the algorithm excluding O_7^{rth} are performed sequentially with a probability of 1. The time of performing logical conditions is equal to 0.3 s. Hence, the ratio of the time for logical conditions to the time of task performance is 0.02. This measure demonstrates that most elements of activity are performed in the same sequence. The performance time of executive components of activity or the performance time of efferent operators (T_{ex})

is equal to zero. The proportion of time for cognitive components of a task (including the perceptual activity) to the performance time for all task ($'N_{cog}$) is equal to one, which means that the first subtask is purely cognitive and there are no motor components involved in its execution.

Let us consider specific measures for this subtask. The time taken for afferent operators can be determined as follows:

$$T_\alpha = \Sigma P_r^\alpha t_r^\alpha$$

where
  P_r^α is the probability of the rth afferent operator
  t_r^α is the performance time of the rth afferent operator

In our example as we have mentioned already, all members of the algorithm are performed in the same sequence every time and they have a probability of 1. By dividing the time required to perform afferent operators by the time of task performance, we receive the measure that shows the fraction of time spent on perceptual actions in the entire task performance:

$$N_\alpha = \frac{T_\alpha}{T} = \frac{1.5}{12.15} = 0.12$$

Receiving information can be complex or simple. For example, receiving weak stimuli is more complicated than when stimuli can be easily perceived.

This task includes members of the algorithm that involve purely mnemonic activity T^μ. Their time can be calculated as follows:

$$T^\mu = O_3^\mu, + O_9^\mu, + O_{15}^\mu = 1.5 + 1.5 + 1.5 = 4.5 \text{ s}$$

The fraction of time for the performance that is mostly based on information extracted from memory can be defined by the following formula:

$$N^\mu = \frac{T^\mu}{T} = \frac{4.5}{12.15} = 0.37$$

Hence, almost 40% of the time, the operator performs the task-based memorized data. This kind of stages of task performance is very undesirable and should be reduced as much as possible.

Thinking actions according to the previously described classification of cognitive actions can be related to recoding thinking actions. The fraction of time for this component is

$$N^{rth} = \frac{T^{rth}}{T} = \frac{5.85}{12.15} = 0.48$$

This means that an operator performs recoding actions that present mental workload for thinking and memory almost 50% of the time. These actions should be reduced as much as possible. This measure of complexity is the most critical one because it demonstrates significant cognitive demands that are required for task performance.

The last two measures are associated with the assessment level of concentration of attention. SSAT offers an ordered scale with five categories of complexity for various elements of activity. The main criteria for such categorization are the level of concentration of attention and the ability to perform some elements of activity simultaneously. According to the level of concentration of attention criterion, the simplest cognitive element is related to the third category of complexity. If during task performance some actions are performed at the same time, the level of concentration of attention can increase. However, we can ignore this factor because all elements of the analyzed task are performed in sequence. According to existing rules, perceptual actions with optimal vision conditions are related to the third category of complexity. Therefore, the fraction of time N_α that characterizes receiving visual information from the radar screen is of the third category of complexity, and the elements of activity associated with the performance of such members of the algorithm as $O_3^\mu; O_9^\mu; O_{15}^\mu$ require the fourth category of complexity according to the level of concentration of attention. Similarly, members of the algorithm $O_2^{rth}; O_5^{rth}; O_7^{rth}; O_8^{rth}; O_{11}^{rth}; O_{13}^{rth}; O_{14}^{rth}$ are also related to the fourth category of complexity. Considering that all elements of activity are performed in sequence, the fraction of time for the fourth category of complexity is

$$N^\mu + N^{rth} = 0.37 + 0.48 = 0.85$$

This indicates that the executed subtask requires a very high level of concentration of attention, meaning that according to all measures of complexity, it requires a lot of mental efforts during its performance.

Algorithmic and time structural descriptions of subtasks 2–4 are presented in Table 12.9. It was developed in a similar way as Table 12.7.

After the development of the time structure of these subtasks, we can evaluate the complexity of the whole navigational–tactical task by using Tables 12.7 and 12.9.

Let us consider the complexity of the whole task (see Table 12.10).

The fraction of time spent performing efferent operators (motor activity) is $'T_{ex} = 0$ because we measure complexity until the point when an operator visually determines and verbally reports the distance between P_i and L_p. The task performance time is equal to 27.25 s. All afferent members of the algorithm are performed in sequence. So T_α can be evaluated as

$$T_\alpha = O_1^\alpha + O_4^\alpha + O_6^\alpha + O_{10}^\alpha + O_{12}^\alpha = 1.5 \text{ s}$$

TABLE 12.9

Algorithmic and Time Structural Description of Subtasks 2–4

Member of the Algorithm	Technological Description of Subtasks	Actions (Psychological Units of Analysis)	Operative Units of Information	Time (s)
Subtask 2. Determining point in time when objects O and X meet				
O_{16}^{th}	Extracting the relationship between the speed of vector O and X lengths	Imaginative thinking action	Image of relationship	0.7
O_{17}^{μ}	Actualization of the angle image with its sides having the same relationship as the vector lengths	Mnemonic action	Image of angle	0.4
O_{18}^{th}	Mental placing of the angle, with one side going through point O and the peak at the extrapolated trajectory of object X	Imaginative thinking action	Image including angle, point O, and trajectory of X	1.4
O_{19}^{μ}	Memorization of the location of the intersection P_i	Mnemonic action	Image of P_i position	1.5
$\overset{2(0.1)}{\downarrow} O_{20}^{th}$	Visual measurement of the distance between X and P_i	Thinking action	Image of line segment X-P_i and vector X	1.6
O_{21}^{th}	Visual measurement of the distance between O and P_i	Thinking action	Image of line segment O—P_i and vector X	1.6
$\overset{2(0.1;0.9)}{l_2 \uparrow}$	Sanctioning of decision about defining P_i or return to O_{20}^{th}	Simple decision-making action from two alternatives	If *yes*, go directly to O_{22}^{μ} ($p = 0.9$). If *no* ($p = 0.1$), perform $O_{20}^{th} + O_{21}^{th}$ and then go to O_{22}^{μ}.	1.4
$\overset{2(0.9+0.1)}{\downarrow} O_{22}^{\mu}$	Repetition (maintenance) in working memory of the position of P_i and time of contact	Mnemonic action	Image of P_i and number of line segment units from object X and O to P_i	0.5
Total performance time of the second subtask 9.1				
Subtask 3. Determining the position–location of object Y at the time of contact				
O_{23}^{μ}	Actualization of the predicted trajectory of object Y	Mnemonic action	Image of line from point Y according to the direction of the trajectory	0.4

(Continued)

TABLE 12.9 (*Continued*)

Algorithmic and Time Structural Description of Subtasks 2–4

Member of the Algorithm	Technological Description of Subtasks	Actions (Psychological Units of Analysis)	Operative Units of Information	Time (s)
O_{24}^{th}	Mental placement of vector Y on the trajectory in time units until the contact of object O and object X	Thinking action	Image of vector, trajectory, and final position (line segment L_p–P_i)	1.8
O_{25}^{μ}	Memorization of location of crossing point L_p		Image of L_p on the trajectory	1.5
Total performance time of the third subtask 3.7				
Subtask 4. Determining the distance between object O and Y when object O and X reach intersection point				
O_{26}^{th}	Visual measurement of the distance between L_p and P_i	Thinking action	Image of points L_p and P_i and the line segment between them	1.6
O_{27}^{rth}	Recoding of imaginative codes into verbal code and pronounces result	Recoding thinking action	One-digit number in verbal code	0.4
Total performance time of the fourth 2.0 s				
Total performance time of the whole task 27.27				

Therefore, the fraction of time for afferent operators (sensory-perceptual activity) in the whole task is

$$N_\alpha = \frac{T_\alpha}{T} = \frac{1.5}{27.25} = 0.05$$

Hence, a perceptual component plays a minor role in the whole task.

The following stage of analysis involves determining a complexity measure that characterizes the demands for memory function for which the performance time for all operators TO^μ has to be calculated.

In the described algorithm of task performance, all its members have a probability of 1 and are performed in sequence. Hence, the time of their performance can be determined as

$$T^\mu = O_3^\mu + O_9^\mu + O_{15}^\mu + O_{17}^\mu + O_{19}^\mu + O_{22}^\mu + O_{23}^\mu + O_{25}^\mu = 8.8 \text{ s}$$

Therefore, the fraction of time for retaining current information in working memory can be defined as follows:

$$N_{wm} = \frac{T^\mu}{T} = \frac{8.8}{27.25} = 0.32$$

TABLE 12.10

Quantitative Evaluation of Complexity of the Whole Navigational–Tactical Task

N of Measure	Measures of Complexity	Value[a]
1	Algorithm (task) execution time (T)	27.25
2	Sum of the performance time of all afferent operators (T_a)	1.5
3[b]	Sum of the performance time of all thinking operators (T^{th}) + (T^{imth})	15
4[b]	Sum of the performance time of all operators that require keeping information in working memory (T^{μ})	8.8
5	Sum of the performance time of all logical conditions (L_g)	1.7
6	Sum of the performance time of all efferent operators (T_{ex})	0
7	Sum of the performance time of all cognitive components of the task (including perceptual activity) (T_{cog})	27.25
8	Sum of time spent on discrimination and recognition of perceptual distinctive characteristics that are approaching threshold of sensory receptors ($'T_a$)	0
9	Fraction of time for afferent operators in the time for the entire task performance (N_a)	0.05
10	Fraction of time for thinking operators in the time for the entire task performance ($^1N_{th}$ + $^2N_{th}$)	0.55
11	Fraction of time for logical conditions in the time for the entire task performance	0.06
14	Fraction of time for retaining current information in working memory in the time for the entire task performance (N_{wm})	0.32
16	Fraction of time for cognitive activity in the time for the entire task performance (N_{cog})	1
21	Scale of complexity (N_{ch}) of the algorithm Almost 87 of cognitive components have the fourth level of complexity	4

[a] Time is given in seconds.
[b] Instead of the symbol T_{wm}, we use here T^{μ}, and instead of the symbol T_{th}, we use here T^{th} and T^{imth}.

Thus, about 1/3 of the task performance time involves retaining information in working memory. This measure demonstrates that there are significant demands for memory during this task performance. The next important measure of complexity evaluation for this specific task is analysis of demands for thinking actions. There are two types of thinking actions: recoding thinking actions and imaginative thinking actions. We first evaluate demands imposed by recoding thinking actions. The performance time for these actions that have a probability of 1 and are performed sequentially is

$$T^{rth} = O_2^{rth} + O_5^{rth} + O_7^{rth} + O_8^{rth} + O_{11}^{rthr} + O_{13}^{rth} + O_{14}^{rth} + O_{16}^{rth}$$

$$+ O_{18}^{rth} + O_{20}^{rth} + O_{21}^{rth} + O_{24}^{rth} + O_{26}^{rth} + O_{27}^{rth}$$

$$= 6.25 \text{ s}$$

We calculate similarly the performance time of imaginative thinking actions that are associated with all O^{th} members of the algorithm:

$$T^{imth} = 8.82 \text{ s}$$

Obtained data allow to calculate two other measures of complexity assessment of the considered task. The first one is the fraction of time for recoding thinking actions during task performance:

$$^1N_{th} = \frac{T^{rth}}{T} = \frac{6.25}{27.25} = 0.23$$

This measure demonstrates that recoding of visually presented information requires approximately a quarter of task performance time. This is an essential factor for cognitive demands during task performance. The next measure reflects cognitive demands that are associated with imaginative thinking actions. It is calculated in the following manner:

$$^2N_{th} = \frac{T^{imth}}{T} = \frac{8.82}{27.25} = 0.32$$

All members of the algorithm associated with O^μ; O_9^{rth}; O^{th} are related to the same category of complexity according to such criterion as the level of concentration of attention. Considering that all of these actions are performed in sequence, we can calculate the fraction of time for the thinking components of activity in the total time of task performance as follows:

$$^1N_{th} + {}^2N_{th} = 0.23 + 0.32 = 0.55$$

Therefore, 0.55 of the time during task execution involves performing the most complex thinking actions with the fourth level of concentration of attention. According to existing SSAT criteria, thinking actions that have the fourth category of concentration of attention are very complex. In stressful conditions, they can be transferred into the fifth category of complexity and become extremely difficult (there is no higher category of complexity in SSAT according to this criterion).

Later, we calculate the total fraction of time for the fourth category of complexity in the whole task:

$$X_r = N_{wm} + N^{rth} + N^{th} = 0.32 + 0.23 + 0.32 = 0.87$$

Therefore, according to existing rules, the whole task belongs to the fourth category of complexity according to the level of concentration of attention.

Finally, it is obvious in the considered example that the fraction of cognitive components in task performance (the fraction of time for cognitive activity in the total task performance time) is $N_{cog} = T_{cog}/T = 1$. Usually, such fraction is significantly smaller when motor activity is included in task performance. This is explained by the fact that usually motor components of task performance require significantly more time than cognitive components.

Analysis of subtask 1 in the context of the whole task clearly demonstrates that recoding thinking actions that are undesirable and should be eliminated dominate in this task.

In order to do that, the speed of the moving object and their direction should be presented not by two numbers but by vectors. Moreover, presenting the predicted trajectory of objects as additional information is desirable. Such changes in task performance require the performance of several additional perceptual actions. However, there will be a significant reduction in thinking, recoding, and mnemonic actions. The main purpose of this study was to demonstrate such method. It is clear that such method can be specifically useful at the analytical stage of design. At the same time, such method can be combined with experimental procedures that can significantly simplify them. Such combination can make experimental procedures more precise.

12.4 Complexity Evaluation and Variability of Task Performance

In automated and semiautomated systems, human work is very flexible and the sequence of cognitive and behavioral actions is not specified. Externally presented information constantly changes and an operator changes his or her sequence of actions based on dynamically presented information. A formalized description of human work activity becomes possible only based on the algorithmic description of human activity. All this implies the development of adequate methods for the quantitative assessment of task complexity.

In ergonomic literature, there is a tendency to contrast the variability of human activity to the possibility of its formal description. This contrast is reflected in the statement *constraints versus instruction* (Vicente, 1999). Constraint-based principle practically rejects the possibility of formalized description of activity and therefore the possibility to design human work activity. Not only is human activity variable. Engineers who design equipment face the problem of variability just like ergonomists. For example, any manufacturing part that was obtained in the machining process is unique in its size and shape. In order to ensure interchangeability of parts in mass production and provide their assembling into a finished product, there is a science known as the *interchangeability*. Mass production is impossible without this science. The fundamental concepts of this science are tolerance, tolerance zone, maximum limit of size, minimum limit of size, etc.

Thus at present, there is a contradiction between the requirements to develop models of human performance and flexible and adaptive human behavior.

We demonstrate that it is possible to develop work activity models even for very flexible and adaptive task performance. A constraint-based approach contradicts design principles because worker's performance is practically unknown in such situation and models of human performance cannot be developed. Interchangeability principles allow us to develop the design model of activity for very variable tasks. Suppose there are 110 versions of performance of the same task. The version of task performance depends on the type of information presented to the operator, his past experience, and individual characteristics. Depending on the required accuracy of the design models of the operator's work activity, a specialist can choose for the analysis of a few basic strategies as standard versions of task performance. Suppose there are three main strategies of task performance. Each strategy can be described algorithmically and the time structure of activity for each strategy can be developed. Each strategy has its own probability and range of variation. If variation in the performance time or other characteristics of selected representative strategies is in the range of tolerance, we can neglect their variation. If the accuracy of the analysis should be chosen, higher than not three but, for example, four strategies as a reference sample can be selected. Hence in any particular situation, when a worker uses various strategies that are described algorithmically and their variation does not exceed the range of tolerance, it can be considered as equivalent to the strategy selected as a reference sample. So even very flexible human activity can be described by using reference strategies as samples of task performance. Each selected strategy can be described algorithmically and their time structure can be developed. If required, it is possible to integrate all described algorithms into one general algorithm of task performance. Therefore, the complexity of separate representative strategies of task performance or the complexity of a whole task can be evaluated.

In Chapter 11, we have described tasks performed in the military when the flexibility of task performance was not significant. In the following example, we in abbreviate manner consider evaluating quantitatively the complexity of a flexible task in laboratory conditions. We have described such tasks in more detail in Bedny and Karwowski (2007). In Section 7.2, we considered the algorithmic description of a task when subjects worked on a specially designed control panel.

First, we will briefly recap this task. An operator may use several or all five controls depending on the information displayed by five indicators (see Figure 7.1).

The first digital indicator can present only numbers 1 or 2. Based on this information, an operator turns the switch to the appropriate position. Then an operator grasps a four-position hinged lever (7) that has a button on the handle and depresses the top button using the thumb. The subject waits 3 s, and then depending on the pointer's position of the second indicator, he or she moves lever 7 into one of four positions. After that, a digital indicator demonstrates one number from 10 possible numbers. The subject moves his or her hand to switch 8 and turns it in the corresponding position. The last action involves pressing the green or red button, depending on whether a green or red indicator on the control panel is illuminated. A general algorithmic description of this task and the time structure of activity during its performance are presented in Section 7.3.

The design of the experimental panel allowed 110 different versions of task performance to be carried out. It is theoretically possible to evaluate the complexity of each of the 110 versions of the algorithm. However, we use the basic concept of mass production known as range of tolerance to evaluate quantitatively such flexible human activity. According to this principle, we can neglect any variation in human activity if this variation is within an accepted range of tolerance. Based on this principle, we select four most representative versions of task performance and consider all other versions of task performance to be practically identical to one of the four selected versions of task performance. The first version of the algorithm considers the scenario when presented on the control panel information requiring all five controls to be used. In the second version, a four-position hinged lever (7) is not used. In the third version of the algorithmic description of task performance, the multiposition switch (9) is not used. In the fourth and final versions, both lever 7 and switch 9 are not used. All selected versions of the task can be described algorithmically. Evaluating the complexity of these versions of the algorithm reflects the complexity of the task in general.

Moreover, it is possible to consider each version of an algorithm in a simple manner. For example, in our task, a four-position hinged lever (7) has four identical directions for movement. Therefore, the direction of lever movement is irrelevant to the algorithmic description of a particular version of the task. In contrast, a rotation angle of multiposition switch 9 clearly influences the time it takes to manipulate it. This is why we specify a certain rotation angle when describing a method of manipulating this control for a particular version of the task.

We selected four versions of task performance that can be described algorithmically and then the time structure of each version of the algorithm is developed. Suppose that the probability of each member of the algorithm is 0.25. The complexity can be calculated in various ways. For example, we can calculate the complexity of each version of an algorithm. Knowing the probability of each version, we can calculate the complexity of the whole task or the following procedure can be utilized.

At the first step, we have to determine the total time of task performance (T). This time can be determined as the mathematical mean of the performance time of each of the four versions of the algorithm, using the following formula:

$$T = 0.25 \times T_1 + 0.25 \times T_2 + 0.25 \times T_3 \times 0.25 \times T_4$$

where
  T_1 through T_4 are the performance times of each version of the algorithm
  0.25 is the probability of occurrence of each version of the algorithm

TABLE 12.11

Members of the Algorithm and Their Performance Time in the First Version of Task (Time in Seconds)

O_1^α and l_1	O_2^ε	O_3^α and L_2	O_4^ε	O_2^α and L_3	O_7^ε	O_8^α and L_4	O_9^ε	O_{10}^α and l_5	O_{12}^ε
0.3	0.62	0.3	0.69	0.3	0.16	0.3	0.70	0.3 (overlapped by motor components)	0.90

Let us calculate the performance time of the first version of the algorithm. All members of the algorithm have a probability of 1 in each version. Therefore, we can determine T_1 as a sum of all listed members of the relevant algorithm. Suppose that only such cognitive members of the algorithm as O_{10}^α and l_5 are overlapped by motor components of activity. The members of the algorithms and their performance time in seconds for this version of the task were extracted from the algorithmic and time structure description of the task and presented as separate tables (see Table 12.11).

Therefore, the total time of task performance for the first version is

$$T_1 = 0.3 + 0.62 + 0.3 + 0.69 + 0.3 + 0.16 + 0.30 + 0.70 + 0.90 = 4.27 \text{ s}$$

The performance time of the three other versions of task is calculated similarly. In our example, this time is $T_2 = 2.87$ s, $T_3 = 3.27$ s, and $T_4 = 1.92$ s. Thus, the mean performance time for the whole task is

$$T = 0.25 \times 4.27 + 0.25 \times 2.87 + 0.25 \times 3.27 + 0.25 \times 1.92 = 3.05 \text{ s}$$

We can also determine the time for receiving information, keeping information in memory, decision making, etc. For example, the total time taken to perform logical conditions, L_g, can be determined as the mathematical mean of the performance time for logical conditions in each of the four versions of the algorithm, using the following formula:

$$L_g = 0.25 \times L_{g1} + 0.25 \times L_{g2} + 0.25 \times L_{g3} + 0.25 \times L_{g4}$$

where

L_g is the total time taken to perform all logical conditions (decision making)
L_{g1}–L_{g4} is the time for logical conditions for the first through fourth versions of the task
0.25 is the probability of each version of the algorithm

Using these data, we can determine L_g, which is computed as follows:

$$L_g = 0.25 \times 0.75 + 0.25 \times 0.45 + 0.25 \times 0.60 + 0.25 \times 0.3 = 0.52 \text{ s}$$

Then we can calculate the fraction of time for decision making. It is $L_g/T = 0.52/3.05 = 0.17$. All other previously considered measures of complexity can be determined as well.

SSAT considers task complexity evaluation as an important stage of the design process. Complexity is a multidimensional characteristic of task and therefore multiple measures of complexity should be utilized in the design process. Such measures provide comparison of physical configuration of the equipment with the structure of activity during task performance. Moreover, task complexity evaluation can be performed at the analytical stage of the design process. Measures of complexity can be compared with each other. They are quantitative indexes that can be used for optimization of the design process. Because an activity is a process, the duration of cognitive and behavioral actions and their operations are used as units of measurement. Not only the duration of elements of activity but also their qualitative characteristics, ability to perform activity elements sequentially and simultaneously, logical organization of activity elements, and the probability of these elements appearing in the structure of activity are taken into consideration. In this chapter, we consider the concept of complexity from morphological analysis perspectives. However, the concept of complexity in unity with the concept of difficulty can be considered as an important mechanism of activity regulation. This is the functional analysis of activity, when activity is considered as a self-regulated system. This aspect of complexity and difficulty is discussed in Bedny et al. (2014).

12.5 Formalized Methods and Complexity Evaluation in the Analysis of Task Safety

Currently, most of the reliability and safety analysis and design are based on cognitive psychology (McCormic and Ilgen, 1985; Reason, 1990; Senders and Moray, 1991). This section presents methods of safety and reliability analysis based on SSAT, an alternative psychological framework to cognitive psychology. Such methods as qualitative, algorithmic, time structure, and complexity analysis incorporating the use of the MTM-1 system to describe motor actions are demonstrated and discussed. These methods, which generate detailed models of human activity during task performance, are particularly useful at the early stages of safety analysis. For ergonomists and safety engineers, evaluating the safety and reliability of a work process is especially challenging in the absence of any real hardware or actual task performance. In such cases, appropriate analytical methods are essential. Later we demonstrate that the SSAT method of morphological analysis of activity that includes an algorithmic description, followed by a time structure analysis and the quantitative evaluation of task complexity, can facilitate the detection of potential danger points in task performance. The value of this approach is that these points are detected not by observation or experiment but by building models of human work activity. This can support a more effective safety analysis and problem-solving in the early stages of the design process.

As an example, we will consider the safety analysis method that is utilized for the analysis of a press operator's transition from leg to two-hand control. Modern mechanical presses used for serial and small-serial production processes typically provide for either two-hand or one-leg control. For example, a press operator in the standing position can start the ram moving down by simultaneously pressing two buttons, one with each hand. Alternatively, the ram can be set in motion by pressing a foot pedal. The operator can switch between these modes of control by manipulating a two-position switch with the right hand. Safety rules determine that the mode of control should be employed: leg control is permissible when the metal sheet feed and removal of blanks and finished parts are mechanized, as such a setup does not require the operator to place his or her hands into the danger zone. An additional safety requirement for the leg-control mode is the use of a guard, which prevents inadvertent entry of the operator's hands into the danger zone. Guard removal is only permitted when the hand-control mode is selected via the two-position switch. Once the press is in the hand-control mode, the movement of the ram is only possible when both left and right hands are pressing their respective buttons, which must be held down until ram movement is complete. Any break in contact with either button immediately interrupts the movement of the ram or slide.

Clearly, this design incorporating mutually exclusive modes of control is well justified from an ergonomic viewpoint. However, closer consideration reveals some drawbacks. The mode of control automatically neither deploys nor withdraws the protective guard; the operator may ignore the safety procedures and remove the protective guard while in leg-control mode, raising the possibility of inadvertently operating the press while their hands are in the danger zone. The risk is heightened by the fact that switching between control modes may not necessarily follow any regular pattern, the operator's choice of control mode being dependent on the technical requirements of the specific work process and their judgment as to the best way to tackle it. This suggests that the press design should be modified so that whenever the operator removes the protection guard, the machine should be automatically switched from the leg-control mode to two-hand control.

At first glance, the foregoing conclusion may seem self-evident and the design flaw easily detected. However, experience shows that such insights are not always nearly so obvious at the equipment design stage; rather, there is often a mismatch between the designer's understanding of how the equipment will be used and what operators will choose to do in practice. One example of such a mismatch is offered by the 1986 disaster at the Chernobyl nuclear plant. Employees at reactor four were highly motivated to test the system within an allotted schedule. This led to the numerous violations of the safety requirements laid down by operating rules and regulations, including the disabling of essential safety systems (UNSCEAR, 2008). Incidents such as this suggest that in order to ensure that safety considerations are fully incorporated at the system design stage, it is vitally important to analyze users' preferred strategies of task performance during the work process, which depend not only on cognitive but also on emotionally evaluative and motivational factors. Let us consider the safety analysis method that is utilized for the analysis of a press operator's transition from leg to two-hand control.

We now present an example of the analytical description of a production operation involving the manual loading of a blank. The first and second stages of task analysis are qualitative and algorithmic safety analysis.

Prior to any design innovation, the performance of this production operation can be described as follows: the press operator works in a standing position; a special table to their left holds uncut metal pieces, while a similar table on their right holds finished pieces. The uncut blanks weigh 10 kg. In order to take a blank from the left table or deposit a finished piece on the right, the operator must make a body rotation of 45°; taking a blank from the left requires a hand movement of 80 cm, while depositing a piece on the right requires a hand movement of 50 cm. The hand movement required to reach the two-position switch (as described earlier) is 30 cm. The calculated distance of the hand movements includes some body motion in the same direction; the effect of the body movement is to diminish the magnitude of the hand movement distance—other distances of movement are considered during a more detailed time structure analysis as described later, which combines SSAT and MTM-1. At this early stage of analysis, a simple narrative description of the work process suffices.

The operator selects two-hand or leg control by turning the two-position switch to the required position with his or her right hand. They then take one blank from the left table with both hands, move it onto the work surface of the press, push it against the stop, and then activate the press by simultaneously pressing the left and right buttons. When the cutting process is complete, the operator releases the buttons, takes the workpiece with both hands, and deposits it on the table on the right. In cases where the operator forgets, or chooses to ignore the safety regulations, it is possible to use the leg control without the protection of the safety guard. Table 12.12 presents an algorithmic description of the production process as outlined earlier.

As can be seen from Table 12.12, the individual members of a human algorithm are designated by special symbols. Each member consists of a set of qualitatively different actions integrated by a common goal, where the possible combinations of actions are constrained by the specificity of their logical organization and the capacity of the operator's short-term memory. The arrows associated with the members of the algorithm in Table 12.12 indicate the transition from one member to another.

Individual members of a human algorithm are designated by special symbols. These are the psychological units of analysis. On the right side of the table, we present technological units of analysis that are simply a verbal description of the members of a human algorithm. An analysis of the algorithm discloses its potential danger points, understood as those cognitive or behavioral actions—or their combination—whose execution could lead to injuries to the operator.

TABLE 12.12

Algorithmic Description of the Production Operation Performed by a Press Operator Involving Transfer from One Mode of Control to Another without Automatic Switching to Guard Protection

Member of the Algorithm	Description of the Algorithm Member
$^{b}O_1^{\mu}$	Recall safety rules or forget/ignore them intentionally.
$l_1^{\mu} \overset{1(1-2)}{\uparrow}$	If safety rules are forgotten or ignored, decide to perform O_3^{ε}; if recalled, decide to perform operator O_2^{ε}.
$^{b}\overset{1(1)}{\downarrow}O_2^{\varepsilon}$	Move the two-position switch to the required position with the right hand (for two-hand control of the press) and remove the protection guard.
$\overset{1(2)}{\downarrow}O_3^{\varepsilon}$	Take a blank from the left table with both hands and put it on the work surface of the press.
O_4^{ε}	Push the blank to the stopper.
$^{c}l_2^{\mu} \overset{2(1-2)}{\uparrow}$	If a safety rule is performed (O_2^{ε} is performed), decide to turn on the press with two-hand control (go to O_5^{ε}). If O_2^{ε} is not performed, decide to use leg control (perform O_6^{ε}) even when protection guard is removed.
$\overset{2(1)}{\downarrow}O_5^{\varepsilon}$	Turn on the press with two-hand control when protection guard is removed and go to ω_1.
$^{a}\omega_1 \overset{\omega}{\uparrow}$	Always-false logical condition (go to $O_6^{\alpha\omega}$).
$_b\overset{2(2)}{\downarrow}O_6^{\varepsilon}$	Turn on the press with leg control even when protection guard is removed; then go to $O_7^{\alpha\omega}$.
$^{b}\overset{\omega}{\downarrow}O_7^{\alpha\omega}$	Wait based on visual control until ram completes its working movement.
O_8^{ε}	Release the two buttons or pedal and move the finished piece to the right-hand position table.

a The *always-false* logical condition is a syntactical device used to indicate the transition from one member of the algorithm to another (go to $O_6^{\alpha\omega}$). It does not represent any actual actions or operations during task performance.
b Symbols in bold designate danger points during the production process.
c Logical condition l_2^{μ} (decision-making) performs checking functions, in the case.

In Table 12.12, such members of the algorithm are designated in bold type and comprise $O_1^{\mu}; O_2^{\varepsilon}; O_6^{\varepsilon}; l_1^{\mu}; l_2^{\mu}; O_7^{\alpha\omega}$. The *active waiting period* of task performance ($O_7^{\alpha\omega}$) is that time during which the operator observes the press in operation—that is, the downward movement of the ram—after having pressed the two-button or leg control. Although the operator does not perform any motor actions during this period, they are required to actively focus their attention on the machine's operation. Risks can emerge during this period, particularly if the operator has ignored the safety instructions and is working in leg/pedal control mode. If this is the case, as distraction can lead to injury, a higher level of focused attention is required, increasing the complexity of the task.

TABLE 12.13

Algorithmic Description of a Production Operation Performed by a Press Operator Involving Transfer from One Mode of Control to Another with Automatic Switching to Guard Protection

Member of the Algorithm	Description of the Algorithm Member
O_1^ε	Take a blank from the left table with both hands and put it on the work surface of the press.
O_2^ε	Push the blank to the stopper.
O_3^μ	Recall safety rules or forget/ignore them intentionally.
$l_1\mu\overset{1}{\uparrow}$	If safety rules are recalled, perform O_4^ε. If safety rules are forgotten or ignored, perform O_6^ε.
$\overset{2}{\downarrow}O_4^\varepsilon$	Move the two-position switch to the required position with the right hand for two-hand control.
O_5^ε	Turn on the press with two-hand control and go to ω_1.
$^a\omega_1\overset{\omega}{\uparrow}$	Always-false logical condition (go to $O_7^{\alpha\omega}$).
$\overset{1}{\downarrow}O_6^\varepsilon$	Turn on the press with leg control even when protection guard is removed and go to l_2.
$l_2\overset{2}{\uparrow}*$	If leg control does not work, then return to O_4^ε.
$\overset{\omega}{\downarrow}O_7^{\alpha\omega}$	Wait based on visual control until ram completes its working movement.
O_8^ε	Release the two buttons or pedal and move the finished piece to the right-hand position table.

[a] ω is always a false logical condition that has only one output. It does not represent any actual actions or operations and is used to indicate the transition from one member of the algorithm to another (go to $O_7^{\alpha\omega}$).

In order to illustrate the design potential of the algorithmic analysis of operator performance strategies under differing conditions, Table 12.13 shows the same production operation after the implementation of a design innovation. In this alternative design, the switch from leg to two-hand control mode is carried out automatically whenever the protective guard is removed. This change means that it is no longer possible to carry out the production operation in violation of safety requirements.

A comparison of the algorithmic descriptions of task performance before (Table 12.12) and after (Table 12.13) the design innovation demonstrates the removal of all dangerous points of the production operation. The decision making associated with l_2^μ in Table 12.1 is eliminated and thus there is no need for later checks. As erroneous operation of the leg control is no longer possible, member $O_7^{\alpha\omega}$ is no longer a potential danger point. In those cases where the operator's task performance strategies are difficult to predict, or where a more in-depth assessment of possible risks and their associated prevention costs is required, further, more detailed stages of SSAT task analysis can be carried out

(Bedny, 2006). In such situation, the time structure analysis is used. It involves a description of the time structure of activity in tabular or graphic form. When the time structure of an activity includes elements that can only be performed sequentially, the tabular form of presentation is usually used. The example discussed here includes relatively simple cognitive actions. The time allotted for the performance of these actions is based on data presented in a standard handbook of engineering psychology (Lomov, 1982; Myasnikov and Petrov, 1976). The performance duration of the basic motions is based on data in the MTM-1 system manual (MTM 1 Analyst Manual, 2001). The MTM-1 system manual contains a full description of the system and comprehensive tables of motion data (the abbreviated descriptions available in general handbooks of time study are inadequate for practical use). The SSAT methodology recommends that MTM-1 analysis should only be performed after carrying out an analysis of the possible strategies of task performance and an algorithmic description of the task. SSAT has developed new principles for the use of MTM in task analysis where there is a complex combination of cognitive and motor components of activity (motor and cognitive actions) and the tasks performed by workers have very flexible strategies. Table 12.14 shows the time structure of activity during performance of the production operation depicted in Table 12.12.

In Table 12.14, members of the algorithm $O_1^\mu; l_1^\mu; l_2^\mu$ include one cognitive action; operator O_3^ε includes two motor actions combined with body movements. For the right hand, the first action is *R80ABA+G1B* and the second is *M80B10/2BA+RL1*. Both hand actions are combined with a body assistance movement *AS30*, where 30 indicates the distance through which the body moves; the operator reaches the metal blank and moves it a distance of 80 + 30 = 110 cm. Each motor action includes motions; the first motor action includes three motions integrated by one action goal: motion *R80ABA* accompanied by body assistance and the grasping of the blank (*G1B—grasp object lying close against a flat surface*). O_6^ε (move leg + press pedal) includes one motor action that is usually performed with the right leg. Thus, it can be seen that SSAT uses MTM-1 differently from other approaches.

The last stage of safety analysis includes quantitative evaluation of task complexity. In quantitative safety analysis, the usual practice is to assess the probability of errors leading to injury; this approach may usually be supplemented with the quantitative methods of task complexity evaluation developed within SSAT. The combination of probability of errors and complexity evaluation methods arguably provides a more comprehensive safety assessment, allowing improved quantification of the degree of danger and thus more accurately targeted risk-reduction measures. Recognizing that nonsubjective methods of obtaining the probabilistic characteristics of human performance are difficult and also tend to be insufficiently accurate, the method presented here combines probabilistic measures derived from subjective expert judgments with objective measurement procedures. It should be noted that what is being assessed using these methods is not the probability of an accident per se but rather the probability of potential danger points in

TABLE 12.14

Time Structure of Production Operation[a]

Members of Algorithm	Description				Mental Components of Activity
	Right Hand	Time (s)	Left Hand	Time (s)	Time (s)
O_1^μ	Extracting information from memory				1.2
I_1^μ	Decision making (yes/no type)				0.3
O_2^ε	R30A + G1A + M2.5 + RL1 + R30E	0.98			
O_3^ε	R80ABA + G1B + M80B10/2BA + RL1 (AS30)	1.86	R50ABA + G1B + M50B10/2 + RL1 (AS30)		
O_4^ε	R5A + G5 + M35A(10/2) × 0.4 + RL2	0.7			
I_2^μ	Decision making (yes/no type)				0.3
O_5^ε	R40A + G5 + AP2	0.79	R40A + G5 + AP2		
ω_1	No time	—		—	
O_6^ε	One step and pressing pedal	1.47			
$O_7^{G\omega}$	Active waiting period	3.0			
O_8^ε	RL2 + R30A + G1B + M80B10/2BA + RL1 (AS30)		RL2 + R30A + G1B + M80B10/2BA + RL1 (AS30)	1.54	

[a] Time analysis for hand actions is performed based on MTM-1 data; other data for time analysis were taken from the *Handbook of Engineering Psychology* (Lomov, 1982).

the work process. Accidents will occur at such danger points only under certain combinations of conditions; quantitative safety measures provide useful tools for identifying and thus better avoiding such conditions.

The first step in the task complexity evaluation is to determine which, if any, elements of the activity under analysis are performed simultaneously; in the task under discussion, all elements are performed sequentially, simplifying the analysis. The next step involves describing the probabilistic characteristics of the task. This allows calculation of the mathematical mean performance time of the various task components, making it possible to calculate the proportion of the overall task performance time associated with potential danger points. In order to do so, we begin by calculating the probability of occurrence of individual members of the algorithm. Initial verbal estimates are translated into numerical values using the data provided by Zarakovsky (1966). A similar approach to obtaining subjective probability measures can be found in Kirwan (1994). In the example under discussion, the accuracy of such estimates can be considered fairly high, as there are only two possible outcomes for each logical condition.

SSAT identifies two types of human algorithm: deterministic and probabilistic. In a deterministic algorithm, the logical conditions have only two possible outputs, each with an equal probability of 0.5. In a probabilistic algorithm, the logical conditions can have two or more outputs, each of which may have a different probability of occurrence. The algorithm in Table 12.12 is thus a probabilistic algorithm. Logical condition l_1^μ has two outputs: the first has a probability of 0.9 and the second a probability of 0.1. Logical condition l_2^μ also has two outputs: the first (go to O_5^ε) has a probability of 0.9 and the second (go to O_5^ε) a probability of 0.1. From this it follows that the probabilities of occurrence of those members of the algorithm of task performance (see Table 12.12) that differ from 1 are the following:

$$P(O_2^\varepsilon) = 0.9; \quad P(O_5^\varepsilon) = 0.9; \quad P(O_6^\varepsilon) = 0.1$$

According to Table 12.14, the performance time of each member of the algorithm can be designated (in seconds) as follows:

$$t(O_1^\mu) = 1.2; \quad t(l_1^\mu) = 0.3; \quad t(O_2^\varepsilon) = 0.98; \quad t(O_3^\varepsilon) = 1.86; \quad t(O_4^\varepsilon) = 0.7$$

$$t(l_2^\mu) = 0.3; \quad t(O_5^\varepsilon) = 0.79; \quad t(O_6^\varepsilon) = 1.47; \quad t(O_7^{aw}) = 3.0; \quad t(O_8^\varepsilon) = 1.54$$

The performance time of the production operation prior to improvement can be determined from the following formula:

$$T = \Sigma P_i t_i$$

where
   P_i is the probability of ith member of the algorithm
   t_i is the time of performing ith member of the algorithm

Then T for performance of the operation can be determined as

$$T = 1 \times t(O_1^\mu) + 1 \times t(l_1^\mu) + 0.9 \times t(O_2^\varepsilon) + 1 \times t(O_3^\varepsilon) + 1 \times t(O_4^\varepsilon) + 1 \times t(l_2^\varepsilon)$$

$$+ 0.9 \times t(O_5^\varepsilon) + 0.1 \times t(O_6^\varepsilon) + t(O_7^{\alpha w}) + t(O_8^\varepsilon)$$

$$= 1.2 + 0.3 + 0.9 \times 0.98 + 1.86 + 0.7 + 0.3 + 0.9 \times 0.79 + 0.1 \times 1.47 + 3.0 + 1.54$$

$$= 10.64 \text{ s}$$

The duration of the executive components of activity T_{ex} (total duration of all efferent operators with symbol O^ε) can be determined as follows:

$$T_{ex} = T = 0.9 \times t(O_2^\varepsilon) + 1 \times t(O_3^\varepsilon) + 1 \times t(O_4^\varepsilon) + 0.9 \times t(O_5^\varepsilon)$$

$$+ 0.1 \times t(O_6^\varepsilon) + 1 \times t(O_8^\varepsilon)$$

$$= 0.9 \times 0.98 + 1.86 + 0.7 + 0.9 \times 0.79 + 0.1 \times 1.47 + 1.54$$

$$= 5.84$$

The time taken for logical conditions, afferent operators, the extraction of information from memory, and the executive (response) components of activity can be determined similarly:

$$L_g = \Sigma P_i^l t_i^l; \quad T_\alpha = \Sigma P_r^\alpha t_r^\alpha; \quad T_\mu = \Sigma P_k^\mu t_k^\mu; \quad T_{ex} = \Sigma P_j^o t_j^o$$

where
  P_i^l, P_r^α, P_j^o, P_k^μ are the probability of ith logical conditions and rth afferent and jth efferent operators
  t_i^l, t_r^α, t_j^o, t_k^μ are the performance time of ith logical conditions and rth afferent and jth efferent operators

The time spent on logical conditions (decision making) associated with potentially dangerous points of the production operation can be calculated:

$$'L_g = 1 \times t(l_1^\mu) + 1 \times t(l_2^\mu) = 0.3 + 0.3 = 0.6 \text{ s}$$

In fact, the faulty execution of logical conditions during task performance may not result in injury. Thus, it is possible to calculate a mathematical mean for the performance time of all logical conditions using a similar formula. On this basis, if necessary, we can determine the fraction of time for the logical (decision-making) components associated with danger, using the following formula:

$$\Omega L_g = \frac{'L_g}{L_g} = 1$$

where
   $'L_g$ is the mathematical mean for performance time of logical conditions associated with danger
   L_g is the mathematical mean for performance time of all logical conditions

Next, we calculate the relationship between the time spent on logical conditions associated with danger points and the time used for the executive components of activity (i.e., the time for efferent operators), or the time for overall task performance:

$$'N_1 = \frac{'L_g}{T_{ex}} \quad \text{or} \quad N_1 = \frac{L_g}{T}$$

where
   $'L_g$ is the time for logical conditions associated with danger
   T_{ex} is the time for response (executive) components of activity
   T is the general time of task performance

We calculate only $'N_1$ as an example. This measure demonstrates the relationship between the logical and executive components of activity, giving the fraction of logical components in task performance associated with potentially dangerous points of the task:

$$'N_1 = \frac{0.6}{5.84} = 0.1$$

When attempting to evaluate the complexity of logical conditions, it is also necessary to distinguish between internally and externally driven decision-making processes. In some situations, the performance of actions and decision-making processes is determined largely by information retrieved from long-term memory. This is a more complex situation than those where the decision-making process is predominantly guided by stimuli or information external to the individual and is therefore particularly undesirable when associated with danger (Bedny, 1987; Konopkin, 1980). The following measure of complexity applies to logical conditions performed on the basis of information extracted from long-term memory:

$$'L_{ltm} = \frac{'l_{ltm}}{L_g}$$

where

$'L_{ltm}$ is the proportion of time for logical components of work activity depending mostly on memory and associated with danger

$'l_{ltm}$ is the mean performance time for logical conditions predominantly governed by memory and associated with danger

In the production operation under consideration, both logical conditions are associated with danger points and both are performed based on information extracted from memory. Hence, l_{ltm} is equal to L_g and therefore $'L_{ltm} = 1$. In our example, $'L_{ltm}$ indicates that the worker must rely on safety rules extracted from long-term memory during the decision-making processes, suggesting a lack of external information or stimuli to provide guidance.

There are no afferent operators (independent perceptual operators) included in the algorithmic description of the production operation. Hence, $T_\alpha = 0$. Instead of T_α, we have T_μ which is equal to 1.2 s (the time taken to actualize or extract the required information from memory). The information to be recalled concerns the rules of task performance, which may or may not be adequate for the specific situation. Such rules are important for safety. Hence,

$$'T_\mu = 1.2 \text{ s}$$

There is only one member of the algorithm that relates to an active waiting period associated with danger:

$$'T^{aw} = 3 \text{ s}$$

It is also possible to calculate the relationship between an active waiting period that can be considered as a potential danger point and overall task performance:

$$'N^{aw} = \frac{'T^{aw}}{T} = \frac{3}{10.64} = 0.28$$

which indicates the fraction of the active waiting period associated with potential danger points.

When the elements of an activity generally follow a habitual sequence, it can be considered as stereotypical. The performance of stereotyped decisions is normally easier. In such situations, decisions always take place in a set order after the performance of particular elements of activity. In contrast, an activity is considered changeable or variable when its elements constantly alter their sequence and some unexpected elements emerge.

The need to perform unexpected decisions, which may appear in the structure of the activity randomly, is more complex than habitual decision. Such decision can cause errors. Potentially dangerous and unexpected decision-making actions are generally undesirable.

Hence, the last measure that we will consider is $'L_{ch}$. This measure demonstrates that logical components (the need to perform decisions) emerge unexpectedly for the worker.

To calculate L_{ch}, we must first determine the time allotted for the variable logical conditions. It is defined by the following formula:

$$'L_{ch} = \Sigma P^l_{chi} t^l_{chi}$$

where

P^l_{chi} is the probability of the appearance of the ith variable logical conditions
t^l_{chi} is the performance time of the ith variable logical conditions

The technological process described involves the variation of machined parts and a random shift from foot to manual control and vice versa. This means that operators have to transition from foot to manual control and vice versa randomly in time, and such requirements for decision emerge unexpectedly for workers. The operator can forget about the need to perform the required decision upon transition from one control regime to the other. This means that the two logical conditions considered in this example and the associated operator's decisions are variable and require constant attention from the operator in order for them to be performed adequately.

Each of the two variable logical conditions has a probability of appearance 1 and time of execution 0.3 sec. The measure of variability of logical conditions is determined according to the following formula:

$$'L_{ch} = \frac{l_{ch}}{L_g}$$

From this it follows that $'L_{ch}$ can be calculated as follows:

$$'L_{ch} = \frac{0.6}{0.6} = 1$$

This indicates that those logical conditions that can be considered as potential danger points are flexible and may occur randomly during transitions from one mode of control to another. Later, we consider another example of the safety analysis.

Control/Ram Movement Relationship and Accident Prevention
Another example is related to work on a different type of press, in this case based on the analysis of a real accident. This example helps us further

demonstrate the systemic-structural activity approach to the analysis of safe work practices and the prevention of accidents using analytical procedures. It can also be considered as an ergonomic analysis of the compatibility principle, which, as the speed of response had no special significance in this case, apparently was not taken into consideration during the design of the control/ram movement relationship despite its clear relevance to safety issues.

In this version of the press design, when the operator moves the control up, the ram moves down, and vice versa. This design caused an accident under emergency conditions. In order to analyze the accident, we use the algorithmic description of the relevant subtask, an analysis of the time structure of activity during subtask performance, and some quantitative measures of task complexity. In contrast to the previous example, where the time structure data were presented in tabular form, in this example, we also use a graphical method for describing the time structure of the considered element of activity. We begin with an algorithmic description of the fragment of the production task associated with the decision to move the ram up or down.

There was emergency situation. The operator incorrectly sets up the blank on the press table and then involuntarily and erroneously moved his right-hand control up while simultaneously attempting to move the blank into the position with his left hand. As a result, the ram moved down causing a hand injury. For the purposes of analysis, it is sufficient to consider only the right-hand movement. An algorithmic description of the subtask *move the control to the required position* is presented in Table 12.15.

The logical conditions have two outputs each with a probability of 0.5. Hence, this is a deterministic algorithm of the performance subtask *move the control*. Let us now consider the time structure of task performance, which is presented in Table 12.16. Extraction of simple information from memory takes 0.3 s, and according to MTM-1, a simple decision requires 0.26 s.

In Table 12.15, O_2^ε and O_3^ε differ only in the direction of movement (up or down). Each member of the algorithm includes two motor actions. The first is *reach the handle of the control and grasp it* ($R40A + G1A$). Its performance time is $R40A + G1A = 0.47$ s. The first motion label $R40A$ means *reach the object in a fixed location*. $G1A$ indicates *easily grasp the handle*. The second motor action is *move the control to the required position and release the handle* ($M20B + RL1$).

TABLE 12.15

Algorithmic Description of the Subtask *Move Control*

Member of the Algorithm	Description of the Algorithm Member
O_1^μ	Extract required information from memory.
$l_1 \overset{1}{\uparrow}$	Decide to perform O_2^ε or O_3^ε.
O_2^ε	Move control down with the right hand.
$\overset{1}{\downarrow} O_3^\varepsilon$	Move control up with the right hand.

TABLE 12.16

Time Structure of Subtask under Normal Conditions

Members of the Algorithm	Description		
	Right Hand	Time (s)	Mental Components Time (s)
O_1^μ	Extraction of information from memory		0.30
l_1	Decision making (yes/no type)		0.26
O_2^ε	R40A + G1A + M20B + RL1	0.92	
O_3^ε	R40A + G1A + M20B + RL1	0.92	

It also consists of two motions: M20B (move the object to the approximate location) and RL1 (release by opening fingers). Here M20B rather than M20A is selected because the direction of this movement is variable. The performance time of the second motor action is 0.45 s. The performance times of two actions for each member of the algorithm are presented in Table 12.16.

This example demonstrates that, in contrast to traditional methods of using MTM-1, the SSAT approach before dividing the task into its constituent motions takes into account the hierarchical organization of activity, showing that the activity is composed of members of the algorithm, each of which represents a cluster of actions combined by a common goal into a subsystem of activity. Here the goal *move the control to the required position* integrates two motor actions, each of which consists of two motions.

Let us now present the time structure of the member of the algorithm O_2^ε or O_3^ε in a graphical form. As both members of the algorithm are identical in their content, we describe only one of them. The time structure of O_2^ε or O_3^ε and the complexity of their elements is shown in Figure 12.2.

The length of the segments in Figure 12.2 indicates the overall duration of the execution of the corresponding elements of activity. Data for the duration of individual movements are drawn from MTM-1, which also provides rules for the combination of motions, both sequentially and simultaneously. SSAT extends the MTM-1 methodology by providing additional rules for combining not only motor but also mental components of activity (Bedny, 1987;

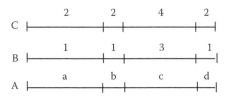

FIGURE 12.2
Complexity evaluation of motor components of activity (O_2^ε or O_3^ε) in normal and stressful situations, where A is the elements of activity and their duration, B the complexity of elements in normal conditions, and C the complexity of elements in stressful conditions. a: *R40A*; b: *G1A*; c: *M20B*; d: *RL1*.

Bedny and Karwowski, 2007). The application of these rules to the algorithmic analysis, which describes the logic of transition from one member of the human algorithm to another, allows the development of a time structure for each separate member of the algorithm and for activity during the performance of the task as a whole. One of the most important requirements when constructing activity time structures in this way is to use psychological rather than technical units of analysis. To illustrate, whereas *move the control into the required position* is considered to be a technological unit of analysis, "R40A + G1A + M20B + RL1" is a psychological unit. The conversion of technological into psychological units of analysis, and their combination as appropriate, allows the very precise description of the structure of activity during task performance.

The lower horizontal line "A" in Figure 12.2 demonstrates the elements of activity and their duration. The upper horizontal line "B" indicates the complexity of the elements, which depends on the level of concentration of attention required during their execution. SSAT orders complexity on a linear scale of 1–5. Elements of activity requiring a minimum level of concentration are related to the first category of complexity; those requiring a medium level of concentration are related to category two; and those elements that require high levels of concentration are related to category three or higher, with five indicating a level at which activity can no longer be performed effectively. According to MTM-1, motor operations (motions) R40A, G1A, and RL1 require a minimum level of concentration of attention; thus, SSAT assigns them to the first category of complexity. Moving up or down requires a choice of direction and therefore is represented in MTM-1 by "MA"; as this movement requires an average level of attention, SSAT initially assigns it to the second category of complexity. However, SSAT rules state that the presence of contradictions between the movement and its effects indicates an additional level of complexity; for example, when the operator moves the control up, the ram is moving down. With this in mind, the movement of the control in a given direction (M20B) is finally assigned to the third category of complexity (Figure 12.2, line B).

In the manufacturing operation illustrated in Figure 12.2, an emergency situation occurs when the operator erroneously presses the pedal in the absence of a protective guard while his or her hands are in the danger zone due to an incorrect positioning of the blank. This emergency situation triggers the emergence of emotional stress on the operator, which leads to an additional mobilization of physical and mental effort and thus to an increase in the operator's level of concentration of attention. According to SSAT, this means that the complexity of elements of activity should be increased by 1–2 levels, depending on an assessment of the level of emotional stress involved. In the situation under consideration, we have chosen to increase the difficulty level of each element of motor activity by one category, as shown in Figure 12.2 (see line C). This high level of complexity indicates that this element should be considered as a potentially dangerous point of the manufacturing operation that might lead to injury. Thus, it can be seen that knowing the probability of occurrence of one or other elements of activity and their duration, as well as the mathematical means of the production operation or task performance, allows a quantitative evaluation of its potential danger points.

The presented material demonstrates that complexity measures can be used not only for discovering and assessment of potentially dangerous points of tasks but also for making recommendations for their reduction. Prediction of human error and the probability of their occurrence are very difficult at the design stage, when real equipment does not exist yet. That is why the development of analytical methods for the analysis of human erroneous actions and their related possible accidents is an important aspect of safety analysis. Material presented in this section demonstrates a new method of safety analysis, which can be used at the analytical stage of equipment development. Quantitative analysis of task complexity is particularly important in such analysis. SSAT developed a unified and standardized approach to the analysis of task performance. At the stage of formalized and quantitative task assessment, the major units of analysis are cognitive and motor actions. Motor actions can be described in terms of standardized motions. SSAT developed new principles of utilizing the MTM-1 system for the analysis of flexible tasks where cognition and motor components of activity can be combined in various manners.

The studies described here demonstrate that a qualitative analysis of activity, including analysis of possible strategies of task performance, an algorithmic description of activity, followed by a time structure analysis and the quantitative evaluation of task complexity, allows the detection of potential danger points in production tasks and operations' performance. The value of this approach is that these points are detected not by observation or experiment but by building models of human work activity. This allows a more effective safety analysis and problem-solving at the early design stage.

Conclusion

In this book, we have presented the systemic-structural activity theory (SSAT), which provides a unified framework for the study of human performance and for work analysis and design in ergonomics (human factors) and work psychology. The main purpose of SSAT is the optimization of human performance and increase of productivity.

Work productivity is an interdisciplinary field studied by labor economists, industrial engineers, ergonomists, and work psychologists. It is a well-known fact established in various fields of science that systemic principles are critically important in the study of human work. However, current ergonomic and psychological approaches are not sufficiently advanced to study human work from the systemic principle perspectives. For example, the behavioral approach considers external human behavior as an aggregation of independent reactions, while the cognitive approach considers cognition as an information processing system that is relatively independent from observable behavior. Further, there is also a separation of cognition and motivation. Such separation of these basic components of human behavior is not productive when studying human work, but cognitive psychology does not facilitate integration of these components.

Integration of cognition, behavior, and motivation in psychology and ergonomics requires a totally different approach that can provide a theoretical foundation for such integration. Such integration is possible only within the framework of activity theory where the principle of unity of cognition and behavior has been introduced. Application of this idea to the analysis of human work is demonstrated in SSAT studies. According to SSAT, the concept of self-regulation is important for the understanding of this relationship. Human motor actions cannot be considered as purely external and not containing cognitive mechanisms. The data obtained in activity theory and in cognitive psychology show that motor components of activity always include cognitive components.

Only an understanding of the relationship between external motor and internal cognitive actions can provide the bases for integration of cognition and behavior. However, the concepts of cognitive and motor actions do not carry rigorous scientific meaning in cognitive psychology. As we have demonstrated in this book, the term action is understood differently by different scientists. SSAT describes cognitive and behavioral actions in a standardized manner and shows that cognitive and behavioral actions are interdependent. These actions are basic elements of human activity. Manual tasks that require high-precision motor actions and their coordination can be very complex; therefore, such behavioral actions require mental efforts for their regulation.

We have shown in this book that manual tasks and tasks with dominantly cognitive components should be studied from a unitary theoretical position.

SSAT introduces various units of analysis and measurement of work activity. The book conceptualizes an application of the technological and psychological unit of analysis. Described in a standardized manner, cognitive and motor actions and operations are used as main psychological units of analysis. Unified methods of activity analysis and their organization into stages and levels of analysis are presented. Morphological SSAT analysis describes the activity structure including algorithmic description of activity and development of the activity time structure.

In contemporary industries, work activity is rather flexible and workers utilize various strategies to perform the same task.

Two methods of task analysis have been developed to analyze flexible human work. One method is qualitative and it describes activity strategies from the perspectives of activity self-regulation. The other method is formalized algorithmic task description of activity during task performance. SSAT offers two types of human algorithms: deterministic and probabilistic. Utilizing deterministic and especially probabilistic human algorithm, one can describe flexible human activity and its mental strategies. It is important to distinguish human algorithm from mathematical or computer algorithms because members of human algorithms consist of cognitive and behavioral actions. Activity is a process and therefore time study is an important aspect of task analysis. The time structure of activity is a new concept in task analysis that has been introduced in SSAT. It demonstrates how cognitive and behavioral elements of activity unfold in time. Comparison of the time structure of activity with configuration of equipment is the critical step of ergonomic design because configuration of equipment determines the structure of activity.

New basic methods of quantitative analysis of human performance have been developed in SSAT. One of the most important quantitative methods of task analysis is the method of task complexity assessment. Complexity is described as a multidimensional phenomenon, which requires multiple measures for its quantitative evaluation. It has been demonstrated that measuring task complexity is useful for a variety of purposes: development of efficient methods of task performance, job evaluation, time study, analysis of productivity, reliability of performance, and optimization of design solution.

Currently, SSAT can be considered as an alternative approach to cognitive psychology in studying human work. It suggests unified and standardized principles to work analysis and design. Cognitive psychology utilizes experiment as the basic method of study and is unable to develop analytical procedures to task analysis and design. At the same time, the main principle of design is developing analytical methods of analysis where an experiment is a supplementary method of study. Cognitive psychology considers cognition as a process and ignores the fact that activity is a complex structure, which makes it impossible to develop human activity models that are necessary for ergonomic design. Their shortcomings make it difficult to use cognitive psychology in applied studies. At the same time, SSAT has utilized and adapted for applied purposes all that is positive in studies of human cognitive psychology. Cognitive psychology data have been utilized when studying separate cognitive processes in task analysis. Another advantage of SSAT is that it studies cognition, behavior, and motivation in unity. For example, the selection and interpretation of information are facilitated not just by cognitive but also by emotionally motivational mechanisms. These mechanisms work in unity. SSAT covers a broad range of issues. It suggests a new understanding of learning and training in industrial settings and provides new findings in analysis of individual style of human performance, in analysis of human safety, etc. Suggested SSAT methods can be applied to traditional and computerized types of human activity.

Bibliography

Aggleton, J. P. and Young, A. W. (2000). The enigma of the amygdale: On its contribution to human emotion. In R. D. Lane and L. Nadel (Eds.), *Cognitive Neuroscience of Emotion*. New York: Oxford University Press, pp. 106–128.

Aladjanova, N. A., Slotintseva, T. V., and Khomskaya, E. D. (1979). Relationship between voluntary attention and evoked potentials of brain. In E. D. Khomskaya (Ed.), *Neuropsychological Mechanisms of Attention*. Moscow, Russia: Science, pp. 168–173.

Al'bukhanova, K. A. (1973). *On the Subject of Mental Activity*. Moscow, Russia: Science Publisher.

Anderson, J. R. (1985). *Cognitive Psychology and its Implications* (2nd edn.). New York: W. H. Freeman.

Andrianov, V. and Dubrovickij, A. (1971). Self-training of pilots. *Aviation and Astronautics*, 1, 13–16.

Angel, R. W. (1976). Efference copy in the control of movement. *Neurology*, 26 (2), 422–424.

Annet, J. (1961). The role of knowledge of result in learning: A survey. Report No. 342-3. New York: US Naval Training Devices Center.

Annet, J. (2000). Theoretical and pragmatic influences on task analysis methods. In J. M. Schraagen, S. F. Chipman, and V. L. Shalin (Eds.), *Cognitive Task Analysis*. Mahwah, NJ: Lawrence Erlbaum Associates, Publishers, pp. 25–37.

Annet, J. and Cunningham, D. (2000). Analysis command team skills. In J. M. Schraagen, S. F. Chipman, and V. L. Shalin (Eds.), *Cognitive Task Analysis*. Mahwah, NJ: Lawrence Erlbaum Associates, Publishers, pp. 401–416.

Annet, J. and Duncan, K. D. (1967). Task analysis and training design. *Occupational Psychology*, 41, 211–221.

Anokhin, P. K. (1962a). *The Theory of Functional Systems as a Prerequisite for the Construction of Physiological Cybernetics*. Moscow, Russia: Academy Science of the USSR.

Anokhin, P. K. (1962b). Anticipatory reflection of reality. *Questions of Philosophy*, 7, 97–111.

Anokhin, P. K. (1969). Cybernetic and the integrative activity of the brain. In M. Cole and I. Maltzman (Eds.), *A Handbook of Contemporary Soviet Psychology*. New York: Basic Books, Inc., Publishers, pp. 830–857.

Arend, I., Colom, R., Botella, J., Contreras, M. J., Rubio, V., and Santacreu, J. (2003). Quantifying cognitive complexity: Evidence from a reasoning task. *Personality and Individual Differences*, 35, 659–669.

Armstrong, M., Cummins, A., and Hastings, S. (2003). *Job Evaluation: A Guide to Achieving Equal Pay*. London, U.K.: Kogan Page.

Arnaut, E. J., Gordon, G. M., Joines, D. H., and Philips, G. M. (2001). An experimental study of job evaluation and comparable worth. *Industrial and Labor Relations Review*, 54 (4), 806–815.

Atkinson, R. C. and Shiffrin, R. M. (1968). Human memory: A proposed system and its control processes. In K. W. Spence and J. T. Spence (Eds.), *The Psychology of Learning and Motivation: Advances in Research and Theory*, Vol. 2. New York: Academic Press.

Atkinson, R. C. and Shiffrin, R. M. (1971). The control of short-term memory. *Scientific American*, 225 (2), 82–90.

Austin, J. T. and Vancouver, J. B. (1996). Goal constructs in psychology: Structure, process, and content. *Psychological Bulletin*, 120 (3), 338–375.

Bakhtin, M. M. (1979). *Aesthetics of Verbal Creativity*. Moscow, Russia: Art Publisher.

Bandura, A. (1989). Self-regulation of motivation and action through internal standards and goal systems. In L. A. Pervin (Ed.), *Goal Concept in Personality and Social Psychology*. Hillsdale, NJ: Lawrence Erlbaum Associates, pp. 19–86.

Bandura, A. (1997). *Self-Efficacy: The Exercise of Control*. New York: W. H. Freeman.

Bandura, A. (1977). *Social Learning Theory*. Englewood Cliffs, NJ: Prentice-Hall.

Banich, M. T. (2004). *Cognitive Neuroscience and Neuropsychology*. Boston, MA: Houghton Mifflin Company.

Barkley Rosser, J. Jr. (2010). Introduction to special issue on transdisciplinary perspectives on economic complexity. *Journal of Economic Behavior and Organization* (Special Issue), 75 (1), 3–11.

Barness, P. M. (1980). *Motion and Time Study Design and Measurement of Work*. New York: John Wiley & Sons.

Bedny, G. and Karwowski, W. (2006). General and systemic-structural activity theory. In W. Karwowski (Ed.), *International Encyclopedia of Ergonomics and Human Factors*, Vol. 3. London, U.K.: Taylor & Francis Ltd., pp. 3159–3167.

Bedny, G., Karwowski, W., and Seglin, M. (2001a). A hart evaluation approach to determine cost-effectiveness an ergonomics intervention. *International Journal of Occupational Safety*, 7 (2), 121–133.

Bedny, G. Z. and Bedny, M. G. (2001). The principle of unity cognition and behavior: Implication of activity theory for the study of human work. *International Journal of Cognitive Ergonomics*, 401–420.

Bedny, G., Karwowski, W., and Sengupta, T. (2006). Application of systemic-structural activity theory to design of human-computer interaction tasks. In W. Karwowski (Ed.), *International Encyclopedia of Ergonomics and Human Factor*, Vol. 1. Boca Raton, FL: CRC/Taylor & Francis, pp. 1272–1286.

Bedny, G., Karwowski, W., and Young-Guk, K. (2001b). A methodology for systemic-structural analysis and design of manual-based manufacturing operations. *Human Factors and Ergonomics in Manufacturing*, 11 (3), 233–253.

Bedny, G. and Meister, D. (1997). *The Russian Theory of Activity: Current Application to Design and Learning*. Mahwah, NJ: Lawrence Erlbaum Associates.

Bedny, G., Polyakov, L., and Stolberg, V. (1996). Russian psychological concept of work activity. *National Social Science Journal*, 8 (2), 145–158.

Bedny, G. and Seglin, M. (1997). The use of pulse rate to evaluate physical work load in Russian ergonomics. *American Industrial Hygiene Association Journal*, 58, 335–379.

Bedny, G. and Seglin, M. (1999a). Individual style of activity and adaptation to standard performance requirement. *Human Performance*, 12 (1), 59–78.

Bedny, G. and Seglin, M. (1999b). Individual features of personality in former Soviet Union. *Journal of Research in Personality*, 33 (4), 546–563.

Bedny, G., Seglin, M., and Meister, D. (2000). Activity theory. History, research and application. *Theoretical Issues in Ergonomics Science*, 1 (2), 165–206.

Bedny, G. Z. (1979). *Psychophysiological Aspects of a Time Study*. Moscow, Russia: Economics Publishers.

Bedny, G. Z. (1981). *The Psychological Aspects of a Timed Study during Vocational Training*. Moscow, Russia: Higher Education Publisher.

Bedny, G. Z. (1987). *The Psychological Foundations of Analyzing and Designing Work Processes*. Kiev, Ukraine: Higher Education Publishers.

Bedny, G. Z. (invited Ed.). (2004). *Preface, Theoretical Issues in Ergonomics Science* (Special Issue), 5 (4), 249–254.

Bedny, G. Z. (2006). Activity theory. In W. Karwowski (Ed.), *International Encyclopedia of Ergonomics and Human Factor*, Vol. 1. Boca Raton, FL: CRC/Taylor & Francis, pp. 571–576.

Bedny, G. Z. and Chebykin, O. Ya. (2013). Application the basic terminology in activity theory. *IIE Transactions on Occupational Ergonomics and Human Factors*, 1 (1), 82–92.

Bedny, G. Z. and Harris, S. (2005). The systemic-structural activity theory: Application to the study human work. *Mind, Culture, and Activity: An International Journal*, 12 (2), 128–147.

Bedny, G. Z. and Harris S. R. (2008). "Working sphere/engagement" and the concept of task in activity theory. Interacting with computers. *The Interdisciplinary Journal of HCI*, 20 (2), 251–255.

Bedny, G. Z. and Harris, S. (2013). Safety and reliability analysis methods based on systemic-structural activity theory. *Journal of Risk and Reliability*, 227 (5), 549–556, Sage Publisher.

Bedny, G. Z. and Karwowski, W. (2003). A systemic-structural activity approach to the design of human-computer interaction tasks. *International Journal of Human-Computer Interaction*, 2, 235–260.

Bedny, G. Z. and Karwowski, W. (2004a). A functional model of human orienting activity. In G. Z. Bedny (invited editor). *Theoretical Issues in Ergonomics Science* (Special Issue), 5 (4), 255–274.

Bedny, G. Z. and Karwowski, W. (2004b). Activity theory as a basis for the study of work. *Ergonomics*, 47 (2), 134–153.

Bedny, G. Z. and Karwowski, W. (2004c). Meaning and sense in activity theory and their role in study of human performance. *Ergonomia*, 26 (2), 121–140.

Bedny, G. Z. and Karwowski, W. (2006). The self-regulation concept of motivation at work. *Theoretical Issues in Ergonomics Science*, 7 (4), 413–436.

Bedny, G. Z. and Karwowski, W. (2008a). Application of systemic-structural theory of activity to design and management of work systems. In W. W. Gasparski and T. Airaksinen (Eds.), *Praxiology and the Philosophy of Technology: The International Annual of Practical Philosophy and Methodology*, Vol. 15. London, U.K.: Transaction Publishers, pp. 97–144.

Bedny, G. Z. and Karwowski, W. (2008b). Activity theory: Comparative analysis of eastern and western approaches. In O. Y. Chebykin, W. Karwowski, G. Z. Bedny, and W. Karwowski (Eds.), *Ergonomics and Psychology: Development in Theory and Practice*. London, U.K.: Taylor & Francis, pp. 221–246.

Bedny, G. Z. and Karwowski, W. (2008c). Time study during vocational training. In O. Y. Chebykin, G. Z. Bedny, and W. Karwowski (Eds.), *Ergonomics and Psychology: Development in Theory and Practice*. London, U.K.: Taylor & Francis, pp. 41–70.

Bedny, G. Z. and Karwowski, W. (2011a). Introduction to applied and systemic-structural activity theory. In G. Z. Bedny and W. Karwowski (Eds.), *Human-Computer Interaction and Operators' Performance: Optimization of Work Design with Activity Theory*. Boca Raton, FL: CRC Press/Taylor & Francis Group, pp. 3–30.

Bedny, G. Z. and Karwowski, W. (2011b). Task concept in production and nonproduction environments. In G. Z. Bedny and W. Karwowski (Eds.), *Human-Computer Interaction and Operators' Performance: Optimization of Work Design with Activity Theory*. Boca Raton, FL: Taylor & Francis/CRC Press, pp. 89–116.

Bedny, G. Z., Karwowski, W., and Bedny, I. S. (2011). The concept of task for non-production human-computer interaction environment. In D. B. Kaber and G. Boy (Eds.), *Advances in Cognitive Ergonomics*. Boca Raton, FL: CRC Press/Taylor & Francis Group, pp. 663–672.

Bedny, G. Z., Karwowski, W., and Bedny, I. S. (2014). *Applying Systemic-Structural Activity Theory to Design of Human-Computer Interaction Systems*. Boca Raton, FL: CRC/Taylor & Francis.

Bedny, G. Z., Karwowski, W., and Jeng, O.-J. (2004). The situation reflection of reality in activity theory and the concept of situation awareness in cognitive psychology. In G. Z. Bedny (invited editor). *Theoretical Issues in Ergonomics Science* (Special Issue), 5 (4), 275–296.

Bedny, G. Z., Karwowski, W., and Sengupta, T. (2008). Application of systemic-structural theory of activity in the development of predictive models of user performance. *International Journal of Human-Computer Interaction*, 24 (3), 239–274.

Bedny, G. Z., Karwowski, W., and Voskoboynikov, F. (2011). The relationship between external and internal aspects in activity theory and its importance in the study of human work. In G. Z. Bedny and W. Karwowski (Eds.), *Human-Computer Interaction and Operators' Performance: Optimization of Work Design with Activity Theory*. Boca Raton, FL: Taylor & Francis/CRC Press, pp. 31–62.

Bedny, G. Z., von Breven, H., and Syntsiya, K. (2012). Learning and training: Activity approach. In N. M. Seel (Ed.), *Encyclopedia of the Science of Learning*. Boston, MA: Springer, pp. 1800–1805.

Bedny, G. Z. and Zelenin, M.P. (1988). *Ergonomic Analysis of Work Safety and the Problem of Safety in Merchant Marine Transportation*, Moscow: Merchant Marine Publisher.

Bedny, G. Z. and Zelenin, M. P. (1989). *Psychological Aspects of Workers' Training for Repair of Ships*. Odessa, Ukraine: Institute of Advance Education for Specialists of Ministry of the Merchant Marine of the USSR.

Bedny, G. Z., Karwowski, W. (2001). Activity theory as a framework of study of human-computer interaction, pp. 342–346, *HCI International Conference*. New Orleans, LA.

Bedny, I. S. (2004). General characteristics of human reliability in system of human and computer. *Science and Education*, 8–9, 58–61, Odessa, Ukraine.

Bedny, I. S. (2006). On systemic-structural analysis of reliability of computer based tasks. *Science and Education*, 1–2 (7–8), 58–60, Odessa, Ukraine.

Bedny, I. S. and Bedny, G. Z. (2011a). Abandoned actions reveal design flaws: An illustration by a web-survey task. In G. Z. Bedny and W. Karwowski (Eds.), *Human-Computer Interaction and Operators' Performance: Optimization of Work Design with Activity Theory*. Boca Raton, FL: Taylor & Francis/CRC Press, pp. 149–185.

Bedny, I. S. and Bedny, G. (2011b). Analysis of abandoned actions in the email distributed tasks performance. In D. B. Kaber and G. Boy (Eds.), *Advances in Cognitive Ergonomics*. Boca Raton, FL: CRC Press/Taylor & Francis Group, pp. 683–692.

Bedny, I. S., Karwowski, W., and Bedny, G. Z. (2010). A method of human reliability assessment based on systemic-structural activity theory. *International Journal of Human-Computer Interaction*, 26 (4), 377–402, CRC Press/Taylor & Francis Group.

Bedny, I. S., Karwowski, W., and Bedny, G. (2012). Computer technology at the workplace and errors analysis. In K. M. Stanney and K. S. Hale (Eds.), *Advances in Cognitive Engineering and Neuroergonomics*. Boca Raton, FL: CRC Press/Taylor & Francis Group, pp. 167–176.

Bedny, I. S. and Sengupta, T. (2005). The study of computer based tasks. *Science and Education*, 1–2 (7–8), 82–84, Odessa, Ukraine.

Bellamy, R. K. E. (1997). Design of educational technology: Computer-mediated change. In B. A. Nardi (Ed.), *Context and Consciousness: Activity Theory and Human-Computer Interaction*. Cambridge, MA: The MIT Press, pp. 123–146.

Bernshtein, N. A. (1966). *The Physiology of Movement and Activity*. Moscow, Russia: Medical Publishers.

Bernshtein, N. A. (1967). *The Coordination and Regulation of Movements*. Oxford, U.K.: Pergamon Press.

Bernshtein, N. A. (1996). On dexterity and its development. In M. L. Latash and M. T. Turvey (Eds.), *Dexterity and Its Development*. Mahwah, NJ: Lawrence Erlbaum Associates, pp. 1–244.

Bloch, V. (1966). Level of wakefulness and attention. In P. Fraisse and J. Piaget (Eds.), *Experimental Psychology*, Vol. 3. Paris, France: University Press of France, pp. 97–146.

Bohlen, G. A. and Barany, J. W. (1966). A learning curve prediction model for operators performing industrial bench assembly operations. *International Journal of Production Research*, 14, 295–302.

Boisard, P., Cartron, D., Gollac, M., Cartron, D., and Valeyre, A. (2003). Time and work: Work intensity. European Foundation for the Improvement of Living and Work Conditions, site web: eurofound.eu.int.

Braha, D. and Maimon, O. (1998). The measurement of a design structural and functional complexity. *IEEE Transaction on Systems, Man, and Cybernetics-Part A: Systems and Humans*, 28 (4), 527–535.

Brecher, C., Muller, S., and Herfs, W. (2013). Design and implementation of a comprehensible cognitive assembly system. In S. Trzcienlinski and W. Karwowski (Eds.), *Advances in Ergonomics in Manufacturing*. Boca Raton, FL: CRC Press/Taylor & Francis, pp. 272–281.

Bruner, J. S. (1964). The course of cognitive growth. *American Psychologists*, 19, 1–15.

Bush, G., Luu, P., and Posner, M. I. (2000). Cognitive and emotional influences in anterior cingulated cortex. *Trends in Cognitive Sciences*, 4, 215–222.

Carlson, N. R. (2004). *Physiology of Behavior*, 8th edn. Boston, MA: Alyn and Bacon.

Carrol, J. M. (Ed.) (1987). *Interfacing Thought*. Cambridge, MA: MIT Press.

Carroll, J. B. (1963). A model of school learning. *Teachers College Record*, 64, 723–733.

Carver, C. S. and Scheier, M. F. (1998). *On the Self-Regulation of Behavior*. New York: Cambridge University Press.

Carver, C. S. and Scheier, M. F. (2005). On the structure of behavioral self-regulation. In M. Boekaerts, P. R. Pintrich, and M. Zeidner (Eds.), *Handbook of Self-Regulation*. San Diego, CA: Academic Press, pp. 42–84.

Chebisheva, V. V. (1969). *The Psychology of Vocational Training*. Moscow, Russia: Education Publishers.

Chkhaidze, L. V. (1970). *Control of Movements*. Moscow, Russia: Physical Culture and Sport.

Cicerone, K., Lazar, R., and Shapiro, W. (1983). Effect of frontal lobe lesions on hypothesis sampling during concept formation. *Neuropsychologia*, 21, 513–524.

Clark, A. (1999a). An embodied cognitive science? *Trends in Cognitive Science*, 3 (9), 345–351.

Clark, A. (1999b). Embodied, situated, and distributed cognition. In W. Betchel and G. Graham (Eds.), *A Companion to Cognitive Science*. Malden, MA: Blackwell Publishing.

Cole, M. and Maltzman, I. (Eds.). (1969). *A Handbook of Contemporary Soviet Psychology*. New York: Basic Books, Inc., Publishers.

Delis, D. C., Squire, L. R., Bihrle, A., and Massman, P. (1992). Componential analysis of problem-solving ability: Performance of patients with frontal lobe damage and amnesic on a new sorting test. *Neuropsychologia*, 30, 683–697.

Daniels, H. (2008). Discourse in activity. In O. Chebykin, G. Bedny, W. Karwowski (Eds.). *Ergonomics and Psychology. Developments in Theory and Practice*, pp. 247–266.

Diaper, D. (2004). Understanding task analysis for human-computer interaction. In D. Diaper and N. Stanton (Eds.), *The Handbook of Task Analysis for Human-Computer Interaction*. Mahwah, NJ: Lawrence Erlbaum Associates, Publishers, pp. 5–47.

Diaper, D. and Lindgaard, G. (2008). West meets East: Adapting activity theory for HCI & CSCW applications? *Interacting with Computers, The Interdisciplinary Journal of Human-Computer Interaction*, 20 (2), 240–286.

Diaper, D. and Stanton, N. A. (2004). Wishing on a sTAr: The future of the task analysis. In D. Diaper and N. Stanton (Eds.), *The Handbook of Task Analysis for Human-Computer Interaction*. Mahwah, NJ: Lawrence Erlbaum Associates, Publishers, pp. 585–602.

Dimerskij, V. Ya. (1965). On applying imaginative actions during training. *Questions of Psychology*, 6, 114–119.

Ditrikh, Y. (1981). *Engineering and Design*. Moscow, Russia: World Publisher.

Drury, C. G. (1995). Method of direct observation of performance. In J. R. Wilson and E. N. Corlett (Eds.), *Evaluation of Human Work: A Practical Ergonomics Methodology*. London, U.K.: Taylor & Francis, pp. 45–68.

Dobrinin, N. F. (1958). Voluntary and involuntary attention. In N. F. Dobrinin (Ed.), *Scientificworks of the Moscow State Pedagogical University*, Vol. 8. Moscow, Russia: Pedagogical Publishers, pp. 34–52.

Donders, F. C. (1969). On the speed of mental processes (trans. W. G. Koster). *Acta Psychologica*, 30, 412–431.

Du, D.-Z. and Ko, K.-I. (2000). *Theory of Computational Complexity*. New York: John Wiley & Sons.

Eisenstadt, S. A. and Simon, H. A. (1997). Logic and thought. *Minds Mach*, 7, 365–385.

Endmonds, B. (1999). What is complexity? The philosophy of complexity per se with application to some examples in evolution. In F. Heylighen and D. Aerts (Eds.), *The Evolution of Complexity*. Dordrecht, the Netherlands: Kluwer, pp. 1–18.

Endsley, M. R. (2000). Theoretical underpinnings of situation awareness: A critical review. In M. R. Endsley and D. J. M. Garland (Eds.), *Situation Awareness Analysis and Measurement*. Mahwah, NJ: Lawrence Erlbaum Associates, Inc., pp. 3–32.

Engeström, Y. (1999). Activity theory and individual and social transformation. In Y. Engeström, R. Miettinen, and R.-L. Punamaki (Eds.), *Perspectives on Activity Theory*. Cambridge, U.K.: Cambridge University Press, pp. 19–38.

Engeström, Y. (2000). Activity theory as a framework for analyzing and redesign work. *Ergonomics*, 43 (7), 960–974.

Fleishman, E. A. and Parker, J. F. (1962). Factors in the retention and relearning of perceptual-motor skill. *Journal of Experimental Psychology*, 64 (3), 215–226.

Fitts, P. M. (1954). The information capacity of the human motor system in controlling the amplitude of movement. *Journal of Experimental Psychology*, 47, 381–391.

Franken, R. E. (2002). *Human Motivation*, 5th edn. Belmont, CA: Wadsworth.

Frese, M. and Zapf, D. (1994). Action as a core of work psychology: A German approach. In H. C. Triadis, M. D. Dunnette, and L. M. Hough (Eds.), *Handbook of Industrial and Organizational Psychology*. Polo Alto, CA: Consulting Psychologists Press, pp. 271–340.

Freud, S. (1916/1917). *A General Introduction to Psychoanalysis*. New York: Pocket Books.

Frolov, N. I. (1976). *Physiology of Movement*. Leningrad, Russia: Science Publishers.

Fuster, J. M. (1985). The prefrontal cortex, mediator of cross-temporal contingences. *Human Neurobiology*, 4, 169–179.

Fuster, J. M. (2002). Physiology of executive functions: The perception-action cycle. In D. T. Stuss and R. T. Knight (Eds.), *Principles of Frontal Lobe Function*. Oxford, U.K.: Oxford University Press, pp. 96–108.

Galactionov, A. I. (1978). *The Fundamentals of Engineering-Psychological Design of Automatic Control Systems of Technological Processes*. Moscow, Russia: Energy Publishers.

Gal'perin, P. Y. (1969). Stages in the development of mental acts. In M. Cole and I. Maltzman (Eds.), *A Handbook of Contemporary Soviet Psychology*. New York: Basic Books, Inc., pp. 249–276.

Gal'sev, A. D. (1973). *Time Study and Scientific Management of Work in Manufacturing*. Moscow, Russia: Manufacturing Publishers.

Galwey, T. J. and Drury, G. G. (1986). Task complexity in visual inspection. *Human Factors*, 28, 95–606.

Gaspaski, W. (1984). *Understanding Design: The Praxiological-Systemic Perspectives*. Seaside, CA: Intersystem Publications.

Gilbreth, F. V. (1911). *Motion Study*. Princeton, NJ: D. van Nostrand Company.

Gil'bukh, U. Z. (1979). *Simulator Devices in Vocational Training*. Kiev, Ukraine: Higher Educational Publisher.

Goldman-Rakic, P. S. and Raric, P. (1984). Experimental modification of gyral patters. In N. S. Gechwind and A. M. Galaburda (Eds.), *Cerebral Dominance: The Biological Foundations*. Cambridge, MA: Harvard University Press.

Gordeeva, N. D. and Zinchenko, V. P. (1982). *Functional Structure of Action*. Moscow, Russia: Moscow University Publishers.

Gray, J. R., Braver, T. S., and Raichle, M. E. (2002). Integration of emotion and cognition in the lateral prefrontal cortex. *Proceedings of the National Academy of Sciences of the United States of America*, 99, 4115–4120.

Green, D. M. and Swets, J. A. (1966). *Signal Detection Theory and Psychophysics*. New York: Wiley.

Grimak, L. P. (1987). *Reserves of the Human Psyche*. Moscow, Russia: Medical Publishers.

Hacker, W. (1986). *Work Psychology*. Bern, Switzerland: Huber.

Hancock, P. A. and Caird, J. K. (1993). Experimental evaluation of a model of mental workload. *Human Factors*, 35 (3), 413–429.

Heath, R. G. (1972). Pleasure and brain activity in man. *Journal of Nervous and Mental Disease*, 154, 3–18.

Heckhausen, H. (1991). *Motivation and Action*. Berlin, Germany: Spring-Verlag.

Heckhausen, H. and Gollwitzer, P. M. (1987). Thought contents and cognitive functioning in motivational versus volitional states of mind. *Motivation and Emotion*, 11, 101–120.

Helander, M. G. (2001). Theories and methods in affective human factors design. In M. J. Smith, G. Salvendy, D. Harris, and R. J. Koubeck (Eds.), *Usability Evaluation and Interface Design. Proceedings of HCI 2001*, Vol. 1. Mahwah, NJ: Lawrence Erlbaum Associates, pp. 357–361.

Henderson, J. M. (1993). Visual attention and saccadic eye movements. In G. D'Ydewalle and J. Van Rensberggen (Eds.), *Perception and Cognition: Advances in Eye Movement Research*. North-Holland, the Netherlands: Elsevier Science Publishers, pp. 37–50.

Herslund, M.-B. and Jorgensen, N. O. (2003). Look-but-failed to-see errors in traffic. *Accident Analysis and Prevention*, 35, 855–891.

Hick, W. E. (1952). On the role of gain of information. *Quarterly Journal Experimental Psychology*, 4, 11–26.

Hilgard, R., Atkinson R. L., and Atkinson, R. C. (1979). *Introduction to Psychology*. New York: Harcourt, Brace, Jovanovich.

Huchingson, R. D. (1981). *New Horizons for Human Factors in Design*. New York: McGraw-Hill Book Company.

Hyman, R. (1953). Stimulus information a determinant of reaction time. *Journal of Experimental Psychology*, 45, 423–432.

Jacko, J. A. (1997). An empirical assessment of task complexity for computerized menu systems. *International Journal of Cognitive Ergonomics*, 1 (2), 137–147.

Jacko, J. A. and Ward, K. G. (1996). Toward establishing a link between psychomotor task complexity and human information processing. *Computers and Industrial Engineering*, 31 (1–2), 533–536.

Jonassen, D. H. (2000). Toward a design theory of problem solving. *Educational Technology Research and Development*, 48, 63–85.

Just, M. A., Carpenter, P. A., and Miyake, A. (2003). Neuroindices of cognitive workload: Neuroimagining, pupillometric and event-related brain potential studies of brain work. *Theoretical Issues in Ergonomic Science*, 4, 56–58.

Kahneman, D. (1973). *Attention and Effort*. Englewood Cliffs, NJ: Prentice-Hall.

Kamishov, I. A. (1968). The methodology of eye movement registration and determination of the operator's eye movement. *Questions of Psychology*, 4, 62–81.

Kandror, I. S. and Demina, D. M. (1978). On the principles and criteria for the physiological classification of work according to their physical effort and intensity. *Physiology of Human Work*, 4 (N1), 136–147.

Kaptelinin, V. (1997). Activity theory: Implication for human-computer interaction. In B. A. Nardi (Ed.), *Context and Consciousness*. Cambridge, MA: The MTI Press, pp. 103–116.

Kaptelinin, V. and Nardi, B. (2006). *Acting with Technology: Activity Theory and Interaction Design*. Cambridge, MA: The MIT Press.

Karat, J. (1993). Software evaluation methodologies. In M. Helander (Ed.), *Handbook of Human-Computer Interaction*. Amsterdam, the Netherlands: North-Holland, pp. 891–903.

Karat, J., Karat, C.-M., and Vergo, J. (2004). Experiences people value: The new frontier for task analysis. In D. Diaper and N. Stanton (Eds.), *The Handbook of Task Analysis for Human-Computer Interaction*. Mahwah, NJ: Lawrence Erlbaum Associates, Publishers, pp. 585–602.

Karger, D. W. and Bayha, F. H. (1977). *Engineering Work Measurement*, 3rd edn. New York: Industrial Press.

Keebler, J. R., Sciarini, L., Fincannon, T., Jentsch, F., and Nicolson, D. (2008). Effect of training modality on target identification in a virtual tank recognition task. *Proceeding of the 52th Annual of Human Factor and Ergonomics Society*, New York.

Kieras, D. E. (1993). Towards a practical GOMS model methodology for user interface design. In M. Helendar (Ed.), *Handbook of Human-Computer Interaction*. Amsterdam, the Netherlands: North-Holland/Elsevier Science, pp. 135–157.

Kim, J. (2008). Perceived difficulty as a determinant of web search performance. *Information Research*, 13 (4), paper 379.

Kirshner, D. and Whitson, J. A. (Eds.). (1997). *Situated Cognition: Social, Semiotic, and Psychological Perspectives*. Mahwah, NJ: Lawrence Erlbaum Associates, Publishers.

Kirwan, B. (1994). *A Guide to Practical Human Reliability Assessment*. London, U.K.: Taylor & Francis.

Kirwan, B. and Ainsworth, L. K. (Eds.). (1992). *A Guide to Task Analysis*. London, U.K.: Taylors & Francis.

Khan, M., Jaber, M. Y., and Plaza, M. (2011). Linking quality to learning—A review. In M. Y. Jaber (Ed.), *Learning Curves. Theory, Models, and Applications*. Boca Raton, FL: CRC Press/Taylor & Francis Group, pp. 211–236.

Kholodnaya, G. N. (1978). *Time Study in Manufacturing*. Moscow, Russia: Economics Publisher.

Klatsky, R. L. (1975). *Human Memory: Structure and Processes*. San Francisco, CA: Freeman.

Kleinback, U. and Schmidt, K. H. (1990). The translation of of work motivation into performance. In Kleinback, V., Quast, H.-H., Thierry, H., Hacher, H. (Eds.). *Work Motivation*, NJ: Lawrence Erlbaum Associates, Inc.

Klochko, V. E. (1978). Goal formation and dynamic of evaluation of problem solving tasks. PhD dissertation, Moscow University, Moscow, Russia.

Konopkin, O. A. (1980). *Psychological Mechanisms of Regulation of Activity*. Moscow, Russia: Science.

Konopkin, O. A. and Zhujkov, Ju. S. (1973). About a person's ability to assess the probabilistic characteristics of alternative stimuli. In D. A. Oshanin and O. A. Konopkin (Eds.), *Psychological Aspects of Activity Regulation*. Moscow, Russia: Pedagogical Publishers, pp. 154–197.

Kosilov, S. A. (1979). *Psycho-Physiological Foundations of Organization of Work*. Moscow, Russia: Economics Publishers.

Kotarbinski, T. (1965). *Praxiology: An Introduction to the Science of Efficient Action*. Oxford, U.K.: Pergamon Press.

Kotik, M. A. (1974). *Self-Regulation and Reliability of Operator*. Tallinn, Estonia: Valgus.

Kotik, M. A. (1978). *Textbook of Engineering Psychology*. Tallinn, Estonia: Valgus.

Kotovsky, K. and Simon, H. A. (1990). Why are some problems really hard: Explorations in the problem space of difficulty. *Cognitive Psychology*, 22, 143–183.

Krilov, I. N. (1970). On the question of the regulation of the pace of simple rhythmic movements. In V. N. Chernigovsky and N. A. Rokotova (Eds.), *Control of Movement*. Leningrad, Russia: Academy of Science, pp. 38–49.

Kuutti, K. (1997). AT as a potential framework for human-computer interaction research. In B. A. Nardi (Ed.), *Context and Consciousness: AT and Human-Computer Interaction*. Cambridge, MA: The MIT Press, pp. 17–44.

LaBerge, D. and Buchsbaum, M. S. (1990). Positron emission tomographic measurements of pulvinar activity during an attention task. *Journal of Neuroscience*, 10, 613–619.

Landa, L. M. (1976). *Instructional Regulation and Control: Cybernetics, Algorithmization and Heuristic in Education*. Englewood Cliffs, NJ: Educational Technology Publication (English translation).

Landa, L. M. (1984). Algo-heuristic theory of performance, learning and instruction: Subject, problems, principles. *Contemporary Educational Psychology*, 9, 235–245.

Lazareva, V. V., Svederskaya, N. E., and Khomskaya, E. D. (1979). Electrical activity of brain during mental workload. In E. D. Khomskaya (Ed.), *Neuropsychological Mechanisms of Attention*. Moscow, Russia: Science Publisher, pp. 151–168.

Lee, T. W., Locke, E. A., and Latham, G. P. (1989). Goal setting, theory and job performance. In A. Pervin (Ed.), *Goal Concepts in Personality and Social Psychology*. Hillsdale, NJ: Lawrence Erlbaum Associates, pp. 291–326.

Lehmann, G. (1962). *Practical Work Physiology*. Stuttgart, Germany: George Theme Verlag.

Leont'ev, A. N. (1972). *The Problem of Psychic Development*. Moscow, Russia: Moscow University Publishers.

Leont'ev, A. N. (1977). *Activity, Consciousness, Personality*. Moscow, Russia: Political Publishers.

Leont'ev, A. N. (1978). *Activity, Consciousness and Personality*. Englewood Cliffs, NJ: Prentice Hall.

Leont'ev, A. N. (1981). The problem of activity in psychology. In J. V. Wertsch (Ed.), *The Concept of Activity in Soviet Psychology*. Armonk, NY: M. E. Sharpe, Inc.

Levinthal, D. and Gavetti, G. (2004). Strategy field from the perspective of management science. *Management Science*, 50, 1309–1318.

Lindsay, P. H. and Norman, D. A. (1972). *Human Information Processing*. San Diego, CA: Harcourt Brace Jovanovich.

Lindsley, D. B. (1957). Psychology and motivation. In M. F. Jones (Ed.). *Nebraska Symposium of Motivation*. Lincoln: University of Nebraska Press, pp. 109–123.

Liu, P. and Li, Z. (2012). Task complexity: A review and conceptualization framework. *International Journal of Industrial Ergonomics*, 42, 553–568.

Lomov, B. F. (1966). Man and Machine: Soviet Radio.

Lomov, B. F. (Ed.). (1982). *Handbook of Engineering Psychology*. Moscow, Russia: Manufacturing Publishers.

Lomov, B. F. (Ed.). (1986). *Foundations of Engineering Psychology*. Moscow, Russia: Higher Education Publishers.

Lomov, B. F. and Surkov, E. N. (1980). *Anticipation in Structure of Activity*. Moscow, Russia: Science Publisher.

Luria, A. R. (1966). *Higher Cortical Function in Man*. New York: Basic Books.

Luria, A. R. (1970). *Traumatic Aphasia*. The Hague, the Netherlands: Mouton.

Luu, P., Flaisch, T., and Tucker, D. M. (2000). Medial frontal cortex in action monitoring. *Journal of Neuroscience*, 20, 466–469.

Maslov, O. P. and Pronina, E. E. (1998). Psychic and reality: Topology of virtual reality. *Applied Psychology*, 6, 41–49.

Maynard, H. B., Stegemerten, G. J., and Schawab, J. L. (1948). *Method-Time Measurement*. New York: McGraw-Hill Book Co.

McCormick, E. J. and Ilgen, D. (1985). *Industrial and Organizational Psychology*, 8th edn. London, U.K.: Unwin Hyman.

Meister, D. (1985). *Behavioral Analysis and Measurement Methods*. New York: John Wiley & Sons.

Meister, D. (1999). *The History of Human Factors and Ergonomics*. Mahwah, NJ: Lawrence Erlbaum Associates, Publishers.

Miller, D. and Swain, A. (1987). Human error and human reliability. In G. Salvendy (Ed.), *Handbook of Human Factors*. New York: Wiley, pp. 360–413.

Miller, G. A., Galanter, E., and Pribram, K. H. (1960). *Plans and the Structure of Behavior*. New York: Holt.

Miller, L., Staney, K., and Guckenberger, E. (1997). Above real-time training. *Ergonomics in Design*, 5, 21–24.

Milner, B. (1982). Some cognitive effects of frontal lobe lesions in man. *Philosophical Transactions of the Royal Society of London*, B298, 211–226.

Moshensky, M. G. (1971). *Time Study and Wages in the West and USA*. Moscow, Russia: Thinking Publisher.

Mowrer, O. H. (1960). *Learning Theory and the Symbolic Processes*. New York: Wiley.

MTM 1 Analyst Manual. UK MTMA (2000) Ltd, 2001.

Muchinsky, P. M. (1990). *Psychology Applied to Work. An Introduction to Industrial and Organizational Psychology*. Pacific Grove, CA: Brooks/Cole Publishing Company.

Murray, H. A. (1938). *Explorations in Personality*. New York: Oxford University Press.

Myasnikov, V. A. and Petrov, V. P. (Eds.). (1976). *Aircraft Digital Monitoring and Control Systems*. Leningrad, Russia: Manufacturing Publishers.

Nardi, A. (Ed.). (1997). *Context and Consciousness: AT and Human-Computer Interaction*. Cambridge, MA: The MIT Press, pp. 17–44.

Nayenko, N. I. (1976). *Psychic Tension*. Moscow, Russia: Moscow University.

Neilson, W. S. and Stowe, J. (2010). Piece rate contracts for other-regarding workers. *Economic Inquiry*, 48, 575–586.

Neumenn, J. and Time, K. P. (1975). *Organization of Work. Psychophysiological Aspects*. Moscow, Russia: Economics (translation from German).

Neumin, Y. G. (1984). *Models in Science and Technic*. Leningrad, Russia: Science Publishers.

Newell, A. and Simon, H. A. (1972). *Human Problem Solving*. Englewood Cliffs, NJ: Prentice-Hall.

Nicholson, N. (1997). Evolutionary psychology: Toward a new view of human nature and organizational society. *Human Relations*, 50, 1053–1078.

Nojivin, U. S. (1974). On psychological self-regulation of sensory motor actions. In V. D. Shadrikov (Ed.), *Engineering and Psychology*, Vol. 1. Yaroslav, Russia: Yaroslav University, pp. 206–210.

Norman, D. A. (1976). *Memory and Attention: An Introduction to Human Information Processing*, 2nd edn. New York: Wiley.

Norman, D. A. (1981). Categorization of actions slips. *Psychology Review*, 88, 1–15.

Norman, D. A. (1986). Cognitive engineering. In D. Norman and S. Draper (Eds.), *User Centered System Design: New Perspectives on Human-Computer Interaction*. Hillsdale, NJ: Erlbaum, pp. 31–61.

Norman, D. A. (1988). *The Psychology of Everyday Things*. New York: Harper & Row.

Novikov, A. M. (1986). *Process and Method of Formation of Vocational Skills*. Moscow, Russia: Higher Education.

Olds, J. and Milner, P. (1954). Positive reinforcement produced by electrical stimulation of septal area and other regions of the rat brain. *Journal of Comparative and Physiological Psychology*, 47, 419–429.

Oviatt, S. (2012). Multimodal interface. In J. Jacko (Ed.), *The Human Computer Interaction Handbook*, 3rd edn. Boca Raton, FL: CRC Press/Taylor & Francis Group, pp. 405–430.

Park, J. (2009). *The Complexity of Proceduralized Tasks*. London, U.K.: Springer-Verlag.

Park, J. (2011). Scrutinizing inter-relationships between performance influencing factors and the performance of human operators pertaining to the emergency tasks of nuclear power plant—An explanatory study. *Annals of Nuclear Energy*, 38, 2521–2532.

Park, J., Jung, W., and Ha, J. (2001). Development of the step complexity measure for emergency operating procedures using entropy concepts. *Reliability Engineering and System Safety*, 71, 115–130.

Park, J. and Jung, W. (2007). A study on the development of a task complexity measure for emergency operating procedures of nuclear power plants. *Reliability Engineering and System Safety*, 9, 1102–1116.

Patrick, J. (1992). *Training Research and Practices*. London, U.K.: Academic Press.

Paus, T., Koski, L., Caramanos, Z., and Westbury, C. (1998). Regional differences in the effects of task difficulty and motor output on blood flow response in the human anterior cingulate cortex: A review of 107 PET activation studies. *Neuro Report*, 9, R37–R47.

Pavlov, I. P. (1927). *Conditioned Reflex*. London, U.K.: Oxford University Press.

Payne, W. (1997). Task complexity and contingent processing in decision-making: An information search and protocol analysis. *Organizational Behavior and Human Performance*, 16, 366–387.

Pervin, L. A. (1989). Goal concepts, themes, issues and questions. In L. A. Pervin (Ed.), *Goal Concepts in Personality and Social Psychology*. Hillsdale, NJ: Lawrence-Erlbaum Associates, pp. 173–180.

Petersen, S. E., Fox, P. T., Posner, M. I., Mintun, M., and Raichle, M. E. (1988). Positron emission tomographic studies of the cortical anatomy of single word processing. *Nature*, 331, 585–589.

Piaget, J. (1952). *The Origins of Intelligence in Children*. New York: International University Press.

Picton, T. W., Alain, C., and McIntosh, A. R. (2002). The theater of the mind: Physiological studies of human frontal lobes. In D. T. Stuss and R. T. Knight (Eds.), *Principles of Frontal Lobe Function*. Oxford, U.K.: Oxford University Press, pp. 96–108.

Pintrich, P. R. and Zeidner, M. (Eds.). (2005). *Handbook of Self-Regulation*. San Diego, CA: Academic Press, pp. 303–341.

Platonov, K. K. (1982). *System of Psychology and Theory of Reflection*. Moscow, Russia: Science Publishers.

Platonov, K. K. (1970). *Work Psychology*. Moscow: Medical Publisher.

Ponomarenko, V. and Bedny, G. (2011). Characteristics of pilots' activity in emergency situation resulting from technical failure. In G. Z. Bedny and W. Karwowski (Eds.), *Human-Computer Interaction and Operators' Performance: Optimizing Work Design with Activity Theory*. Boca Raton, FL: Taylor & Francis, pp. 223–255.

Ponomarenko, V., Bedny, G. Z., and Makarov, R. N. (2006). *Activity Pilot in Flight and Its Imaginative Components.* Maastricht, the Netherlands: Triennial International Congress.

Ponomarenko, V. A. (2006). *Psychology of Human Factor in Dangerous Profession.* Krasnoyarsk, Russia: International Academy of Human Factor in Aviation and Astronautics.

Ponomarenko, V. A. and Zavalova, N. D. (1981). Study of psychic image as regulator of operator actions. In B. F. Lomov and V. F. Venda (Eds.), *Methodology of Engineering Psychology and Psychology of Work of Management.* Moscow, Russia: Science, pp. 30–41.

Ponomarenko, V. A. and Lapa, V. V. (1975). Impact of the intellectual assessment of the situation on nature of emotional reactions in pilots. *Space Biology and Medicine*, 1, 66–79.

Preece, J., Rogers, Y., Sharp, H., Benyon, D., Holland, S., and Carey, T. (1994). *Human-Computer Interaction.* Reading, MA: Addison-Wesley.

Pribram, K. (1971). *Languages of the Brain.* Englewood Cliffs, NJ: Prentice-Hall.

Pushkin, V. N. (1978). Situational concept of thinking in activity structure. In A. A. Smirnov (Ed.), *Problems of General and Educational Psychology.* Moscow: Pedagogy, pp. 106–120.

Rassmussen, J. (1983). Skills, rules, and knowledge; signals, signs, and symbols, and other distinctions in human performance models. *IEEE Transactions on System, Man and Cybernetics*, SMC-15, 234–243.

Rauterberg, M. (1996). How to measure of cognitive complexity in human-computer interaction. In R. Trappl (Ed.), *Cybernetic and Systems*, Vol. 2. Vienna, Austria: Austrian Society for Cybernetic Studies, pp. 815–820 (compl HCI).

Reason, J. T. (1990). *Human Error.* New York: Cambridge University Press.

Reber, A. S. (1985). *Dictionary of Psychology.* New York: Penguin Books.

Rescorla, R. A. (1972). Informational variables in Pavlovian conditioning. In G. H. Bower (Ed.), *The Psychology of Learning and Motivation*, Vol. 6. New York: Academic Press, pp. 1–46.

Reykovski, J. (1979). *Experimental Psychology of Emotion.* Moscow, Russia: Progress (translation from Poland).

Robinson, P. (2001). Task complexity, task difficulty, and task production: Exploring interaction in a componential framework. *Applied Linguistics*, 22 (1), 27–57.

Rodgers, S. H. and Eggleton, E. M. (Eds.). (1986). *Ergonomic Design for People at Work*, Vol. 2. New York: Van Nostrand Reinhold.

Rosser, Jr., J. B. (2010a). Introduction to special issue on transdisciplinary perspectives on economic complexity. *Journal of Economic Behavior and Organization, Elsevier*, 75, 1–2.

Rosser, Jr., J. B. (2010b). Is a transdisciplinary perspectives on economic complexity possible? *Journal of Economic Behavior and Organization, Elsevier*, 75, 3–13.

Rozenblat, V. V. (1975). Principle of physiological assessment of heavy physical work based on heart rate. In V. V. Rozenblat (Ed.). *Function of Organism During of Work Process*, pp. 5–22. Moscow, USSR: Economic Publisher.

Rubakhin, V. F. (1974). *Psychological Foundation of Human Information Processing.* Moscow, Russia: Science Publishers.

Rubinshtein, S. L. (1922/1986). Principles of creative independent activity. *Questions of Psychology*, 4, 101–107 (reprinted from the first publication).

Rubinshtein, S. L. (1934). Problems of psychology in works of Markx. *Psychotechnics*, 1, 3–8.

Rubinshtein, S. L. (1935). *Principles of General Psychology*. Moscow, Russia: Pedagogical Publisher.

Rubinshtein, S. L. (1957). *Existence and Consciousness*. Moscow, Russia: Academy of Science.

Rubinshtein, S. L. (1958). *About Thinking and Methods of Its Development*. Moscow, Russia: Academic Science.

Rubinshtein, S. L. (1959). *Principles and Directions of Developing Psychology*. Moscow, Russia: Academic Science.

Salmon, P. M., Stanton, N. A., Walker, G. H., Jenkins, D., Baber, C., and Mcmaster, R. (2008). Representing situation awareness in collaborative systems: A case study in the energy distribution domain. *Ergonomics*, 51 (3), 367–384.

Salvucci, D. D. and Goldberg, J. H. (2000). Identifying fixations and saccades in eye-tracking protocols. In *Proceedings of the Eye Tracking Research and Applications Symposium 2000*. New York: ACM Press, pp. 71–78.

Sandesrs, M. S. and McCormick, E. J. (1993). *Human Factors in Engineering and Design* (7th edn.). New York: McGraw Hill.

Sapir, E. L. (1956). Language and environment. *American Anthropologist*, 1912. Reprinted in B. L. Wholrf (Ed.), *Language, Thought and Reality*. Cambridge, U.K.: Cambridge University Press.

Schacter, S. and Singer, J. E. (1962). Cognitive, social, and physiological determinants of emotional state. *Psychological Review*, 69, 379–399.

Schiller, P. H., Sandell, J. H., and Maunsell, J. H. (1987). The effect of frontal eye field and superior colliculus lesions on saccadic latencies in the rhesus monkey. *Journal of Neurophysiology*, 57, 1033–1049.

Schneider, W. and Shiffrin, R. M. (1977). Controlled and automatic human information processing: I. Detection, search, and attention. *Psychological Review*, 84, 1–66.

Schultz, D. and Schultz, S. (1986). *Psychology and Industry Today: An Introduction to Industrial and Organizational Psychology*. New York: Macmillan Publishing Company.

Seel, N. M. (Ed.). (2012). *Encyclopedia of the Sciences of Learning*, 1st edn. Boston, MA: Springer, pp. 1800–1805.

Senders, J. W. and Moray, N. P. (1991). *Human Errors: Cause, Prediction, and Reduction*. Hillsdale, NJ: Lawrence Erlbaum Associates, Publishers.

Sengupta, T., Bedny, I., and Bedny, G. (2011). Microgenetic principles in the study of computer-based tasks. In G. Z. Bedny and W. Karwowski (Eds.), *Human-Computer Interaction and Operators' Performance: Optimization of Work Design with Activity Theory*. Boca Raton, FL: Taylor & Francis/CRC Press, pp. 117–148.

Sengupta, T., Bedny, I. S., and Karwowski, W. (2008). Study of computer based tasks during skill acquisition process. *Second International Conference on Applied Ergonomics Jointly with 11th International Conference on Human Aspects of Advanced Manufacturing*, Las Vegas, NV, July 14–17, 2008.

Shallice, T. (1988). *From Neuropsychology to Mental Structure*. Cambridge, U.K.: Cambridge University Press.

Shannon, C. E. and Weaver, W. (1949). *The Mathematical Theory of Communications*. Urbana, IL: University of Illinois Press.

Shchedrovitsky, G. P. (1995). *Selective Works*. Moscow, Russia: Cultural Publisher.

Shuchman, L. A. (1987). *Plans and Situated Actions: The Problem of Human-Machine Interaction.* Cambridge, U.K.: Cambridge University Press.

Siegal, A. I. and Wolf, J. J. (1969). *Man-Machine Simulation Models.* New York: Wiley.

Simon, H. A. (1999). *The Sciences of the Artificial,* 3rd rev. edn. Cambridge, MA: MIT Press, p. 4.

Simonov, P. V. (1982). Need-informational theory of emotions. *Problem of Psychology,* 5, 44–56.

Singh, K. (1997). The impact of technological complexity and interfirm cooperation on business survival. *Academy of Management Journal,* 40 (2), 339–367.

Smidtke, H. and Stier, F. (1961). An experimental evaluation of validity of predetermined elemental time system. *The Journal of Industrial Engineering,* 3, 182–204.

Smirnov, A. A., Leont'ev, A. N., Rubinshtein, S. L., and Teplov, B. M. (Eds.). (1962). *Psychology.* Moscow, Russia: Pedagogical Publishers.

Sokolov, E. N. (1969). The modeling properties of the nervous system. In M. Cole and I. Maltzman (Eds.), *Handbook of Contemporary Soviet Psychology.* New York: Basic Books, Publishers, pp. 671–704.

Sperling, G. (1960). The information available in brief visual presentations. *Psychological Monographs,* 74, 1–29 (Whole No.11).

Sternberg, S. (1969). The discovery of processing stages: Extension of Donder's method. *Acta Psychological,* 30, 276–315.

Sternberg, S. (1975). Memory scanning: New findings and current controversies. *Quarterly Journal of Experimental Psychology,* 27, 1–32.

Sternberg, S. (2008). Identification of neural modules. In O. Y. Chebykin, G. Bedny, and W. Karwowski (Eds.), *Ergonomics and Psychology: Development in Theory and Practice.* Boca Raton, FL: CRC Press/Taylor & Francis, pp. 135–166.

Suh, N. P. (1990). *The Principles of Design.* New York: Oxford University Press.

Taylor, F. W. (1911). *The Principles of Scientific Management.* New York: Harper and Brothers.

Taylor, S. F., Welsch, R. C., Wager, T. D., Luan, P. K., Fitzgerald, K. D., and Gehring, W. J. (2004). A functional neuroimaging study of motivation and executive function. *Neuroimage,* 21, 1045–1055.

Thaompson, J. D. (1968). *Organizations in Actions.* New York: McGraw-Hill.

Thelen, E. (1995). Time-scale dynamics in the development of an embodied cognition. In R. Port and T. van Gelder (Eds.), *Mind in Motion.* Cambridge, MA: MIT Press.

Thelen, E. and Smith, L. (1994). *A Dynamic Systems Approach to the Development of Cognition and Action.* Cambridge, MA: MIT Press.

Thomas, J. C. and Richards, T. (2012). Achieving psychological simplicity measures and methods to reduce cognitive complexity. In J. A. Jucko (Ed.), *The Human-Computer Interaction Handbook: Fundamentals, Evolving Technologies, and Emerging Application.* Boca Raton, FL: CRC Press/Taylor & Francis Group, pp. 491–513.

Thut, G., Schultz, W., Roelcke, U., Nienhusmeir, M., Missiner, J., Maguire, R. P., and Leenders, K. L. (1997). Activation of human brain by monetary reward. *Neuro Report,* 8, 1225–1228.

Tikhomirov, O. K. (1984). *Psychology of Thinking.* Moscow, Russia: Moscow University.

Tolman, E. C. (1932). *Purposive Behavior in Animals and Men.* New York: Century.

Turvey, M. T. (1996). Dynamic touch. *American Psychologist,* 51 (11), 1134–1152.

UNSCEAR. (2008). The Chernobyl accident: UNSCEAR's assessment of the radiation effects. Report. Vienna, Austria: The United Nations Scientific Committee on the Effects of Atomic Radiation.

Vancouver, J. T. (2005). Self-regulation in organizational settings. A tale of two para-digms. In M. Boekaerts, P. R. Pintrich, and M. Zeidner (Eds.), *Handbook of Self-Regulation*. San Diego, CA: Academic Press, pp. 303–341.

Van Santen, J. H. and Philips, N. Y. (1970). Method and time study of mental work. *Work Study and Management Services*, 14 (1), 21–27.

Vargas-Baron, E. (2013). Building and strengthening national systems for early child-hood development. In P. Britto, P. L. Engle, and C. M. Super (Eds.), *Handbook of Early Childhood Development Research and Its Impact on Global Policy*. New York: Oxford University Press, pp. 443–466.

Veikher, A. A. (1978). *Complex Labor*. Leningrad, Russia: Science Publisher.

Vekker, L. M. (1976). *Psychological Processes*, Vol. 2. Leningrad, Russia: Leningrad University Press.

Vekker, L. M., Allen, J. A., and Yufik, Y. (1993). Human-computer interaction: Applications and case studies. In M. J. Smith and G. Salvendy (Eds.), *Proceedings of the Fifth International Conference on Human-Computer Interaction (HCI International' 93)*, Vol. 1, Orlando, FL, August 8–13, 1993.

Vekker, L. M. and Paley, I. M. (1971). Information and energy in psychological reflec-tion. In B. G. Ana'ev (Ed.), *Experimental Psychology*, Vol. 3. Leningrad, Russia: Leningrad University, pp. 61–66.

Vencel', E. S. and Ovcharenko, L. A. (1969). *Theory of Probability*. Moscow, Russia: Science Publisher.

Vicente, K. J. (1999). *Cognitive Work Analysis: Toward Safe, Productive, and Healthy Computer-Based Work*. Mahwah, NJ: Lawrence Erlbaum Associates, Publishers.

Visser, W. (2006). *The Cognitive Artifacts of Design*. Mahwah, NJ: Lawrence Erlbaum Associates, Publishers.

Vinogradov, M. I. (Ed.). (1969). *Foundation of Work Physiology*. Moscow, Russia: Medical Publisher.

Vygotsky, L. S. (1956). *Selected Psychological Research*. Moscow, Russia, Academy of Pedagogical Science.

Vygotsky, L. S. (1960). *Developing Higher Order Psychic Functions*. Moscow, Russia: Academy of Pedagogical Science of the RSFSR.

Vygotsky, L. S. (1971). *The Psychology of Arts*. Cambridge, MA: MIT Press.

Vygotsky, L. S. (1978). *Mind in Society: The Development of Higher Psychological Processes*. Cambridge, MA: Harvard University Press.

Waugh, N. C. and Norman, D. A. (1965). Primary memory. *Psychological Review*, 72, 89–104.

Weimer, W. B. (1977). A conceptual framework for cognitive psychology. In R. Shaw and J. Bransford (Eds.), *Perceiving, Acting and Knowing*. Hillsdale, NJ: Lawrence Erlbaum.

Wickens, C. D. and Hollands, J. G. (2000). *Engineering Psychology and Human Performance*, 3rd edn. New York: Harper-Collins.

Wickens, C. D. and McGarley, J. S. (2008). *Applied Attention Theory*. Boca Raton, FL: Taylor & Francis.

Wilson, M. (2002). Six views of embodied cognition. *Psychonomic Bulletin and Review*, 9 (4), 625–636.

Wojeiulik, E. and Kanwisher, N. (1999). The generality of parietal involvement in visual attention. *Neuron*, 23, 747–764.

Wood, R. E. (1986). Task complexity: Definition of the construct. *Organizational Behavior and Human Decision Processes*, 37 (1), 60–82.

Wright, T. P. (1936). Factors affecting the cost of airplanes. *Journal of the Aeronautical Sciences*, 3 (4), 122–128.

Yarbus, A. L. (1969). *Eye Movement and Vision*. New York: Plenum.

Zabrodin, Yu. M. and Chernishev, A. P. (1981). On the loss of information in describing the activity of human-operator by the transfer function. In B. F. Lomov and V. F. Venda (Eds.), *Theoretical and Methodological Analysis in Engineering Psychology, Psychology of Work and Control*. Moscow, Russia: Science Publisher, pp. 244–250.

Zalkind, M. S. (1966). Influence of the skills on difference threshold of speed perception. *Questions of Psychology*, 1, 27–38.

Zarakovsky, G. M. (2004). The concept of theoretical evaluation of operators' performance derived from activity theory. In G. Bedny (Ed.). *Theoretical Issues in Ergonomics Science*, 5 (4), 313–337.

Zarakovsky, G. M., Korolev, B. A., Medvedev, V. I., and Shlaen, P. Y. (1974). *Introduction to Ergonomics*. Moscow, Russia: Soviet Radio.

Zarakovsky, G. M. and Magazannik, V. D. (1981). Psychological criteria of complex decision making processes. In B. F. Lomov and V. F. Venda (Eds.), *Methodology of Engineering Psychology and Psychology of Work Management*. Moscow, Russia: Science, pp. 63–79.

Zarakovsky, G. M. and Medvedev, V. I. (1971). Psychological evaluation of efficiency of man-machine system. In *Third Conference of Operator Reliability*. Leningrad, Russia: Psychology Society.

Zarakovsky, G. M. and Medvedev, V. I. (1979). Classification of operator errors. *Technical Aesthetics*, 10, 5–7.

Zarakovsky, G. M. and Pavlov, V. V. (1987). *Laws of Functioning Man-Machine Systems*. Moscow, Russia: Soviet Radio.

Zavalova, N. D., Lomov, B. F., and Ponomarenko, V. A. (1971). Principle of active operator and function allocation. *Problems of Psychology*, 3, 3–12.

Zavalova, N. D. and Ponomarenko, V. A. (1980). Structure and content of psychic image as mechanisms of regulation of actions. *Psychological Journal*, 1 (2), 5–18.

Zener, K. (1937). The significance of behavior accompanying conditioned salivary secretion for theories of the conditioned response. *American Journal of Psychology*, 50, 384–403.

Zinchenko, P. I. (1961). *Involuntary Memorization*. Moscow, Russia: Pedagogy.

Zinchenko, T. P. (1981). *Identification and Coding*. Leningrad, Russia: Leningrad University Publishers.

Zinchenko, V. P. (1995). Vygotsky's idea about units for the analysis of mind. In J. Wertsch (Ed.), *Culture, Communication and Cognition: Vygotskian Perspectives*. Cambridge, U.K.: Cambridge University Press, pp. 94–118.

Zinchenko, V. P., Munipov V. M., and Gordon, V. M. (1973). Study of visual thinking. *Questions of Psychology*, (2), 57–66.

Zinchenko, V. P., Velichkovsky, B. M., and Vuchetich, G. G. (1980). *Functional Structure of Visual Memory*. Moskow, Russia: Moskow State University Publishers.

Zinchenko, V. P. and Vergiles, N. Y. (1969). *Creation of Visual Image*. Moscow, Russia: Moscow University.

Index

reticular system, 39
routine sequencing, 36
self-regulation mechanism, 37
short-term memory, 38
thalamus, 41
tonic arousal, 39
verbalization functions, 41
wakefulness, 38
working memory, 36
Goals, operators, methods, and selection
      rules (GOMS) system, 78
Graph theory, 330
Gross errors, 290
Gymnastic skills, 147

H

Habits *vs.* skills, 120
Handbook of Engineering Psychology,
      292–293
Handle-welding operation, 351, 355
Hierarchical task analysis (HTA)
      technique, 103, 105–106
   CL2, 109–111
   drawback, 113
   human reliability assessment, 109
   operation, 113
Homeostasis, 275
Human–computer interaction (HCI),
      activity time structure, 220–221

I

Imaginative actions, 199
Inattentional blindness phenomenon, 149
Individual style of activity, 237
Individual style of performance,
      133–134
Innovative learning, 88
Input–action–feedback (I-A-F)
      structure, 107
Instrumental skills, 121
Intrinsic feedback, 123
ISCAN eye tracking system, 165

J

Job evaluation, analytical *vs.*
      nonanalytical, 311

L

Learning curve
   approach, 127
   Pavlov's study, 127
   power law, 128
   shape, 128
   and skill formation process, 137–139
   temporal parameters, 129
   as tool for analyzing
      formation of skills, 132–133
      human strategies, 130
      skill retention, 129
Long-term memory structure, 118

M

Marxist philosophy, AT, 92
Memorization, 84
Memory
   actions, 199
   workload, task complexity, 313
Mental act, 80
Mental development, 88
Mental operations, 89
Mental tool-sign, 99
Methods–time measurement (MTM-1)
      system, 321–323, 332–333
   critical analysis, 239
   decision-making, 227
   manual tasks, 238–239
   microelements, 207
   motor activity duration, 217
   pace of performance, 249
   problem-solving, 380–381
   time structure
      activity, 226
      analysis, 385–387
   units of analysis, 210
   utilization, 210
Microgenetic method study, 131
Mnemonic actions, 199
Motions, 80
Motivation; *see also* Goal
   anti-goal/work avoidant goal, 29
   emotion, 26, 29
   energy concept, 26
   FSS approach, 32
   homeostatic process, 33

computer-based
 (*see* Computer-based tasks)
creative, 54
deterministic-algorithmic, 54
difficulty, assessment, 310
directness of activity, 52
execution time, 339–340
goal acceptance, 52
HCI, 53
intermittent goals, 55
means of work, 50
mechanization of work, 49, 51
mental representation, 52
nonalgorithmic, 54
performance, 164–165, 319
physical efforts, 54
probabilistic-algorithmic, 54
problem-solving, 53
production and operational-
 monitoring processes
 computerization, 72
 ergonomic design approach, 73
 human activity, 70
 logically organized problem-
  solving tasks, 72
 mechanization stage, 71
 product design, 68
 sensory-perceptual and thinking
  components, 71
 structure of, 69
 technological process, 70
 work process, 70
production operations, 50
safety
 accident prevention, 391–395
 active waiting period, 383, 390
 algorithmic analysis, 394
 cognitive psychology, 380
 complexity of logical conditions,
  389–390
 control/ram movement
  relationship, 391–395
 decision making, 384
 design innovation, 382
 deterministic algorithm, 387–388
 emergency situation, 392, 394
 human error and probability, 395
 manual control, 391
 mode of control, 381–384

motor components of activity,
 393–394
MTM-1 system, 380
probabilistic algorithm, 387–388
quantitative analysis, 385, 395
requirements, 381–382
subtask move control, 392–393
time structure analysis, 385, 393
self-formulated, 55
semiautomatic and automatic
 systems, 50
situation-bounded activity, 51
skill-based behavior, 53
structure of activity, 51
training development, 49
Task analysis
algorithmic analysis, 203
comparative analysis, 90
concept, 85
deny fire solution, 107–109
description/identification stage, 105
element, 114
Fill tanker with CL2, 109–111
graph theory, 105
hierarchical system, 111–112
homeostatic regulation principle, 112
HTA, 105–106
human–machine system, 103
identification, 103
logical condition, 204–205
malfunctioning, 105
man–machine systems, 104
mental actions, 204
operation, 107, 113
operators' responsibilities, 104
performance, 103
purpose, 102
self-regulation (*see* Self-regulation,
 task analysis)
special symbol, 203–204
SSAT, 187–188
system design stage, 104
thinking actions, 204
time structure of activity
 chronometric measurement, 226
 cognitive actions, 225
 decision-making, 227, 229–230
 design, 221
 development, 221